Fred K. Geitner, Ronald G. Eierman
Process Machinery

Also of Interest

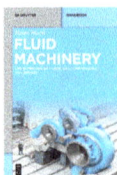

Fred K. Geitner, Ronald G. Eierman

Process Machinery

——

Commissioning and Startup – An Essential Asset
Management Activity

DE GRUYTER

Authors
Fred K. Geitner
Process Machinery Technology Services
1989 Kathleen Avenue
Brights Grove ON N0N 1C0
Canada
fredgeitner@gmail.com

Ronald George Eierman
Eierman Machinery Excellence LLC
10220 Memorial Drive
Houston 77024, TX
USA
eierman09@yahoo.com

ISBN 978-3-11-070097-8
e-ISBN (PDF) 978-3-11-070107-4
e-ISBN (EPUB) 978-3-11-070113-5

Library of Congress Control Number: 2021940897

Bibliographic information published by the Deutsche Nationalbibliothek
The Deutsche Nationalbibliothek lists this publication in the Deutsche Nationalbibliografie;
detailed bibliographic data are available on the Internet at http://dnb.dnb.de.

© 2022 Walter de Gruyter GmbH, Berlin/Boston
Cover image: loonger/E+/Getty Images
Typesetting: Integra Software Services Pvt. Ltd.
Printing and binding: CPI books GmbH, Leck

www.degruyter.com

To the fourth generation, Chloë and Lena Geitner
and
to Mary Doris Harris Eierman, mother, mathematician and mentor.

Foreword

As machinery engineers with many decades of applicable experience, we have observed plant startup and commissioning routines that ranged from deeply flawed and disappointing, to superb and pacesetting. We have come to know that a smooth startup requires planning and management commitment. That said, process machinery commissioning and startup involve well-structured precursor events that represent essential asset management activities. Collectively, motivation, commitment and expert implementation form a bridge from inanimate engineering plans to a construction site and, ultimately, a properly staffed and trained workforce in an exemplary plant. With sound leadership at the helm, sets of drawings are turned into a valuable facility with a happy and competent workforce. Its staffers are value-adders; they operate a safe and highly profitable plant.

Of course, enterprise requirements for capital project execution have evolved. The result is: a significant portfolio of project execution strategies, supplier surveillance, scheduling tools, staffing plans and key performance indicators are used for tracking progress. Together, they include well-defined practices for the development of "Lessons Learned." These valuable lessons are then incorporated into the operating companies' design practices. Others are woven into experience-backed and well-founded exceptions or additions (read "Upgrades") to industry standards of practice.

Understandably, facilities differ, as do management styles. You may not feel comfortable handing this text to novices and instructing them to just follow the book. So then, on the people side of the equation, project team members will include staffers from the engineering, operating, maintenance and enterprise management teams throughout the project's duration, for example, from the pre-front-end engineering and design phase to successful early operation. Success will result if training is carefully executed by subject matter experts. It stands to reason that machinery reliability training includes the designated operation and maintenance teams that will soon be tasked with custody and care of the new assets.

It has been shown that all of these activities result in lower project capital expense, lower asset operating expense, shorter commissioning and start-up schedule and a shorter time to safe and full production. Experience shows that by application of these methods and by using the "tools" they will find in this book, readers and stakeholders can expect longer run-times between turnarounds (i.e., planned shutdowns for maintenance and repair). Moreover, higher asset reliability and production capacity are virtually guaranteed if the reader follows this roadmap. Readers who absorb and then carry out our guidelines will find surprising details in the text that follows.

https://doi.org/10.1515/9783110701074-202

Acknowledgements

It would have been quite impossible to write this text without the help and cooperation of many individuals and companies. These contributors have earned our respect
and gratitude for allowing us to use, adapt, paraphrase, or otherwise incorporate
their work in this text: Shiraz A. Pradhan, B.Sc. MSME, Hurlel G. Elliott, Global Turbomachinery Solutions LLC, who helped us with the review of selected Chapters (both
professionals contributing their experience from major machinery startups). Merv
Behm, Operating Engineer (Canada), Chris Frankcom, P.Eng., Mammoet Crane and
Rigging Supervisor, Ms. Jamie Pogue, CES Energy Services LP (Rigging and Lifting).
Robt. L. Rowan & Assoc. Inc., Texas and Theo de Kok of EMHA, the Netherlands
(Machinery Foundation Installation and Design). Hilary Banda, Olympus Boroscopy.
Flowserve Corporation (Metallic Seal and Commissioning), Chuck Sakers/Koppers
Company (Torque Metering Couplings), Joe Cannatelli, P.E. and Elio Comello, P. Eng.
(Electric Motors). LMF- Leobersdorfer Maschinenfabrik GmbH (API 618 Reciprocating
Compressors), Frank Hoegler of TTS (Gas Turbine MHI concepts), Paul Birdi, Eng.,
AERZEN Blowers and Compressors of Canada Inc. (Helical Screw Compressors and
Blowers), George Talabisco, Siemens Energy, Inc. (Turbocompressors) and Kate Cope-
land of Siemens AG (Steam Turbines).

We are indebted to our close personal friend Heinz P. Bloch who devoted much
of his personal time to a detailed review of selected Chapters. Heinz counseled us
on technical relevance and other concerns.

We hope this text will allow readers to find new and better ways to do their
jobs, broaden their perspective as engineers, and contribute to a fund of knowledge
which—if properly tapped—will bring benefits to everyone.

Fred K. Geitner
Ron G. Eierman

https://doi.org/10.1515/9783110701074-203

Contents

Part II: **Execution**

Chapter 6
Shaft Sealing Systems —— 175

Chapter 7
Commissioning and Startup (C&S) of Electric Motors —— 185

About the Authors

Fred K. Geitner, is a registered professional engineer in the Province of Ontario, Canada. He is the principal engineer of PMTS (Process Machinery Technology Services), an independent consultant and expert litigation witness in the area of process machinery reliability. From 1993 to 1995 he worked for a major gas transmission company in Germany where he was in charge of process machinery technology transfer between the German firm and gas transmission companies in the newly independent states of the former Soviet Union.

Prior to a start-up assignment as senior machinery advisor with Exxon Chemicals in France from 1989 to 1992, Mr. Geitner was employed for 18 years in the refinery and chemical operations of Imperial Oil, the Canadian affiliate of the Exxon Mobile Corporation, USA. During this time, he was a Senior Machinery Engineer and then Engineering Associate involved in design, operation, maintenance and reliability assurance of all types of process machinery in petrochemical refining, ethylene, plastics and other downstream production units.

From 1962 to 1972 Mr. Geitner was employed by Cooper Bessemer of Mt. Vernon, Ohio, a leading manufacturer of large internal combustion engines, gas turbine packages, reciprocating and turbo-compressors. With Cooper, he worked in machinery design, field service engineering, production test and manufacturing management in the United States and Canada.

Mr. Geitner holds a degree (Dipl.-Ing.) from the Technical University Berlin, Germany, and has done graduate studies at the University of Cincinnati, Ohio. Together with H.P. Bloch he co-authored several books on process machinery management and reliability assurance.

Ronald G. Eierman, is a Principal Machinery Advisor with Eierman Machinery Excellence, llc and a Rotating Equipment and Machinery Advisor with Becht Engineering. He holds three degrees from the School of Engineering and Applied Sciences at the University of Virginia; Bachelor of Science Mechanical Engineering with Distinction, Bachelor of Science Applied Mathematics with Distinction and Master of Science Mechanical Engineering specializing in the Dynamics of Turbomachinery. He is a member of the professional organizations; American Society of Mechanical Engineers, Society of Petroleum Engineers, Tau Beta Pi and Sigma Xi. His professional career began in 1974 with Union Carbide Corporation in South Charleston, West Virginia. He retired as a Senior Staff Engineer with Exxon after three decades of service in the worldwide Chemicals and Upstream business sectors of ExxonMobil. He has numerous technical publications spanning the economic application of sealless pumps for chemical plant services to the complex computer modeling of high-performance turbomachinery. He was a member of the first API Taskforce that developed API Standard 685 Sealless Centrifugal Pumps for Petroleum, Petrochemical and Gas Industry Process Service. He is currently providing consulting services to the Hydrocarbon, Petrochemical and Gas Processing Industries for the troubleshooting, application, commissioning, startup, training, and management of machinery systems.

https://doi.org/10.1515/9783110701074-205

Part I: **Laying the Groundwork**

Chapter 1
Introduction and Definitions

1.1 Introduction

There exists an old adage in the process industries that says "As the plant machinery runs so runs the plant." In view of this, the authors chose to focus on the commissioning and startup of process plant machinery because a successful and timely startup of machinery equipment contributes to laying the groundwork for a profitable facility.

The commissioning of process machinery in the hydrocarbon processing industries, for example, and related facilities is a complex endeavor requiring detailed attention to safety and risk mitigation. Moreover, a significant team approach is necessarily required for success.

A way to optimize the safety of commissioning efforts and increase the probability of a successful machinery startup has been presented in this book.

The reader will encounter:
- Definition of process equipment
- Discussion of process equipment criticality
- How the topic fits into asset management (AM)
- A risk assessment method relating to machinery commissioning and startup
- Discussion of capital project specific requirements
- Discussion of commissioning progress management in face of a potentially large machinery population
- Detailed discussion of commissioning and startup phases
- A collection of detailed risk-assessed procedures and checklists for process machinery commissioning and startup.

The authors' goal is to show how addressing the above-mentioned topics will ensure that process equipment and closely related systems are ready for startup. Our thrust is safety and loss prevention.

Not discussed are commissioning activities pertaining to associated equipment other than process machinery. The assumption is that project phases such as front-end engineering design, reliability and maintainability studies, hazard and operability reviews have been successfully completed, manufacturing QA/QC inspections have all been accepted, any deviation notifications have been reviewed and mitigated. We assert further that operator training, instrument calibrations, alarm and trip systems and advanced process controls are being or have been commissioned and tested and that critical spare parts are on order or on hand before the commissioning of process machinery begins.

https://doi.org/10.1515/9783110701074-001

This text presents the techniques and procedures which should be followed to help assure successful machinery startups. Guidelines are given for selected and exemplary types of process machinery, ranging from small non-critical units to large critically important units. Since the startup activity includes the effects of machine shipping and handling, contractor installation, checkout by the startup advisors and run-in under the supervision of the machine vendor's representative and/or the owner's startup advisors, a considerable number of occasions exist for creating a delay in startup. These delays can become expensive; they must be avoided by adequate startup preparation, irrespective of the machine design being satisfactory. Although strict adherence to this guide cannot guarantee to eliminate all delays due to machinery startup problems, it will eliminate redundant mistakes as the sources of delay.

1.2 Definitions

1.2.1 Process Machinery

Process machines, in our context, are prime movers and machines driven by them conveying gases and liquids in hydrocarbon and chemical processing plants. One of the most important machines are compressors and their drivers; they are frequently among a group classified *major machinery* because of their significance to the process and the capital investment required and committed. Their loss or unavailability usually will have a large business impact.

Figure 1.1 attempts to list the family of process machinery in one display.

1.2.1.1 Prime Movers

Prime movers applied in the processing industries are also frequently referred to and listed as *mechanical drives* to differentiate them from the usually much larger prime movers used in power generation. As shown in Figure 1.1, we find the following mechanical drive prime movers in the processing industries:

- Electric motors
- Steam turbines
- Gas turbines
- Internal combustion engines
    - Integral engine compressors
    - Coupled gas engines
    - Coupled diesel engine

Power ranges of prime movers for major process machinery are shown in Figure 1.2.

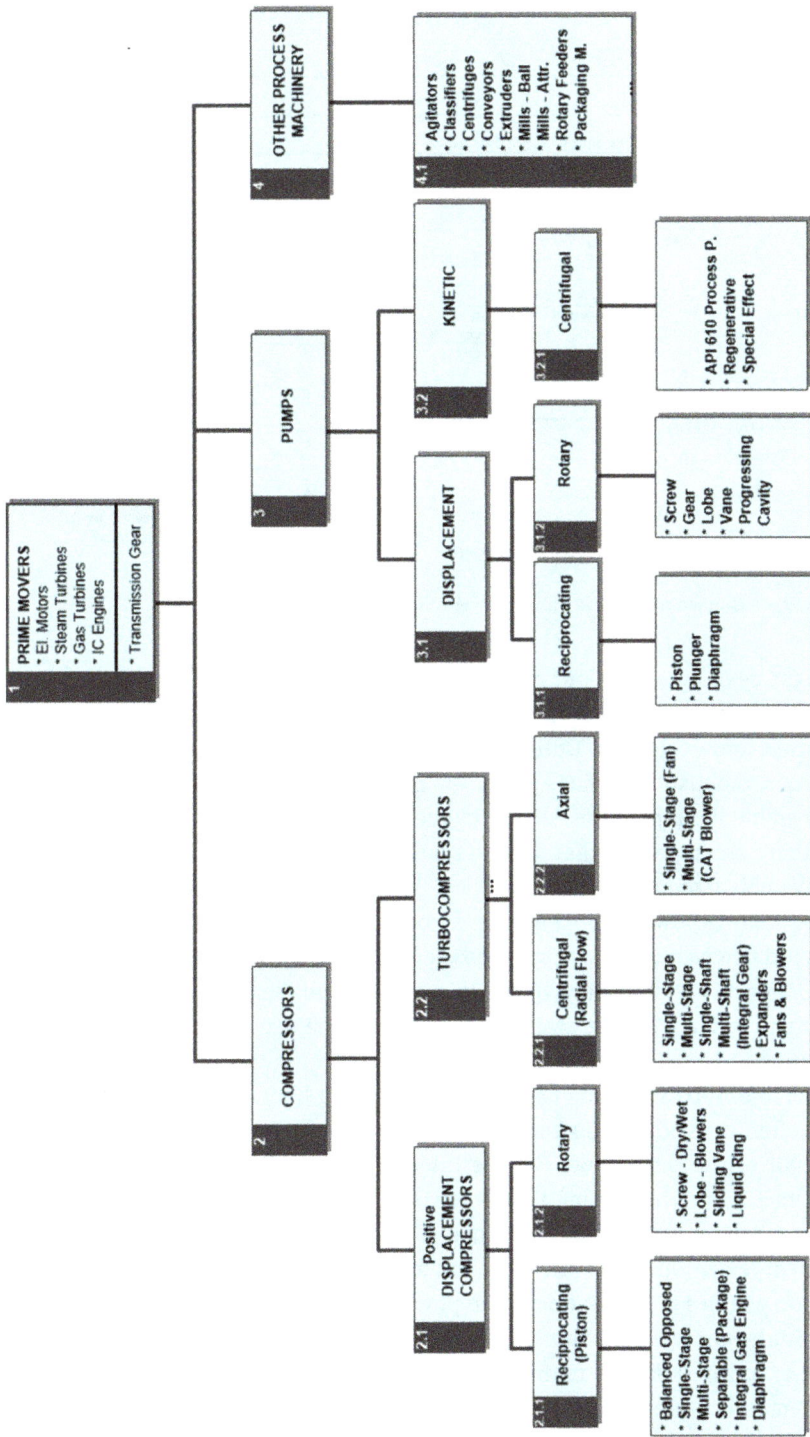

Figure 1.1: Common process machinery.

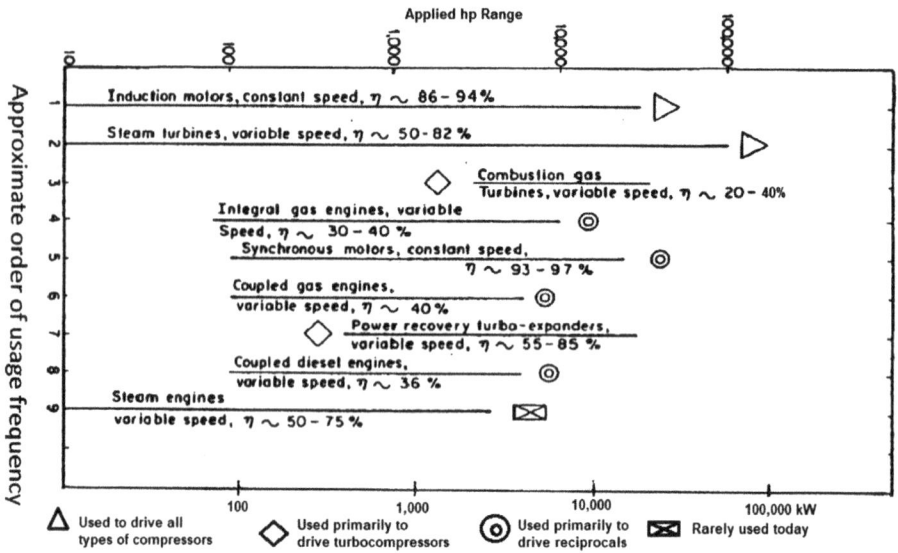

Figure 1.2: Process machinery prime mover power ranges.

It might be of interest to our readers that there are two prime movers that differ from other machinery drivers; this must be taken into account when they are initially started and evaluated. Unlike motors and steam turbines, gas turbines (GT) and internal combustion engines produce a maximum power rate that is dependent upon the ambient temperature defined by their rating lines.

It is therefore important that the site's power requirements are considered across the entire ambient temperature range. The GT startup leader must know that, in most cases, the limiting power will be at high ambient temperatures and hence particular focus should be given to operation in these conditions. The example rating curves in Figure 1.3 are shown for a site with a maximum ambient temperature of 38 °C. It can be seen that, with a GT, significant extra power can be generated at lower ambient temperatures.

Rating lines are defined by mechanical or thermal limits. For the GT, maximum power is often defined by maximum component temperatures in the turbine – the sloping part of the rating line. For the internal combustion engine, the power is often defined by the maximum allowable cooling water temperature. Engines and GTs are rated for various altitudes above sea level, that is, barometric pressures, and ambient temperatures. A rule of thumb for derating naturally aspirated internal combustion engines is 3.5% reduction in power for each 1,000 ft or 300 m above the rating altitude, and 1% reduction for every 10 °F or 5.5 °C above the rating temperature. For exact de-ration of naturally aspirated engines, or for turbocharged engines, the manufacturers must be consulted.

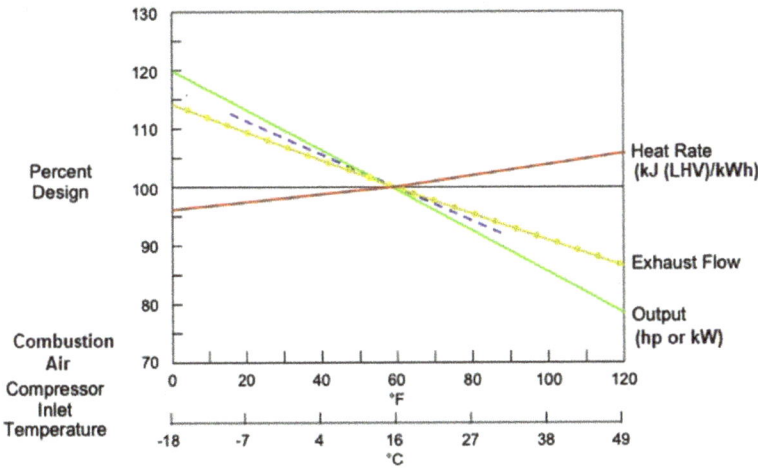

Figure 1.3: Gas turbine rating (Source: Siemens AG).

The foregoing is also a reminder that these machines are intimately connected with the environment taking combustion air in and exhausting it in the form of gas. Both types of prime movers are therefore, highly regulated by environmental protection agencies in almost all countries of the world. Startup personnel must be familiar with state and local environmental regulations regarding their GTs and internal combustion engines.[1]

1.2.1.2 Process Machinery Trains

Once prime movers are connected to the equipment they are driving, they become part of a machinery train where the individual components – prime mover, power transmission components and driven machines – form a virtual chain in that, if one component fails, the whole unit has failed.

Figure 1.4 tabulates compressor prime mover combinations driving major machines in the process industries.

1.2.2 Commissioning Explained

Commissioning is defined as a systematic process of verifying the design, performance of facilities, systems, assemblies and components that are ready to go into service to ensure trouble-free outcomes and mitigate the risk of unplanned delays, outages or downtime. Project managers, contractors, construction engineers, technical specialists and technicians perform pre-commissioning, inspections, tests and safety reviews to confirm that equipment installation and function follow client specifications.

	COMPRESSOR PRIME MOVER COMBINATIONS			
	Steam Turbines	Gas Turbines	Motors	IC Engines
Axial Flow	MANY		SOME WITH speed increaser gearing	FEW IF ANY
Centrifugal	Mostly direct drive - some with high-speed increaser gearing		MANY	FEW Speed increaser gear and special low speed coupling system required
Recip.	SOME Reduction gearing and special low speed coupling system required		MANY Direct drive	MANY Direct drive - separate or integral drive (Midstream of HCPI)
Rotary Screw	SOME Reduction gearing and direct drive - both used	FEW - IF ANY	MANY Speed increaser gearing and direct drive - both used	FEW IF ANY (Upstream of HCPI) MANY (Down & Midstream of HCPI)

Note: Brackets indicate predominant area of application.

Figure 1.4: Prime movers for major compression equipment.

- **Pre-commissioning**. There are regional and industry differences when it comes to a common understanding of pre-commissioning and commissioning. Pre-commissioning is sometimes referred to as cold commissioning, static commissioning or mechanical completion; on the other hand, commissioning is sometimes referred to as hot commissioning, live/dynamic commissioning or startup. Therefore, commissioning teams experience difficulty in determining the proper procedure. As a result, it is imperative to define these terms and their modalities – what, when, where, why, how and who – for every project in order to avoid confusion. Here is an example of what could be the description of what needs to be done during pre-commissioning and commissioning, respectively – after which machinery, systems and process can be started up:
  1. Piping and instrumentation diagram (P&ID) check, also known as system check or walkdown, should be performed by the commissioning team to identify engineering and construction errors as well as presence of specified features. Create a P&ID short or "Punch" list and address identified issues before pre-commissioning.
  2. Verifying pressure testing and line flushing, N_2 purging and removal of blanks.
  3. Pre-commissioning activities start from mechanical completion, where running-in of equipment such as control system sequence tests (dry commissioning),

water or solvent introduction to closed-loop pumps (wet commissioning) and other operating scenarios where process fluids are NOT YET used.

4. The pre-startup safety review (PSSR) is a defining milestone. PSSR is a thorough safety inspection of a new or modified facility to be conducted before any initial startup. This is the point where pre-commissioning and commissioning ends. Correct any system failure, resolve all safety risks and re-take the PSSR to comply with health and safety regulations prior to commissioning.

- **Commissioning:** This is the overall performance testing of the facility and its systems such as process machinery, HVAC systems, piping, static process equipment and lighting. Outstanding issues or "punch (list) points" should be resolved before routine operation. The objectives are:

1. System completion and handover to the owner and operator – see Figure 1.5.
2. Process conditions are established and process fluids – the actual raw materials to be used for routine production – are introduced into the system.
3. Initial operation, or the first production run, aims to determine if the production process produces results that meets output design requirements. Identify non-conformance and allow for adjusting the system or optimize the process prior to routine production.
4. Performance testing involves operating the asset and performing a series of defined tasks. The objective is to measure the performance of the new facility and equipment against the contract, design and nameplate designations. This is a very specialized subject of high variability depending on equipment type with little room for generalization. It is not included in this text.

1.2.3 Asset Management (AM)

ISO 55000[2] provides an overview of AM and AM systems – i.e., management systems for the management of assets. It identifies the "commissioning process" as one of some 25 AM activities. These activities are a solid component of an AM plan as shown in Figure 1.6. They are, for example, data management, condition monitoring, risk management, qualification and assessment of personnel and life cycle costing to name a few.

The target audience of ISO 55000 is persons and entities:
- considering how to improve the realization of value for their organization from their asset base;
- involved in the establishment, implementation, maintenance and improvement of an AM system;
- involved in the planning, design, implementation and review of AM activities; along with service providers.

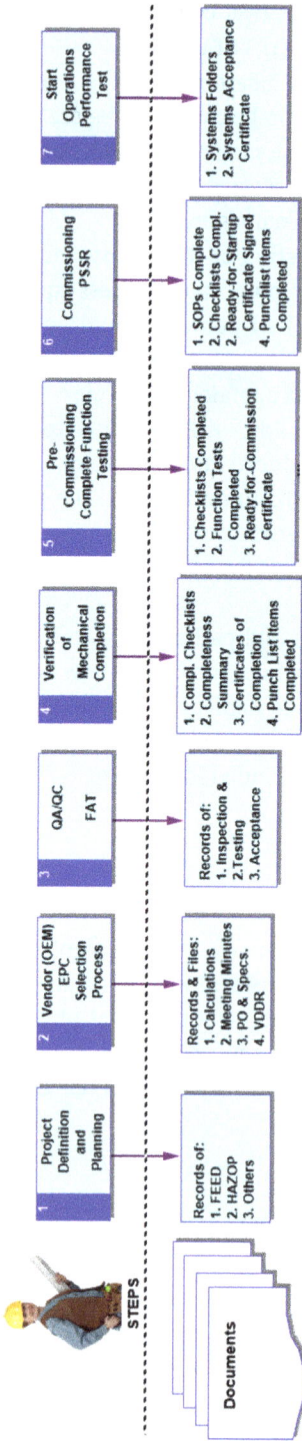

Figure 1.5: System completion execution process (adapted from reference[3]).

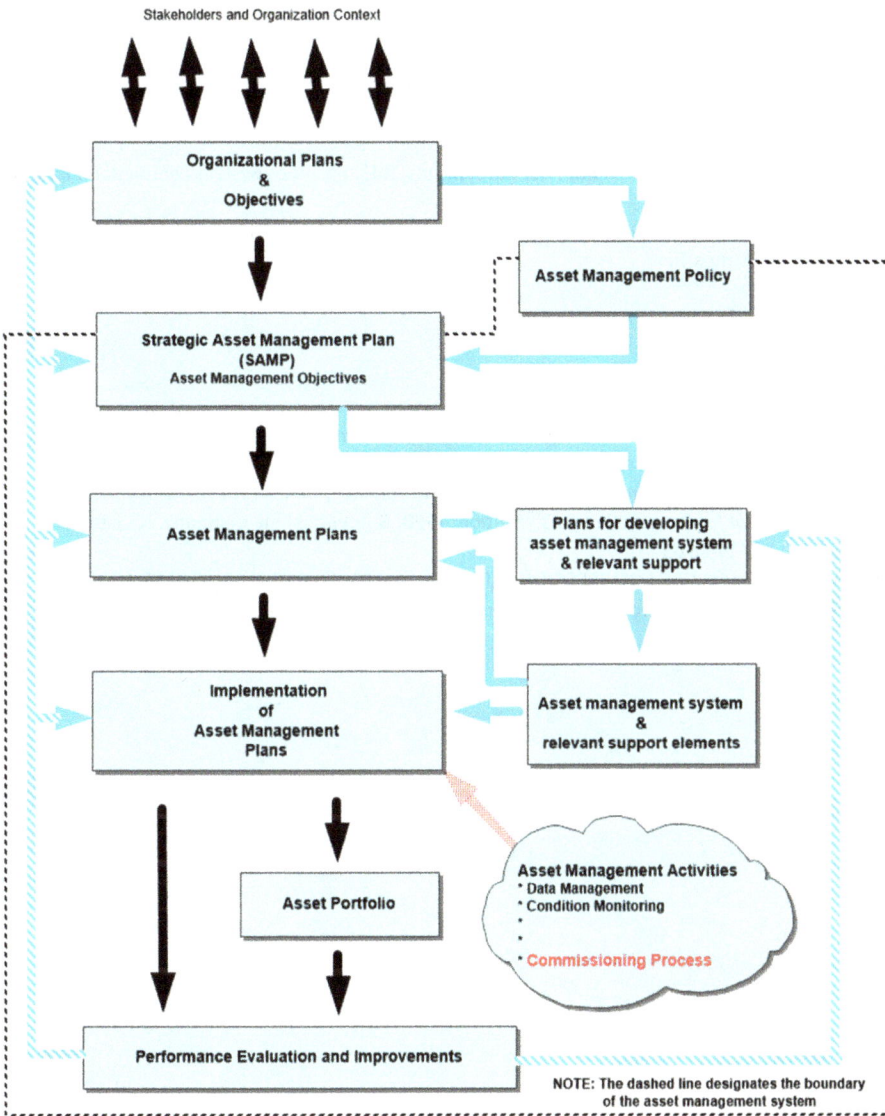

Figure 1.6: How asset management activities fit into the asset management system (adapted from reference[4]).

The following benefits are being claimed:
- The adoption of this International Standard enables an organization to achieve its objectives through the effective and efficient management of its assets.
- The application of an AM system provides assurance that those objectives can be achieved consistently and sustainably over time.

References

1 http://step2compliance.com, a firm offering emission compliance tools and services.
2 INTERNATIONAL STANDARD ISO 55001, First edition 2014-01-15, Reference number ISO 55001:2014(E)
3 API Recommended Practice 1FSC, First Edition, July 2013, Facilities System Completion Planning and Execution.
4 INTERNATIONAL STANDARD ISO 55002, First edition 2014-01-15, Reference number ISO 55002:2014(E), Annex B.

Bibliography

[1] Bloch, H.P. & C. Soares, *Process Plant Machinery*, Butterworth-Heinemann, Woburn, Mass., USA, a member of the Reed Elsevier Group, 2nd Edition, 1998, ISBN: 0-7506-7081-9
[2] Heinz P. Bloch, *Improving Machinery Reliability,* Volume I, Series *Practical Machinery Management for Process Plants*, 3rd Edition, 1998, Gulf Publishing Company, Houston TX, ISBN: 0-88415-661-3
[3] W. Norm Shade, GAS COMPRESSION: A PRIMER ON COMPRESSION EQUIPMENT AND TECHNLOGY, 2021, Third Coast Publishing Group L.L.C., Houston TX, ISBN 978-1-7330413-1-7

Chapter 2
Project Development and Risk

2.1 Project Phases

2.1.1 Project Development

Figure 2.1 describes the stages of a major capital project. The first phase usually consists of preliminary process and economic studies. These studies serve to determine if the project appears viable from both process and economic points of view. Such investigations are usually conducted by the owner's technical and economic staff without contractor involvement. If, after these preliminary studies are completed, the project is still considered viable, the first elements of very detailed process and economic studies get under way. During this stage of project development several objectives must be accomplished:

- Prepare a cost estimate by addressing project strategy and schedule. Questions must be asked such as: Is it advantageous to go out for fixed-price bid or a reimbursable cost arrangement for the detailed design and construction contract? What is the labor situation likely to be? Should the project be split into individual bid packages, or be awarded as one large package? How much "in-house" manning is available?
- Years ago, equipment and overall plant availability consistent with life cycle cost considerations were discussed as leading factors. Today, forces driving a project are primarily safety, ROI, CAPEX, OPEX and the concept of "fit for purpose." To satisfy the latter, the project financing plan frequently provides for a general fund to help the future operating unit to deal with problems that were identified during project execution and not solved.
- Schedule: overall timing must be looked at in detail because financial requirements and project returns have a strong bearing on project schedule and economics.
  An important stage in the project development and definition phase centers around the preparation of the process design specification. This stage may involve the following activities:
  - Conduct enough design work to produce a more realistic cost estimate. Cost estimates that reflect only the lowest possible pricing are worthless.
  - Start interfacing with outside design firms, if the project scope is beyond in-house capabilities.

Enter into contract negotiations with the outside firm or firms selected to do the principal design work and/or the principal construction work. These firms are often engineering, procurement and construction (EPC) companies specialized in certain processes such as oil refining, that is, the upstream business of the oil and gas industries, or their midstream and/or downstream sectors. If necessary, engage an engineering company

https://doi.org/10.1515/9783110701074-002

that provides "owner's engineer services" to augment the existing skill set of the asset owner.

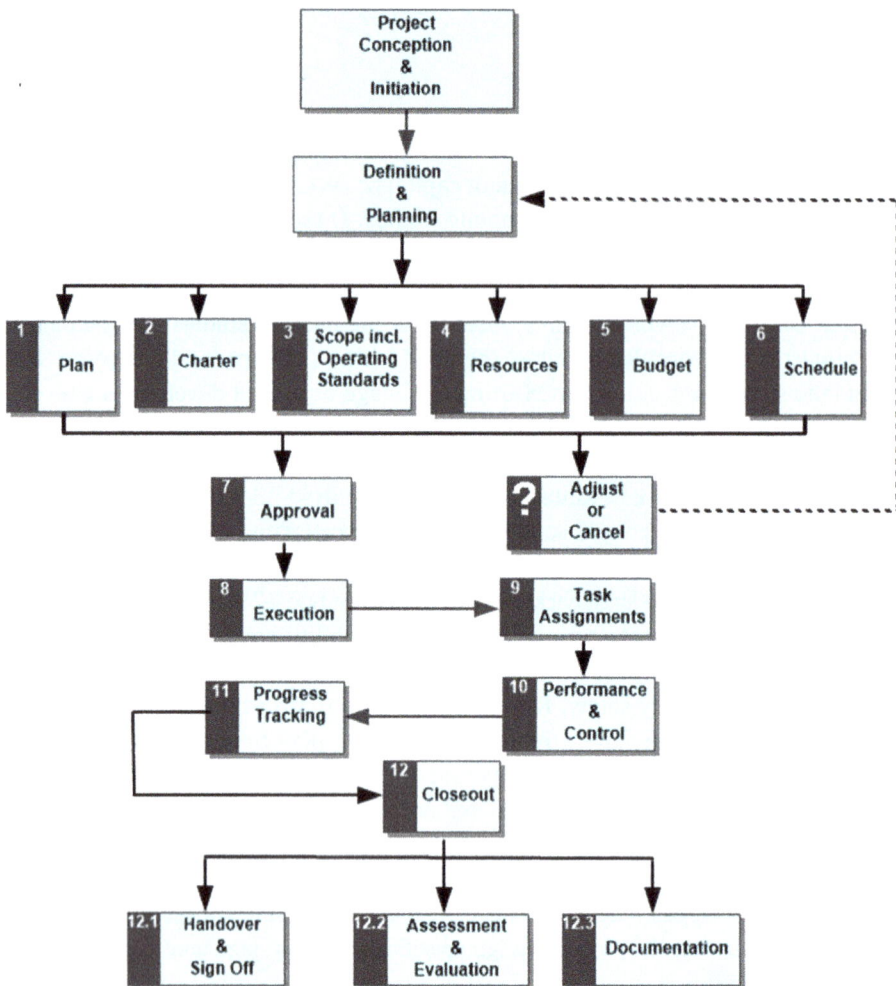

Figure 2.1: Typical capital project phases.

The objective at this point in time is to arrive at a front-end engineering and design (FEED) package. While typically accounting for between 0.8% and 2% of the total capital cost, FEED is where owners have the largest opportunity to impact capital cost reduction.[1] The design, fabrication and procurement decisions made in FEED shape a project in either one that will not go forward due to high costs, or into one that will pass hurdle rates and deliver long-term value.

2.1.2 Approaches to Project Execution

The two main approaches for project execution at this stage are:
1. Enter into a contract with an EPC to do only FEED work. Once FEED is completed, the client company uses the FEED documents and solicits bids to do the remaining design procurement and construction part of the project – hopefully for better pricing and for projecting a reliability focus.
    a) *Advantages:*
    -   The overall CAPEX of the completed project can result in less cost, if the owner is using judiciously this strategy to bid on each portion of the project.
    -   This works better if the EPC doing FEED is the licensor of the technology being implemented in the project. Schedules can be improved since this type of EPC can simply recycle project documents from past projects in a timely manner.
    b) *Disadvantages:* The contractor that is successful to do the design procurement and construction part of the project has to enter into an NDA with the FEED contractor. This can to be problematic – at times – if they are technology competing licensors.

2. Client company enters into a full-blown design engineering procurement construction (DEPC) contract. By doing a DEPC contract, the EPC will do FEED and transition seamlessly into the design procurement and construction parts of the project:
    a) *Advantages:*
    -   Seamless transition from FEED to other phases of project execution
    -   Long lead equipment can be determined early in the project
    -   Continuity of project personnel
    b) *Disadvantages*
    -   Usually costlier than option 1
    -   Non-disclosure agreements (NDA) with technology licensors can prove challenging if the DEPC contractor is asked to implement the technology of someone else.

The task now becomes:
-   Prepare mechanical flowsheets
-   Determine the major, critical long-delivery equipment
-   Obtain quotations on long-delivery equipment
-   Place orders for major long-delivery equipment, subject to cancellation, as the project is still in its design phase. Normally obtaining of quotations and placing of orders is done by the contractor. However, there have been many occasions, where critical plant equipment, such as non-spared machinery, has been pre-ordered by the owner organization and then handed over to a contractor for installation.

- Develop final realistic time and cost estimates.
- Refine the project economics based on the cost estimate.

It is during these last stages of the project definition phase where the foundation of the interfaces between owners and contractors during project execution is laid.

2.1.3 Charter – Operating Standards Are Needed to Define Asset Availability Goals

The purpose of all interfacing between owner and contractor should be the attainment of asset service factor (availability) goals. These goals must be set by appropriate reliability and maintainability input during the project development and definition phase. A prerequisite for this input is the concept of operating standards as part of the project charter.

Operating standards are a set of rules based on an operating philosophy incorporated into the project during its planning and definition phase. As the basis for all agreements between the parties involved it usually begins with a vision statement such as: "We want to be quality leaders," "We want to be the best" and "We want to be an excellent manufacturing operation."

All these catch words could well be meaningless, if they are not followed by detailed guidelines reflecting reliability and maintainability goals. An enlightened senior management would, for example, issue the following introductory statements:

A. "Our prime objectives: We want to build and operate a plant to achieve excellence in everything we do. This means that we shall start up efficiently and on time. The facility will then be operated reliably and predictably. It will perform following strict, pre-established standards and procedures. Our plant will seldom experience upsets or shutdowns from avoidable causes.

B. Our equipment reliability and safety goals: We shall require a high service factor. All equipment will be selected to enable our plant to run for at least three years between shutdowns. We will make extensive use of predictive maintenance tools such as machinery monitoring – IoT based if possible – corrosion and leak detection, state-of-the-art instrumentation to measure force, temperatures and electrical values. Refer to our "Reliability and Safety Guidelines."[2]

C. Our work environment: We want to minimize all routine and low skill work when operating our plant. Our goal is to have our designers consider this in the detailed design of the plant. Here is our list of requirements, "Maintainability and Serviceability Guidelines."[3]

D. Each of these points will have, of course, an economic limit, and in most cases, there will have to be good judgment applied. We shall appoint sponsors and change agents[4] to help in the decision processes.

E. Our turnover and startup procedures: The facility will be completed and turned over for commissioning during a 4- to 6-month time frame. The turnover will be sequential. There will be problems of lack of skilled manpower resources and of interference with ongoing construction work. These problems can be minimized by planning and designing during the detailed engineering phase. We shall define all "systems" as early as possible. Our goal is to engineer by systems in order to make construction and commissioning easier."

While working on operating philosophies it would be well to consider the concept of intermeshing project components expressed in Figure 2.2.

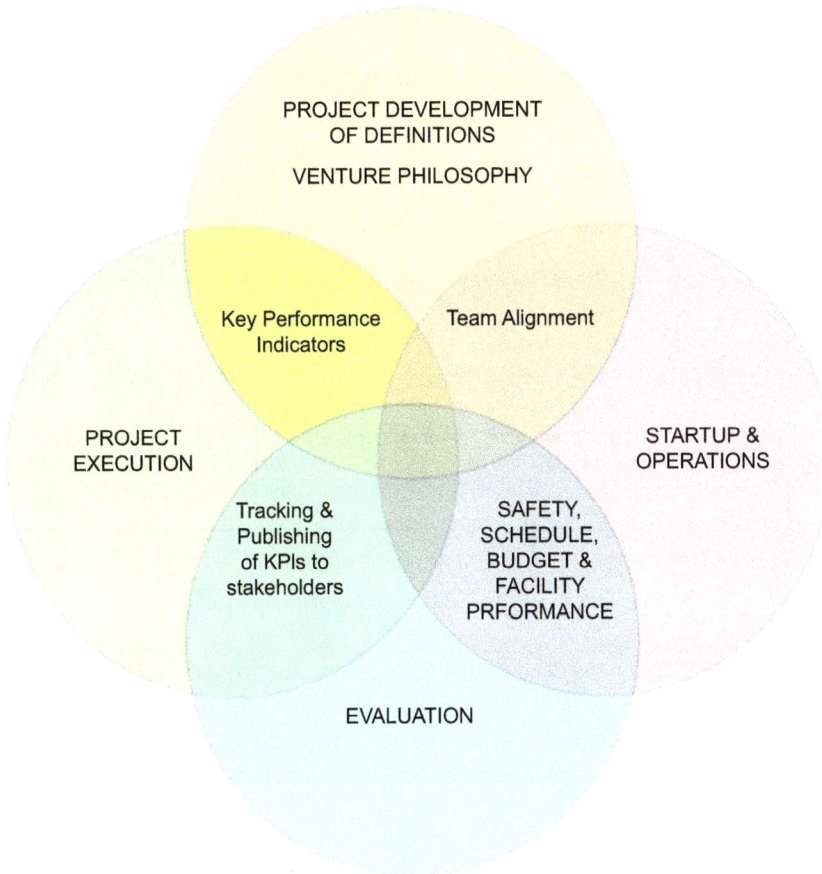

PROJECT DEVELOPMENT
OF DEFINITIONS

VENTURE PHILOSOPHY

Key Performance
Indicators

Team Alignment

PROJECT
EXECUTION

STARTUP &
OPERATIONS

Tracking &
Publishing
of KPIs to
stakeholders

SAFETY,
SCHEDULE,
BUDGET &
FACILITY
PRFORMANCE

EVALUATION

Figure 2.2: Project factors.

It can be costly not to begin project activities with well-thought-out operating philosophies. We must remember, "For every one dollar it costs to fix a problem at the conceptual stage it will cost:

10 $ at the flowsheet stage
100 $ at the detailed design stage
1,000 $ after the plant is built and over
10,000 $ to clean up the mess after an accident."[5]

As a corollary, the opportunities for reliability and maintainability input will diminish with the advancement of the project as shown in Figure 2.3.

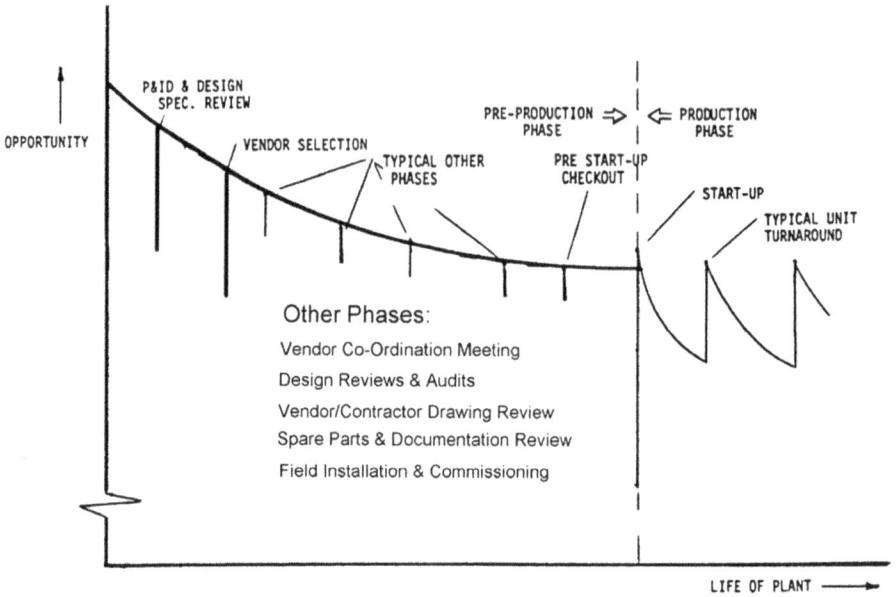

Figure 2.3: Loss of improvement opportunity as a function of project advancement.

Once operating standards have been issued and communicated, they need to be interpreted and constantly reinforced throughout the duration of the project. This is done by daily interfacing with contract personnel, such as designers, project engineers, construction planners, specialists, construction superintendents, field engineers, vendors and subcontractors.

2.1.4 Project Execution

The following relate to several successful, "few surprises" projects where the authors were members of the team. Upon appropriation, which concluded the project development and definition phase, the project went into its implementation phase. The design was now completed in detail. During this phase a detailed engineering model of the facility was as usual constructed. On a three-eighth-inch model all equipment such as compressors, exchangers, furnaces, towers, vessels, valves and other important components were shown to scale. The scale model was an essential tool to assure plant maintainability and avoid future maintenance load. This physical model has now become a thing of the past. Virtual 3D models are now being developed by Intergraph™ software on screen throughout each phase of the project for reviews. One can view these on a large screen during team reviews or on one's desktop computer individually. There are 30%, 60%, sometimes 90%, team reviews before the final 100% review. With these 3D model reviews on screen one can walk through the plant and operators can check ergonomics – based on the average height of the site-specific local population.

During the project execution phase, the necessary steps towards design completion were carried out: Plot plans, foundations, sewer systems, pipe spools and other details were finalized on paper. After site preparation, plant construction was begun. Frequently 35 to 50 percent of the mechanical design work was completed at this time. A typical simplified project time line is illustrated in Figure 2.4.

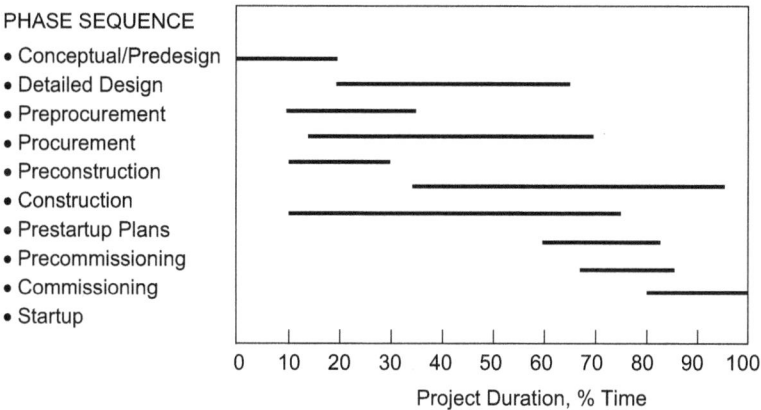

PHASE SEQUENCE
- Conceptual/Predesign
- Detailed Design
- Preprocurement
- Procurement
- Preconstruction
- Construction
- Prestartup Plans
- Precommissioning
- Commissioning
- Startup

Project Duration, % Time

Figure 2.4: Project execution components.

The major steps of equipment commissioning and startup (C&S) are illustrated in Figure 2.5.

The final stage of the project execution phase was mechanical completion. Mechanical completion usually terminates the contractor's involvement and leads to startup and operation of the new facility by the owner's operating team.

2.1.4.1 Evaluation

In all human endeavor, we just wanted to know how we did. "Feedback is the breakfast of champions," someone once said. A contracting firm worth its name would be clearly

1	Preparation & Planning
2	Mechanical Completion & Integrity Checking
3	Pre-Commissioning
4	Commissioning & Initial Operational Testing
5	Start-Up & Initial Operation
6	Performance & Acceptance Testing
7	Post-Commissioning

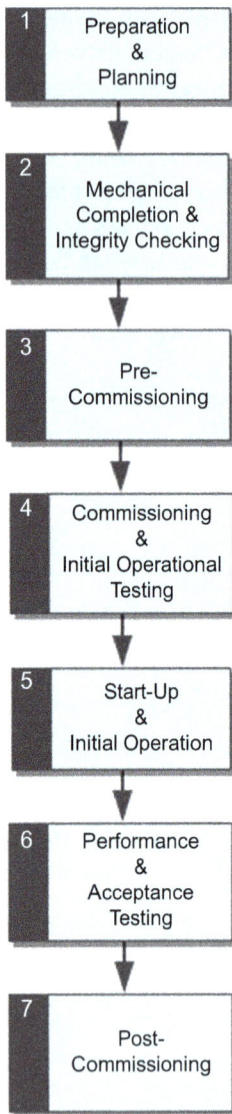

Figure 2.5: Process machinery commissioning sequence.

interested in this topic. The company would want to know how it fared. Table 2.1, a contractor questionnaire, will give us an idea of how such feedback could be accomplished. It also reflects the potential problems that can be encountered in a major project. The effort to look back and to ask the question "How can we do better next time?" is a good beginning for continuous improvement.

Table 2.1: Contractor rating questionnaire.

to no extent		to some extent		to a great extent	
1	2	3	4	5	*(please circle number)*

1. The engineering contractor field team (ECFT) made every effort to clarify its roles and interfaces with our organization.	1 2 3 4 5
2. It was clear to me how ECFT work fitted with my individual work.	1 2 3 4 5
3. ECFT members were consistently sensitive to the needs of others they worked with on a day-to-day basis.	1 2 3 4 5
4. It often seemed as though ECFT was an integrated part of the owner's project administration team (PAT).	1 2 3 4 5
5. I believe that integration between ECFT and owner's PAT was a desirable aim.	1 2 3 4 5
6. ECFT were rarely over-defensive on behalf of the contractor's design work.	1 2 3 4 5
7. The major decisions in the field were made by the owner's PAT.	1 2 3 4 5
8. The ECFT worked well together as a team.	1 2 3 4 5
9. The leadership of ECFT encouraged initiative and personal empowerment.	1 2 3 4 5
10. I felt that the response time to queries and questions was good.	1 2 3 4 5
11. I felt that any member of the ECFT would be able to handle any query I had.	1 2 3 4 5
12. I felt high personal trust in ECFT members' ability to deliver to my requirements.	1 2 3 4 5
13. ECFT encouraged an informal personal approach, which I appreciated.	1 2 3 4 5
14. I would be happy to work with such an ECFT again on a future project.	1 2 3 4 5
15. Overall ECFT completed their assignment to the satisfaction of my organization.	1 2 3 4 5

Any other additional comments or feedback:

2.2 Process Plant Machinery Project Phases

Process equipment is usually moved through distinct steps as shown in Figure 2.6 within the framework of Figure 2.1:

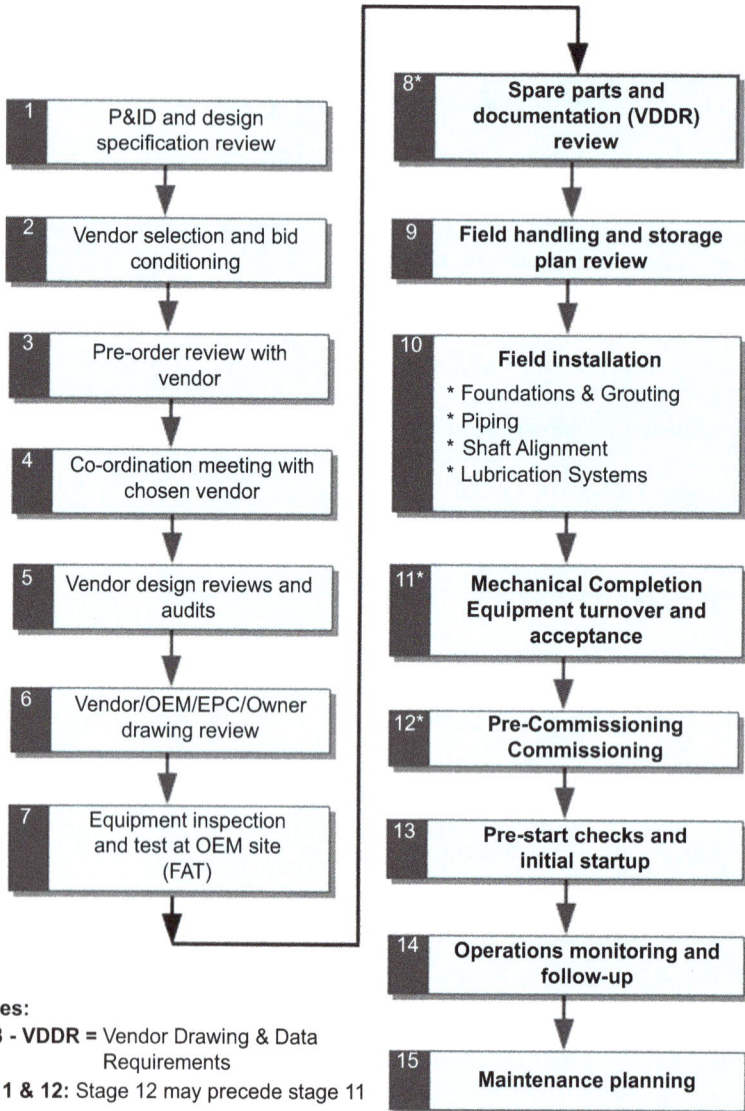

1 P&ID and design specification review	**8*** Spare parts and documentation (VDDR) review
2 Vendor selection and bid conditioning	**9** Field handling and storage plan review
3 Pre-order review with vendor	**10** Field installation * Foundations & Grouting * Piping * Shaft Alignment * Lubrication Systems
4 Co-ordination meeting with chosen vendor	**11*** Mechanical Completion Equipment turnover and acceptance
5 Vendor design reviews and audits	**12*** Pre-Commissioning Commissioning
6 Vendor/OEM/EPC/Owner drawing review	**13** Pre-start checks and initial startup
7 Equipment inspection and test at OEM site (FAT)	**14** Operations monitoring and follow-up
	15 Maintenance planning

*Notes:
1. #8 - VDDR = Vendor Drawing & Data Requirements
2. #11 & 12: Stage 12 may precede stage 11

Figure 2.6: Process machinery project phases.

2.2.1 Commissioning and Startup

Steps 7–15 of Figure 2.6 are the phases connected and leading to commissioning activities involving three to six months of calendar time to complete and typically representing an investment of several million dollars. If the commissioning or startup is delayed, additional costs will be incurred every day. The cumulative total may rapidly become prohibitive.

2.2.2 Reducing the Delay Potential

In order to reduce the delay potential which can statistically be attributed to the critical machinery required in each project, purchasers or owners have intensified scrutiny of critically important non-spared machinery during the procurement phase. This includes expanded "Machinery Quality Assessment" or MQA[6] of machine trains before purchase commitment. This activity would commence during Phase 2 and continue through Phase 7 listed in Figure 2.6. MQA is the central element of future reliability. It is what made BiCs (best-in-class) performers into BiCs. MQA continues throughout the phases listed in Figure 2.7 and beyond. It entails not only vendor selection but also OEM design reviews and audits by the purchaser or his/her agents/ representatives. This effort can definitely reduce the delay attributable to machinery problems. However, regardless of the care which has gone into the machine design, very expensive delays can still occur during the startup phase if proper pre-startup checks and startup procedures are not followed.

2.2.3 Determining Vendor Responsibility

As part of the vendor selection, we should consider single-point responsibility for major machines. Equipment purchasers or owners are always interested in meeting the cost and schedule targets for projects. There could, however, be conflicting goals when purchasing drivers and/or auxiliaries from vendors other than the primary manufacturer or OEM. While cost targets might initially be met, missed schedules and future warranty disputes present risks that must be considered. Using compressors as an example, it is worth noting how experienced owner-operators avoid these types of conflicts.

There must be an agreement on 'train responsibility'. Whenever BiC companies separate the procurement of a major machine unit such as a turbocompressor from the procurement of its driver, one of the two suppliers is given train responsibility. The owner-operator or purchaser/end-user pays in this case the responsible vendor – usually the major pump or compressor manufacturer – a modest sum for accepting single-

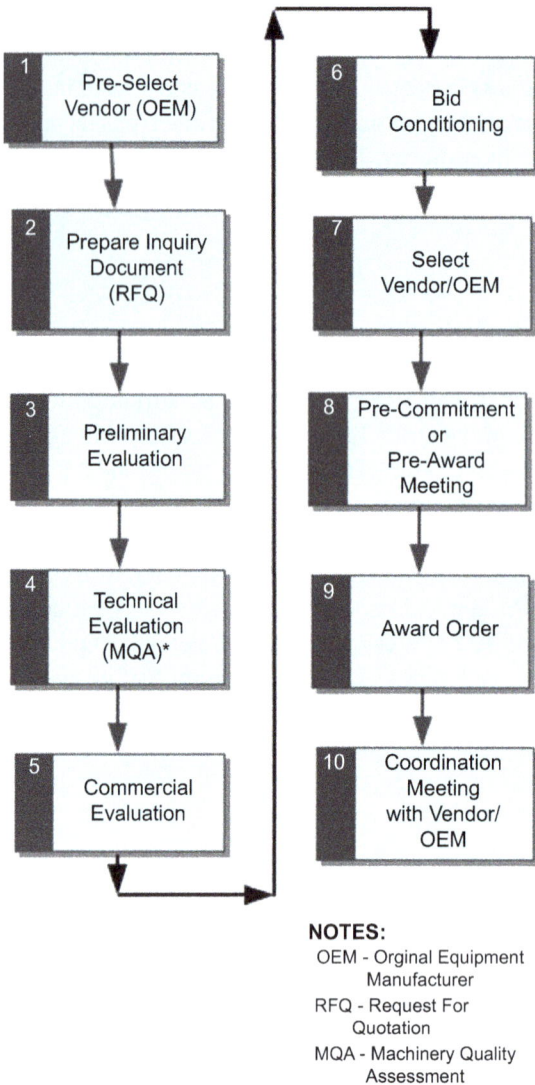

Figure 2.7: Equipment vendor selection steps – process industries.

point train responsibility. Although this practice dates back quite a few years, some purchasers choose not to follow it.

It should be noted that the monetary savings of buying the driver from a detached bidder often vanish when a single additional startup-delay day results from dividing responsibilities.

2.2.4 Exceptions to the Rule

Of course, there are always exceptions to the rule in that giving a vendor single-point responsibility does not mean that all risk has been removed. As a case in point, we recall some reciprocating compressors purchased from a manufacturer that had recently acquired another company's product line. Unfortunately, the purchaser/end-user operation quickly encountered glitches in spare parts identification and timely availability of spares.

Regrettably, many interesting failure experiences and relevant failure-avoidance steps are frequently hidden from our view. Litigants often sign protective non-disclosure agreements because neither side wants their reputation blemished. All involved are reluctant to publish detailed articles or papers about these events, and lessons learned turn into lessons forgotten.

2.2.5 Reducing Risk by Using a Package Concept

Much less risk exists when a purchaser opts for entire machinery trains from top-of-the-line manufacturers. Such manufacturers would have a proven record of long-term quality and as-promised performance. There are modern process machinery packages where train responsibility would lie with the packagers. In the upstream business of the petrochemical industry, for example, these machinery trains are built in accordance with API standards. A typical case is the integrally geared centrifugal compressor, conceived to meet the demands of the petroleum, chemical, power and gas industries. It has been said that they are typical for a single-source machine unit. Drivers, sealing and lubrication systems may, perhaps, be provided by different vendors, but the responsibility stays with the compressor manufacturer, and the package is "single-sourced."

However, when we look at machinery packaging in the midstream markets of the hydrocarbon processing business, we must apply caution, as lesser standards govern. Here everyone is saying they want high quality, but no one really wants to pay for a high-quality package.[7]

Priorities for the review and inspection effort must be set and kept in mind; it will become evident that not all equipment items can be addressed to the same extent. Priorities are intuitively determined by categorical considerations expressed in Figure 2.8. Other simple approaches to determine attention priorities are shown in Figure 2.9 and Table 2.2.

A more thorough approach is taken by performing an equipment criticality assessment. Such an assessment is done by a team of engineers over typically a four-day period similar to a HAZOP study[8] prior to committing to equipment acquisition. The team consists of machinery engineers (SMEs), process design engineers, instrumentation engineers, reliability engineers and operations personnel. A matrix as shown in Figure 2.10

Cat. 1   • General Purpose, spared
         • Non-critical
         • Example: process pumps

Cat. 2   • Critical spared or unspared
         • High IOM&R costs
         • Example: utility pumps, air compressors, blowers

Cat. 3   • Unspared turbomachinery, reciprocating comp.
         • Forced outage has significant business impact
         • Example: process refrigeration compressor

Figure 2.8: Process machinery categories explained.

Table 2.2: Machinery commissioning resource commitment.

1.0 HOW CRITICAL IS THE MACHINE?		_____
1.1	PROCESS CAN OPERATE WITHOUT IT FOR SHORT PERIODS WITHOUT CAPACITY LOSS?	_____
1.2	PROCESS CAN OPERATE WITHOUT IT FOR SHORT PERIODS WITH CAPACITY LOSS?	_____
1.3	PROCESS CANNOT OPERATE WITHOUT IT	_____
2.0 WHAT DOES A MACHINE SHUTDOWN COST?		_____
2.1	ONE DAY	_____
2.2	THREE DAYS	_____
2.3	SEVEN DAYS	_____
3.0 HOW MUCH EFFORT (RESOURCES) SHOULD BE PUT INTO COMMSSIONING AND STARTUP?		_____

is used to perform a risk assessment by weighing severity and consequence and probability of incident occurrence. Risk levels – extreme, high, medium, low – are arrived at and actions taken. Generally, there are three responses to any risk:
- Avoidance – eliminate the possibility of occurrence
- Mitigation – reduce the impact or consequences
- Acceptance – live with the consequences of occurrence

Ten percent of the equipment that falls within the highest consequences are considered to be critically important. However, a final pass at priority setting must be made by continually evaluating risk throughout the project progress. This approach is shown in the following chapters.

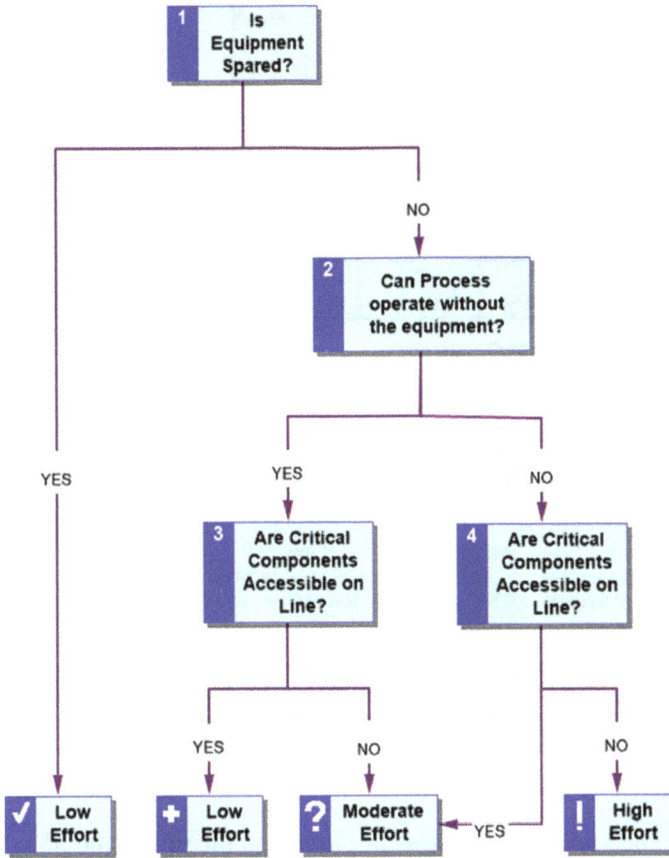

Figure 2.9: Determining process machinery review and inspection efforts.

2.3 Risk – The Function of Probability and Loss Severity

2.3.1 Probabilities

Looking at the overall picture of what we call the project, pre-commissioning, C&S represent important phases in the lifecycle of a plant – see again Figure 2.3. This is universal for all projects – regardless of the nature of any underlying organizational or contractual modalities. While the opportunity for plant reliability enhancements has diminished when commissioning commences, the curve must be kept flat in

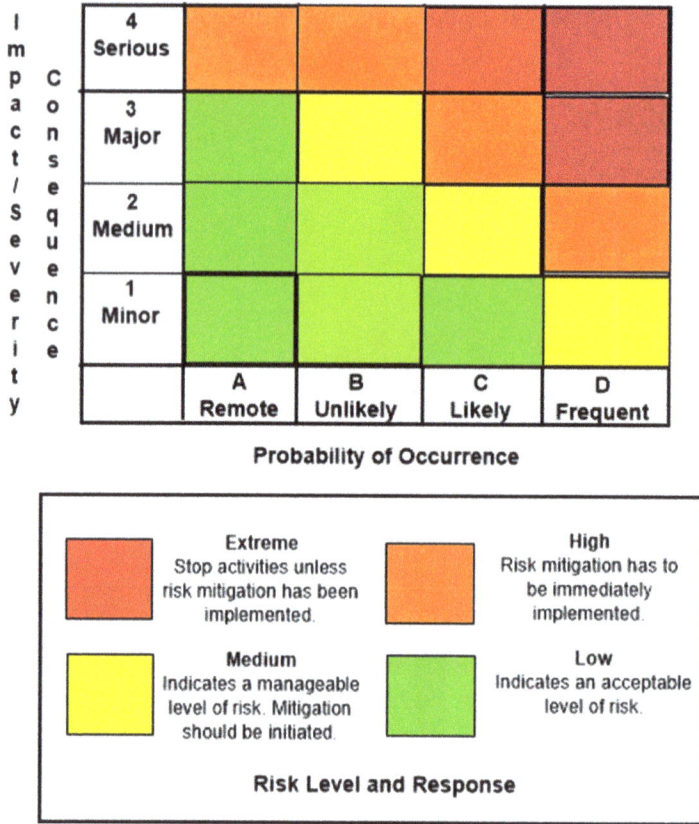

Figure 2.10: Risk matrix.

order to not spoil the good record. A flawless transition into the run-in and initial operating phase has to be achieved.

What then are the probabilities of being unsuccessful? We believe that the risk surrounding pre-commissioning and subsequent C&S activities is defined by insidious early failure probabilities of equipment. This statistical phenomenon has been referred to in the technical literature as "infant mortality."[9] Figure 2.11 illustrates its characteristics. It simply describes the fact that equipment tends to have an elevated initial failure rate attributable to seven types of potential deficiencies cited in the top part of Figure 2.12 below.

A fair amount of intuitive or heuristic knowledge attests to this fact in that we undertake equipment trials prior to equipment release for operation. Examples are factory acceptance tests (FATs), trial runs of ships after they have been launched, tradesmen or technicians attending to the initial startup and run-in of a process pump after having replaced the mechanical seal. We conclude, that infant mortality is inherent in C&S activities and must therefore, be mitigated if not eliminated by appropriate measures.

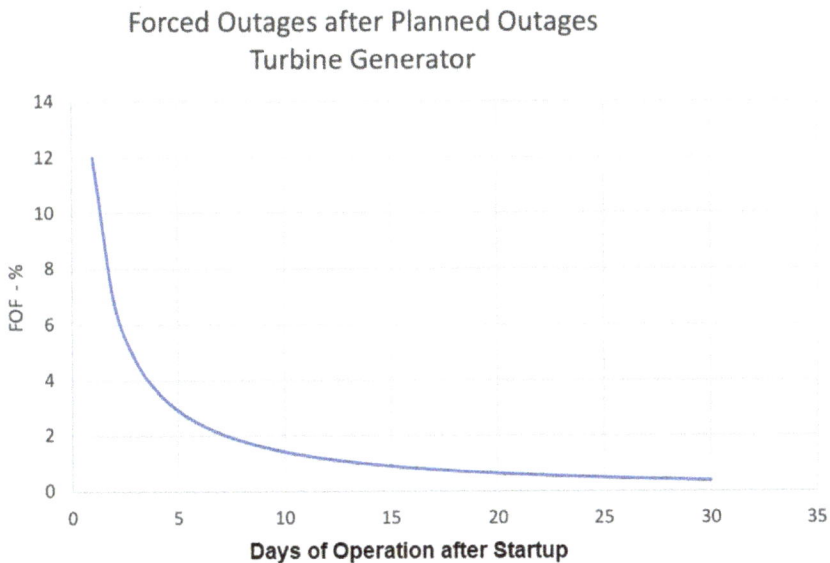

Figure 2.11: Example of equipment infant mortality effect (adapted from references[10,11]).

2.3.2 Risk

Equipment C&S is a human endeavor subject to errors and omissions. Many things can go wrong. No matter how good the plan was, there is always another thing to do[1] before we push the button. People involved in C&S can get careless and ignore their training.

C&S efforts are uniquely tied to one particular project or situation and have no continuum. Once we are started up, we will monitor, care for and feed the equipment for a while, but then walk away to perhaps plan and prepare the next equally unique C&S situation.

We conclude that we are faced with an enterprise that is encumbered with risks of failure. Fundamentally, these risks derive from causes identified in Figure 2.12.

Risk in our context is a function of probability and the consequences of an unsuccessful startup. Whereas the impact of consequences or severity is often not difficult to determine because they are determined by the owners' business environment, probabilities – of a startup gone awry – are subjective and difficult to determine. It is because of this uncertainty that asset managers continue to intuitively assign engineers, technicians, mechanics and other staff to the C&S effort of their process equipment.

---

1 Major Warden in "*The Bridge on the River Kwai*": ". . . following the theory of there is always another thing to do . . ."

**Seven Basic Risk Causes for
Process Machinery**
(Bloch/Geitner)

1. Maintenance Deficiencies
2. Assembly/Installation Defects
3. Off-Design Service
4. Improper Operation
5. Fabrication or Processing Errors
6. Faulty Design / Not fit for Purpose
7. Material Defects

General Risk Causes

FC **Work Process, Asset:** Original
Planing and Design unfit for purpose,
poorly programmed; difficult to
operate and maintain

DO **Instructions, Documentation:**
Incorrect, missing, incomplete,
unintelligible, impractical, difficult to
retrieve

MT **Maintenance:** Assets in poor shape,
damaged, poorly monitored, cared
for, improperly repaired; Maintenance
Department not well organized

CO **Communication:** Missing; not
executed, transmitted, received;
unintelligible, incorrect, ineffective

OR **Organization:** Goals and Objectives
poorly developed: Values, Tasks,
Responsibilities, Strategies, Priorities
not defined

QP **Personnel Qualifications:**
Deficiencies in Knowledge,
Experience, Instructions,Training;
Individuals not suited for the job

General Risk Causes

HS **Health and Safety:** Poor Protection,
Warning, Alarms, Rescue Service;
First Responders and Responsible
CARE* not established

HK **Housekeeping:** Workplace filthy,
cluttered, impractical; Floors and
Machinery Foundations dirty and
oily

EN **Danger Promoting Environment:**
Heat/Cold, Noise, Light, Dust,
Vapors; Unmotivated and
endangered Personnel

PA **Paradigms - Conflicting Goals
and Objectives:** Contradictory
Orders, Operating and Work
Conditions; Time Pressures,
Financial Resource and Manpower
Constraints

AV **Availability, Condition, Quality of
Facilities and Tools:** Poor, not
suitable for current Demands,
incomplete

*Note:Safety and Health Program
sponsored by the Chemistry Industry
Association of Canada (CIAC)

Trouble/Failure Incident

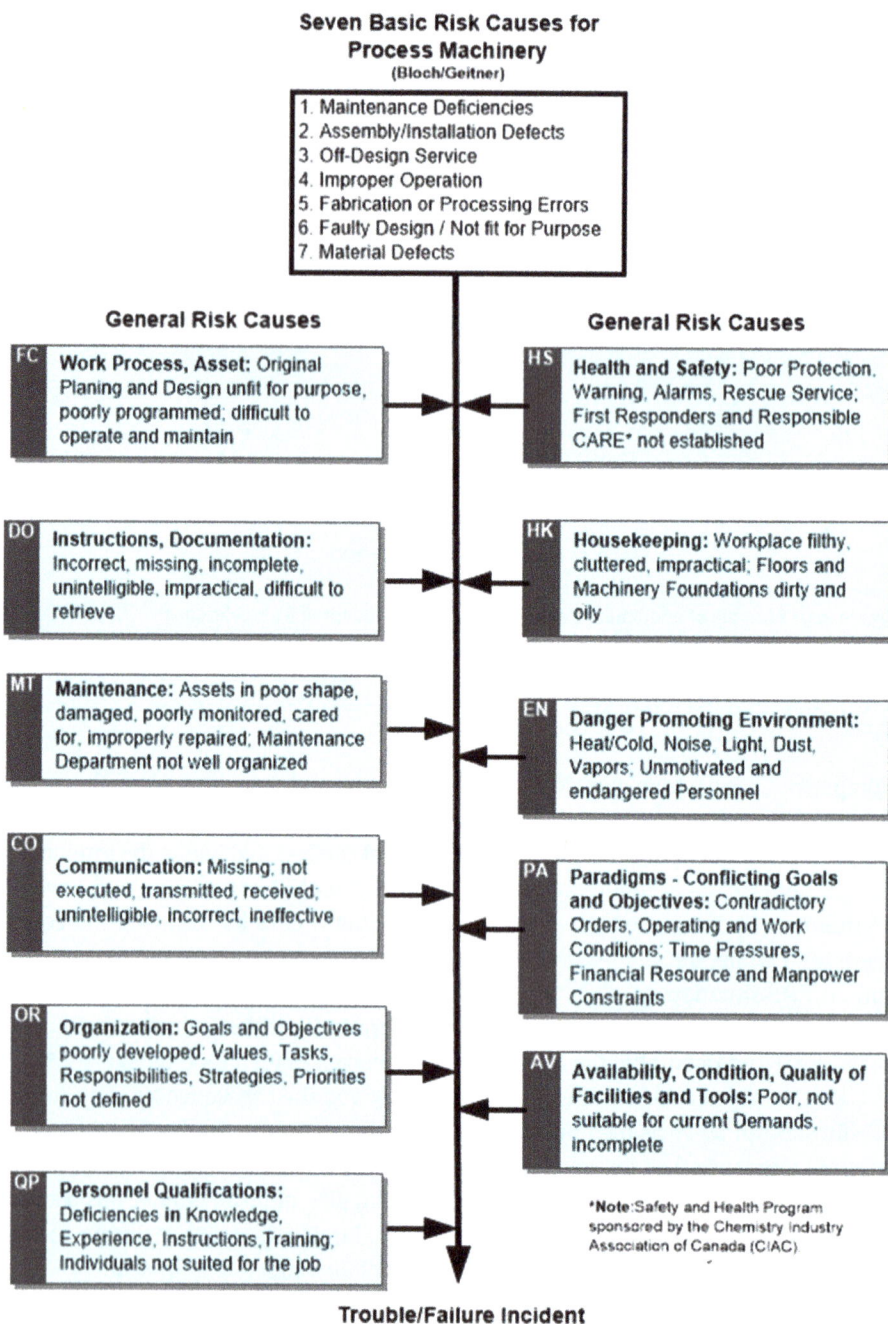

Figure 2.12: Risk causes.

2.3.2.1 Risk Determinants
We see risk connected to three distinct circumstances. They are:
- Arrangements the owner enters into with the contractor, frequently an EPC company.
- Equipment complexity translated into intrinsic reliability and probability of failure.
- Project culture.

2.3.2.1.1 Contract Modalities
Before ground is broken on a construction project, the owner must decide what kind of construction contract will be signed. The most common options available to an owner are the lump sum contract and the cost-plus-fee contract. Both of these contracts have advantages and disadvantages from an owner's perspective. Generally, the owner must keep his/her financial interest in mind, while at the same time, allowing the contractor to make a reasonable profit and build a quality project. This is where MQA[12] helps.

2.3.2.1.1.1 The Lump Sum Contract
For an owner who has very tight budget constraints or lacks experience in the construction industry, the lump sum contract is often ideal. The lump sum contract is the most basic form of an agreement between an owner and contractor and is fairly easy to manage. It is often known as "turnkey contract" because the EPC takes a project from initial inception to completion. For this kind of contract to be effective, the owner must have sufficiently detailed and complete drawings and specifications. The owner must have a contract that penalizes the contractor for buying equipment that does not conform with the mutually agreed purchase specifications, for example, equipment that requires excessive maintenance, does not allow for easy access and has other features that generate excessive operating costs. The construction documents must be well defined at the time of the bid to allow the bidders to properly estimate the cost of labor and materials.

Under a lump sum agreement, the contractor is responsible for completing the project within the agreed-upon fixed cost defined in the contract. If the contractor completes the project under the fixed total cost, then the contractor keeps the difference and makes a profit from the work. The owner is not entitled to any savings if the project is completed below the fixed total cost. A lump sum contract is generally a closed-book arrangement, so the contractor does not have to report the cost of labor and materials to the owner.

Advantages. There are several advantages for an owner to enter into a lump sum contract. It is a huge benefit to an owner that the contract is easy to manage. Payments are progress payments. Made to the contractor, they are based on the percentage of completed work. Generally, the payment schedule is created by the contractor and reviewed by the owner's construction manager. Payment to the contractor is done after all parties agree to the completed work, endorsed by the owner's construction manager and SMEs during engineering and construction.

Disadvantages. There are disadvantages an owner must consider under a lump sum agreement. Perhaps the biggest concerns are cost overruns and lack of flexibility. In the event that the contractor exceeds the fixed total cost of the project, the theory goes that the contractor is responsible for any cost within the scope of the work that exceeds the agreed-upon total. In practice, however, the contractor may stop working on the project and blame the owner and others for cost overruns. It is best to write a contract that addresses how cost overruns will be dealt with.

If the owner wants to make changes to the project during the course of construction, he or she may find that the lump sum contract does not provide for much flexibility. Making changes while construction is in progress can be fraught with difficulty because the contractor bid on the project according to completed plans, not ever-changing plans. Changes may be costly and difficult for the owner to obtain. Since the project's true cost is locked in a closed book, the owner must specify the materials that the contractor is required to use during construction; otherwise, the contractor is likely to use the lowest suitable grade materials to save money. More than a detailed specification must be invoked.

In order for an owner organization to protect itself, it must work closely with the design team during the planning stages of the project. This will ensure that everything it wants is documented in the final plans. C&S lack the necessary details to assure success as these activities are frequently covered by a few sparse lines such as "the EPC shall be responsible for commissioning." The responsibility then rests on the assigned owner's engineer, often a machinery technology subject matter expert and his or her colleagues. They will have to contend with the EPC representatives and the owner's project manager to muster the resources necessary for a proper commissioning strategy. The EPC wants to do the minimum and wants money for any activity deemed to be extra. The project manager, on the other hand, will put the pressure on the owner's engineer and resist additional activities necessary for proper commissioning. Frequently, no standard or specs are agreed up front. For example, a standard such as API RP 686[13] might be invoked, but no standard acceptance targets are being agreed to, for example, on the degree of cleanliness after flushing lubrication systems.

It would be a good idea for the owner's organization to put its project manager on notice that he/she will be assigned to be the plant manager of the new facility for a period of five years or until overall asset availability exceeds 98.5% for at least five years.

Any EPC organization has an interest in attaining mechanical completion milestones while many commissioning activities actually can only be completed after mechanical completion. So once mechanical completion is accomplished, EPC forces tend to leave not having completed commissioning. Mechanical seal system commissioning is one case in point. Piping according to seal plans is a field activity and is normally not very well executed. Seal pot cleanliness requirements are frequently not adhered to during construction. The owner's dilemma is whether or not to do cleaning of seal pots and seal piping. The EPC is reluctant to undo the piping for inspection after it has been completed. If these activities are not accomplished the consequences are risk of seal failures and attendant costs. Another example would be piping hydro-testing and the related issues resulting in rust in machinery piping. This can be a significant issue. No proper drying of piping and static equipment is done leading to machinery full of rust and scale. Examples are reciprocating and screw compressors, major pump piping and heat exchangers in the line. If attention is not paid during hydro-testing, liquid puddles are left in machine casings with the potential for rust formation. Oil flushing is another example of an activity that straddles the mechanical completion stage and commissioning activities.

This type of contract results in considerable risks for the successful C&S of process machinery. General experience has been that the EPC contract regarding pre-commissioning is flawed. There are examples where owner organizations released their contractors at around 90–95% completion of their projects assuming C&S self-directed with their own forces.

One way around the issue with clean pipes, for example, is to insert a clean-pipe clause into the contract. Many owners are now doing this and getting good results, including oil flushing of large systems, and blowing of steam lines.

2.3.2.1.1.2 The Cost-Plus-Fee Contract

For an owner who has experience in the construction industry, or for an owner who cannot initially define or sufficiently detail the scope of the work, the cost-plus-fee contract – known in the industry simply as the "cost-plus" contract – is the best solution. In a cost-plus agreement, the contractor is reimbursed by the owner for the actual cost of performing the work. The contractor is not supposed to make a profit on any phase of the construction and the project is open book.

Before the project begins, the owner and contractor agree on a fee (often a set monthly fee or a fee based on a percentage of the cost of the work) that the contractor will retain for profit and overhead. The idea of the cost-plus arrangement is for the owner to pay the cost of the actual work without markups, plus a set fee for the contractor's profit. To avoid disputes, the owner and contractor should specify early on in the process what is a reimbursable expense to the contractor (for his/her general conditions such as employees on the project and insurance), and what is considered a cost to the owner.

Advantages. The cost-plus agreement is ideal for fast-tracked projects or for situations where the contractor becomes involved before the construction documents are substantially completed. The cost-plus contract allows an owner to have more flexibility to change designs and materials as the project proceeds. Furthermore, this agreement usually requires the contractor to obtain several competitive bids for each trade, allowing the owner to review the bids and secure the lowest cost. And, since the project is open book, the owner is entitled to know the cost of materials and labor at each phase of the construction process. If the owner and contractor have a good working relationship, the cost-plus contract allows for a flexible and efficient project execution.

In this case, the risk of failure to commission and starting up successfully is substantially reduced because the owner routinely assigns an owner's engineer who works with the contractor organization. Here the owner's engineer has control of the standards and practices of pre-commissioning, C&S. This means the owner decides how much commissioning to conduct and he determines the standard of cleanliness and many other details.

Disadvantages. The major disadvantage of the cost-plus contract is that the costs can rise quickly. Under the cost-plus contract the owner must verify hundreds and often thousands of claimed costs; such an arrangement is ripe for a dishonest contractor to defraud the owner.

2.3.2.1.1.3 The GMP Add-On

While the traditional cost-plus agreement does not have a fixed budget, an owner and contractor often agree to cap the price once the project's design is substantially complete. This is known as a guaranteed maximum price (GMP) provision. Under a GMP agreement, a contractor who exceeds the capped amount is responsible for the difference, and if the total cost of the project is below the capped cost, the owner and contractor often agree to a "shared savings" benefit.

Even with a GMP contract, the owner must remain alert and ensure that the contractor does not set the GMP too high or use loopholes in the agreement to get around the GMP cap. Alternatively, some contractors set the GMP higher than need be so that the total cost of the project comes in below the capped cost. In so doing, the contractor can try to receive "extra profit" through the shared savings clause. Again, MQA needs to be in place.

If owners are well informed and take an active role in the development of the project, they can avoid the pitfalls of both the lump sum and cost-plus contracts, and enjoy the fruits of a well-constructed, final project that was completed within budget.

2.3.2.1.2 Equipment Complexity Translated into Intrinsic Reliability and Probability of Failure

An assessment of intrinsic process equipment reliability and failure probability would logically give consideration to its complexity. By making a machinery complexity assessment – see Table 2.3 – and using a severity score concept – see Table 2.4 – we can arrive at a relative risk index number (RIN) that would help the C&S team focus on their priorities. An example would be an un-spared steam turbine–driven compressor train designed to current industry standards such as API 617:[14]

- Power 13,400 hp or 10 MW
- Speed 12,000 rpm
- Typical train configuration shown in Figure 2.14
- Natural gas
- No known train complexities or technological step-outs

From Table 2.3

Train complexity no. $= \sum$ Complexity by (configuration + size + speed
$+$ casing count + application + miscellaneous)

$$= \sum 7 + 4 + 8 + 1 + 5 + 0 = 25$$

From Table 2.4

Train severity score $= \sum$ Consequence (health + environmental
$+$ public impact + business impact)

$$= \sum 5 + 8 + 8 + 4 = 25$$

The relative RIN = train complexity no. × train severity score = 625, which could be considered relatively *medium* compared to the risk presented by equipment trains with different complexities and severity of failure consequences. Figure 2.13 represents a spread sheet summary of the above calculations.

Table 2.3: Machinery train relative complexity numbers.

1.0	Complexity by process machinery train configuration	Rank
1.1	Motor-gear-driven turbocompressor train	6
1.2	Motor-variable speed train (mechanical speed variation component/fluid coupling)	9
1.3	Motor-driven reciprocating compressor	8
1.4	Motor-driven helical screw compressor – oil free	6
1.5	Motor-driven helical screw compressor – oil injected	4
1.6	Steam turbine–driven screw compressor – oil injected	5
1.7	Steam turbine–driven turbocompressor	7
1.8	Gas turbine–driven turbocompressor	10
1.9	IC engine–driven turbocompressor	2
1.10	Integral gas engine compressor – see Figure 2.A1	5
1.11	IC gas engine–driven reciprocating compressor – detachable	4
1.12	Gas or steam turbine–driven generators	6

Table 2.3 (continued)

2.0	**Complexity by size, category III (special-purpose) equipment**	
2.1	Over 15,000 hp or 11 MW	10
2.2	5001–15,000 hp or 3.7–11 MW	4
2.3	501–5,000 hp or 375–3,700 kW	2
2.4	1–500 hp or <1–375 kW	1
3.0	**Complexity by size, pumps – rating given to entire pump population in a given plant**	
3.1	Predominantly large pumps – least complex, over 100 hp or 75 kW	10
3.2	Predominantly medium size, 25–100 hp or 18–75 kW	20
3.3	Predominantly small pumps, <25 hp or 18 kW	30
4.0	**Complexity by operating speed (highest train speed)**	
4.1	Over 20,000 rpm	10
4.2	10,001–20,000 rpm	8
4.3	5,001–10,000 rpm	6
4.4	1,0001–5,000 rpm	4
4.5	0–1,000 rpm	2
5.0	**Complexity-driven casings count**	
5.1	Straightforward summation of casings, excluding prime movers	. . .
6.0	**Complexity by application – compressors**	
6.1	Corrosive/flammable/explosive/toxic	10
6.2	Easily liquefied gas	9
6.3	Low MW gas	8
6.4	Gas mixture/water	7
6.5	Diatomic gas/natural gas	5
6.6	Air	1
7.0	**Complexity by application – pumps**	
7.1	Corrosive/flammable/explosive/toxic	10
7.1	Hot services where the pumping temperature is above 350 °F or 175 °C	9
7.2	Cryogenic services	9
7.3	Low NPSH	8
7.4	Auto-start services, loading services, emergency services, loading services, viscous Services	7

Table 2.3 (continued)

7.5	Other services	1
8.0	**Miscellaneous complexities**	
8.1	Major scale-up of past design experience	5
8.2	Technological step-out, one component	4

See also: [i]Bloch, H.P. & Geitner, F.K., *An Introduction to Machinery Reliability Assessment*, Second Edition, Gulf Publishing Co., Houston, Tokyo, London, Spring 1994, and preceding edition, Pages 148 to 167.

Table 2.4: Relative severity scores.

2.4.1 Score	Health and safety – consequence of failure (severity)
10	Catastrophic: permanent injuries and fatalities
8	Critical: permanent injuries
5	Moderate: injured person temporarily loses his/her ability to work
3	Low: the injured person temporarily does not lose his/her ability to work
1	No personal injuries
2.4.2 Score	**Environmental – consequence of failure**
10	Catastrophic
8	Critical
5	Moderate
3	Low
1	No impact
2.4.3 Score	**Public impact of a failure**
10	Extreme
8	High
5	Moderate
2	Low
1	No impact
2.4.4 Score	**Repair/replacement cost/business loss (USk)**
8	>1,000 (over 1 million dollars)
4	>500 ≤ 1,000
2	>100 ≤ 500
1	≤100

1	2
	13,400hp/10MW
	Set of Steam Turbine /
	CentCmp at 12,000rpm
Rel. Risk #	625
Rel.Risk Level:	MEDIUM
Complexity	25
1.7	7
2.2	4
4.2	8
5.1	1
6.5	5
8.0	0
Severity	25
Table 2.4.1	5
Table 2.4.2	8
Table 2.4.3	8
Table 2.4.4	4

Figure 2.13: Determining RIN for a steam turbine–driven turbocompressor train.

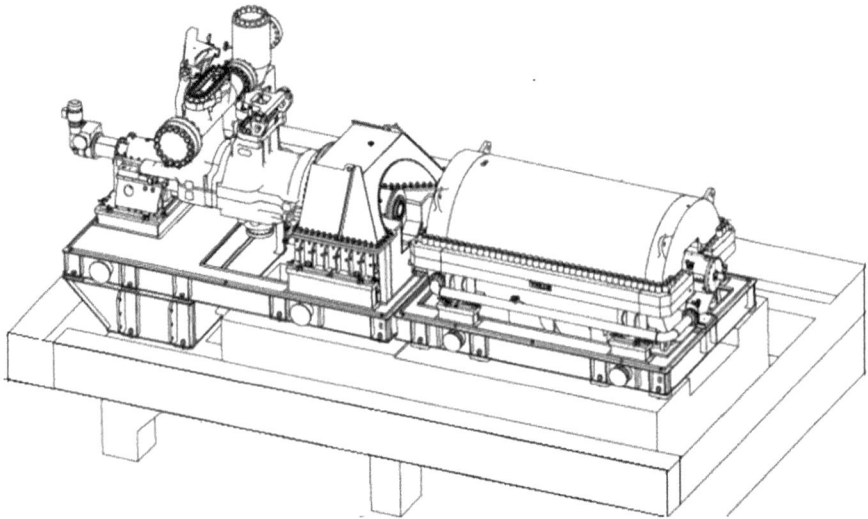

Figure 2.14: Typical steam turbine–driven process compressor train (Elliott).

2.3.2.1.3 Project Culture

The third risk determinant is a project's culture manifested by the degree of adherence to well-understood rules and regulations, stated in this text and elsewhere, pertaining to pre-commissioning, C&S activities. A good test would be to continuously find answers to the familiar old question, "What can go wrong?" As an aid, we should consider Table 2.5, a worksheet for potential risk analysis. In the following chapters, we shall revisit this form and use it.

Table 2.5: Form: machinery potential risk analysis record.

Phase	Risk	Most probable cause[1]	Suggested actions/mitigating measures
Rigging and lifting	•	•	•
Handling, staging and storage protection	•	•	•
Foundation and grouting	•	•	•
Piping and tubing	•	•	•
Shaft alignment	•	•	•
Lubrication system	•	•	•
Startup and initial operation	•	•	•

[1] Refer to Figure 2.12

Finally, project management and company culture determine how individual risks are recognized and dealt with. In the following we are going to lay the groundwork for the task at hand.

Appendix 2.A

During the period from 1945 to the end of the 1970s the integral-angle gas engine compressor – see Figure 2.A1 – went through an extensive development which saw its thermal efficiency increase from around 25% to some 37%.[15] This was mainly achieved by moving from an engine-driven "scavenging" arrangement to a pure turbocharged design. High efficiencies of engine compressors tended to delay the acceptance of competing gas turbine–driven centrifugal compressor packages. These engines also kept up with the times; in 1978, in response to the increasing pressure being brought on by the EPA, it was the first gas engine to adopt "Clean Burn" combustion.

When this engine was finally replaced by gas turbine packages, particularly in the gas transmission field, it had enjoyed a remarkable 55-year production run. The assembly line style manufacture of this product by one OEM in Ohio, for example, provided a high level of employment.

Figure 2.A1: Four power-cylinder integral gas engine compressor (Cooper-Bessemer).

Total production was:

Produced by C-B in the USA and Canada	2,825
Produced by licensees*	225
Produced by Soviet Union**	1,566
Total	4,616

* One OEM's production locations included Canada, the United Kingdom, France, Germany, Italy, Mexico and Japan
** Produced at Gorky Works.

References

1 J. Britain, *Reshaping the industry toward innovation, certainty and efficiency*, Hydrocarbon Processing, January 2017, P. 23.

2 Bloch, H.P. & Geitner, F.K., *Maximizing Machinery Uptime*, First Edition, Houston, Tokyo, London, 2006 at www.elsevier.com, ISBN: 0-7506-7725-2, 672 pages, Pages 287 to 288.

3 John S. Mitchell, Physical Asset Management Handbook, Fourth Editions, 2007 Clarion Technical Publishers, Houston TX, ISBN 0-9717945-4-5, Pages 361 to 374.

4 Change agents could be internal consultants promoting change in relationships between people that have to interface to work successfully.

5 Attributed to Trevor Kletz. Trevor Asher Kletz, OBE, FREng, FRSC, FIChemE (1922–31 October 2013) was a prolific British author on the topic of chemical engineering safety. He is credited with introducing the concept of inherent safety, and was a major promoter of HAZOP.

6 MQA – Machinery Quality Assessment. A rigorous and multifaceted engineering work effort by the future equipment owner aimed at uncovering risks and vulnerabilities that exist in low-cost or prototype-like machinery offers. See also:

   1. Heinz P. Bloch, *Improving Machinery Reliability*, Volume I, Series *Practical Machinery Management for Process Plants*, 3rd Edition, 1998, Gulf Publishing Company, Houston TX.
   2. H.P. Bloch and F.K. Geitner, *COMPRESSORS – How to Achieve High Reliability & Availability*, Mac Graw-Hill, New York – Toronto, 2012, Pages 117 to 128.

7 Klaus Brun and Sarah Simons, *RECIP PACKAGING – WHAT HAPPENED TO THE REST OF MY INSTRUMENTATION?* gascompressionmagazine.com, SEPTEMBER 2017, Pages 8 to 9.

8 HAZARD AND OPERABILITY ANALYSIS

9 Robert B. Abernethy, *The New Weibull Handbook*, 5th Ed. 2004, ISBN-10 0-9653062-3-2.

10 V. Schüle, ABB Power Generation.

11 N. Pinto, *Applying tools to strengthen safety and risk management*, December 2020, Hydrocarbon Processing, Pages 66 to 69. www.hydrocabonprocessing.com.

12 See reference 6.

13 API RP 686, 2nd Edition, 2009, Recommended Practice for Machinery Installation and Installation Design.

14 API 617 – Axial and Centrifugal Compressors and Expander-compressors, API 692 – Dry Gas Sealing Systems for Axial, Centrifugal, and Rotary Screw Compressors and Expanders, API 671 – Special-Purpose Couplings for Petroleum, Chemical, and Gas Industry Services, Fifth Edition, API 613 – Special Purpose Gear Units for Petroleum, Chemical and Gas Industry Services, and several others.

15 Geitner, F.K., Contributing Editor, *Nostalgia, the era and role of integral gas engine compressors reviewed*, Pipe Line & Gas Technology, Hart Energy Publishing LLLP, New York, NY., November 2006.

Chapter 3
Planning

3.1 The Task

Commissioning and startup (C&S) planning should be approached with the adage: Expect the unexpected and plan accordingly. In the oil and gas industry, for example, the cost of any extended forced downtime can run into millions of dollars. With many external factors impacting C&S, a startup team must assume that problems will be encountered. On the one hand, any delay in upstream commissioning is frequently expected to be compensated for by a quick and trouble-free startup of machinery. This creates pressure for an early startup. On the other hand, building a contingency into the plan will prevent the C&S effort from being derailed when the inevitable hurdle appears.

Despite a high commitment to the general quality idea, capital project organizations in the processing industries often fail to stay focused on what ought to be their main goal, namely asset availability and its two components, reliability and maintainability. The reasons are schedule compression resulting in unforeseen delays, frustrations and managerial errors. In order to try to escape these problems, project professionals frequently turn to sophisticated techniques that range from critical path diagrams to integrated project-management software. In the meantime, little energy is directed toward the interfaces and relationships with contractors, such as engineering, procurement and construction companies (EPC), original equipment manufacturers (OEMs), equipment vendors and other suppliers to the project. The quality of these relationships may well have the most important influence on the outcome of C&S efforts and ultimately on the service factor of a new plant. The various players in the commissioning have to be working together as a team. From an organizational point of view, owner–contractor interfaces during the project execution, and frequently during the latter part of the definition phase, are systematized by a project coordination procedure. As an example, refer to the table in Appendix 3. A. It lists the main points of this agreement between owner and contractor. In a simplified view this procedure regulates the different project stages as illustrated in Figure 3.1.

3.2 Organization

C&S require large teams, and the makeup of the group is a crucial driver of success. Mature organizations in hydrocarbon processing, for example, such as Shell, Exxon or BP have worldwide access to teams with a wide array of experience in executing these events across their organizations. However, the vast majority of businesses

https://doi.org/10.1515/9783110701074-003

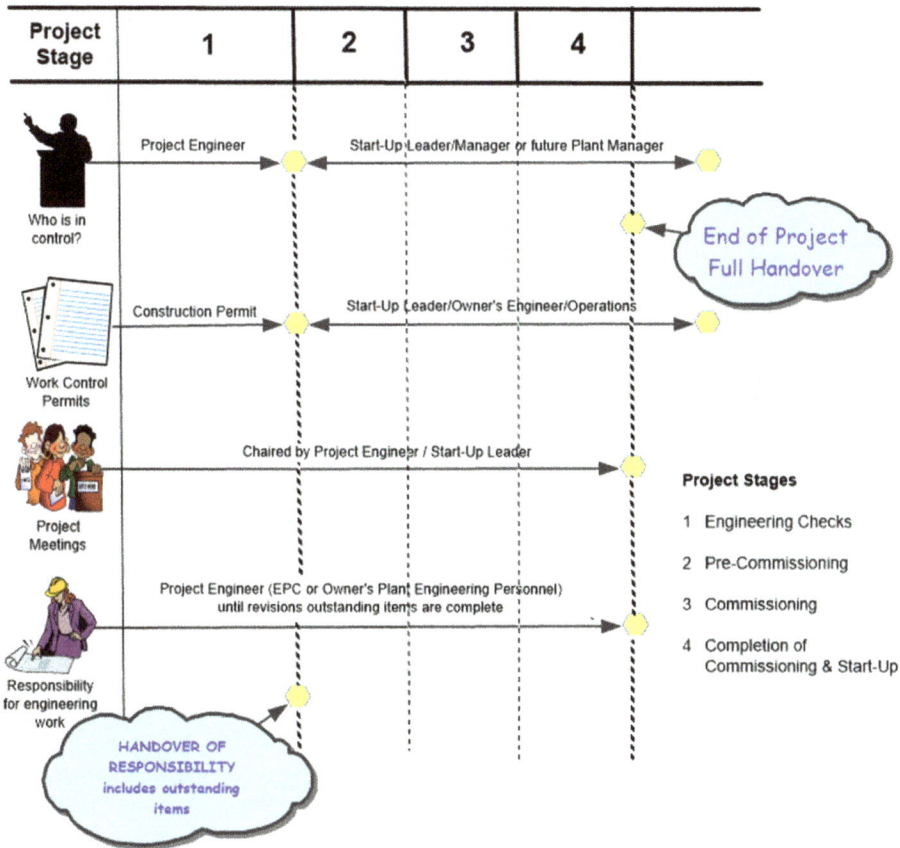

Figure 3.1: Project agents versus project stages.

lack this knowledge base or personnel power, so adding external expertise is vital. An integrated team consisting of highly skilled specialists who help plan, direct and support the internal team is the right model to adopt.

As to selecting external resources, owners often pick the cheapest option without ensuring that they are bringing in the right set of skills and expertise. Another way they cut costs is to limit the timing of the specialist contractors, keeping them out of the critical planning stage or the project evaluation step, where key learnings are determined. In the long term, adopting these approaches will drive up costs. External expertise should be approached as a strategic investment rather than a project overhead.

C&S responsibilities can be effectively discharged by two owner organizations. One, a mechanical section and the other a technical section as shown in Figure 3.2. The technical startup section is staffed by engineers and technicians with prior startup experience. Their task is to define the necessary steps leading to successful commissioning, startup and satisfactory long-term operation of rotating and reciprocating machinery. The mechanical startup section is supervised by a machinery specialist and by mechanical first-line supervisors – often abbreviated FLS – foremen, and machinists[1] whose principal task is to execute all necessary steps leading to the same results.

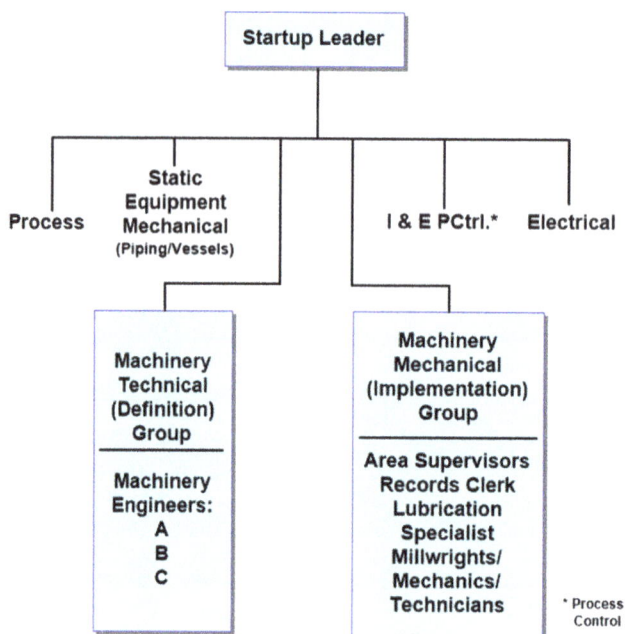

Figure 3.2: Machinery commissioning and startup team.

As can be seen from the organization chart in Figure 3.2, the two section supervisors are preferably reporting to the same startup leader or manager. As an alternative, the reporting structure indicated in Figure 3.3 has been employed and can also be made to work. However, the latter's effectiveness can only be assured if the two section supervisors share mutual background and trust and respect each other. If their relationship is lacking in these elements, a game of "one-upmanship" and finger pointing may result. Startup progress and effective utilization of available resources may suffer from this kind of relationship.

3.3 Startup Management Tools and Resources

Successful startups, as in any work activity requiring interfaces with several areas of expertise – process, electrical, instrument, mechanical, contractor, and other specialists – must begin with an explicit plan of attack. Schedules must be developed as

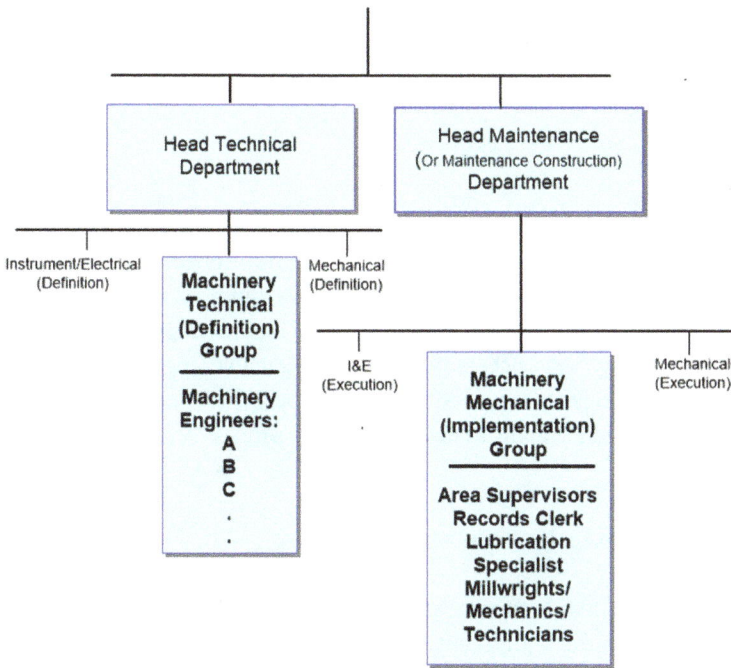

Figure 3.3: Machinery commissioning and startup team – alternative.

soon as practical and updated as changing situations require. Contact personnel must be designated for each area of responsibility required for the machinery startup. Most of all, agreement as to the overall responsibility and the "chain of command" required for decision making in the event of conflicting opinions must be established at the outset.

Following or coincident with the startup planning stage, the startup advisors must become familiar with the specific equipment involved in the project. Of course, the overall startup schedules will determine the machinery startup planning priorities. Concentrated effort must first be directed toward the large horsepower and critical machines of a plant – Category III, Figure 2.8. This requires detailed review of the information supplied in the design specification and in the vendor's literature as well as field review of the machinery installation. Startup schedules including the details of drum inspections, lube oil flushing, cold alignments, instrument and electrical checks, mechanical checks and others should be determined and included in the original plant schedule for acceptance by the responsible parties.

Noncritical or spared equipment – Categories II and I – should come next on the agenda. Hot service pumps, for example, should be reviewed critically in terms of alignment capabilities, piping supports, pump materials, warmup facilities and run-in alternatives. Other spared equipment must be similarly reviewed. However, emphasis is first given to equipment with higher risk of startup problems.

3.4 Planning

As alluded to earlier, a startup must be approached with a plan in mind. Due to the initial makeup of a startup team or task force, process startup schedules incorporating contractor turnover dates, process startup requirements and product commitment dates are usually available. However, they often omit commissioning requirements for machinery. Consequently, it is necessary for the responsible engineer to obtain process startup schedules as soon as possible, superimpose the machinery startup schedule, and make contingency plans to meet or alter the process schedule.

3.4.1 Typical Schedules Are Easily Established

To arrive at a reasonable schedule for machinery startup in a given process plant, it is necessary to define all of the equipment involved by keeping an equipment list, determine the contractor's schedule within his/her limits of responsibility, determine process priority requirements and establish electrical/instrumentation completion and process control – SCADA,[2] for example – dates. The machinery startup must take place within the framework established by these other schedules.

For the case of a major petrochemical processing plant, as a basis for arriving at a machinery startup schedule, for example, we assume that a contractor-process-product commitment schedule similar to the schedule shown in Figure 3.4 exists. With this as our guide, pumps and blowers must be operable first. Compressors are required last for the final phase of the process startup and for product delivery. However, a detailed review of the machinery startup requirements for the entire plant should clearly show that the compressor work must begin first in order to meet the final product delivery date. In addition, the initial process schedule may have to be rearranged to make the process system around the compressor available for compressor run-in.

It is essential to understand detailed contract modalities, meaning whether or not the EPC contract involved "clean pipes" at mechanical completion or not. But suppose the steam piping for the turbines was specified as clean after flushing with no allowance for steam blowing time. Assume further that it remains questionable whether or not the electrical facilities will be available when required due to staggered completion dates on the substations. In that case and quite obviously, a problem in scheduling exists and resolution will require the cooperation of all parties involved.

The machinery startup schedule is frequently actually superimposed on the product commitment schedule without explanation. It would always be advantageous to

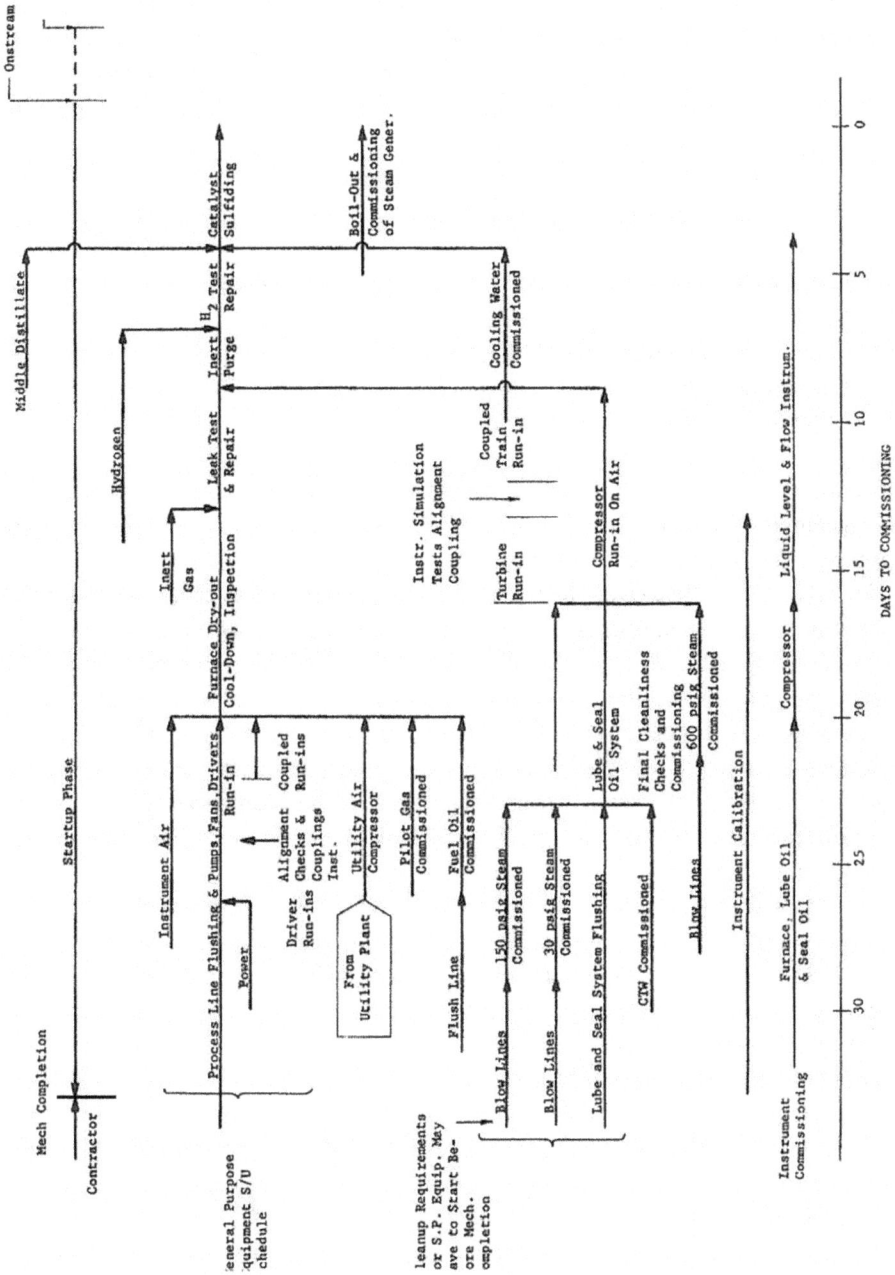

Figure 3.4: Contractor commitment schedule – Exxon.

the startup engineer, however, if some detailed justification for the startup schedule were laid out.

Beginning with the pump requirements, we would consider the following work necessary: Subsequent to mechanical acceptance from the contractor, the pumps are completely installed and

1. Final cold alignment with pipes connected done
2. Spring hangers blocked in cold condition
3. Temporary strainers installed
4. Check valves in correct direction following pipe hydrotest
5. Bearings flushed and lubricated
6. Coupling spacers fitted but not installed
7. Run-in plates mounted on driver coupling halves
8. Other details clearly spelled out

In addition, process and electrical/instrument and process control systems have been completed by the contractor. In the following chapters we shall provide appropriate procedures and pertinent checklists.

Construction contractors or EPCs executing, for instance, large petrochemical projects are using a multitude of written guidelines dealing with the installation of rotating equipment. Many of these procedures describe future plant maintenance tasks and should, therefore, be assembled as soon as possible. Many other good references[3] and aids exist or will be generated by the technical section, from general cautionary advice to detailed procedures and checklists which determine C&S sequences. However, they will stay in a theoretical realm – on paper – if the project startup leader, or the owner's engineer, does not recognize that these tools become useful only if the team does take them seriously.

It has been our experience that when one or several skilled and dedicated machinists or millwrights are charged with the task of field inspection, they will pursue this task and its documentation in form of completion of checklists in a serious and conscientious manner. The authors found that the best person to interact with the contractor's personnel as mechanical field inspector is an experienced first-line maintenance supervisor or lead-hand mechanic. While the owner's specialist field engineers are capable of "pinch hitting," they are far too busy to spend all their time in the field. Mediocre results may be achieved, if startup engineers are not supported by experienced maintenance personnel. Also remember, contractors often know very little about maintenance!

During equipment startup, maintenance millwrights, together with process operators, become the final line of defense against any remaining risks. We must not forget that the operator becomes eventually intensely involved and a few days before actual first-run he or she can see what transpired or has been overlooked.

The work of the startup team is accomplished by daily interfacing with contract personnel, such as designers, project engineers, construction planners, specialists, construction superintendents, field engineers, OEM FSRs (vendor field service representative) and subcontractor personnel. This can only happen when management has seen to it that all disturbances or obstacles to this process have been recognized and removed. Justified attention to detail – see Figure 3.5 – by the owner's field inspectors will frequently lead to interface problems with the contractor's personnel. This should be anticipated, and interface meetings should be held during which contentious points are aired. Senior specialists will usually chair these meetings.

Figure 3.5 content:

PETRONTA
Refinery
Process Unit: Front End

| Tag No.: All Pumps – Front End | Machine: P-502 | |
| Subject: Grouting of pumps | *Notification:* P- 011 | 15/05/73 |

To BUSY CONTRACTOR: Attention Jack Golden

Dear Jack, this is the second time I am raising this subject.
Your company personnel are not adhering to the agreement we made concerning the Grouting of our Pumps. Please refer to the sketch below. We have checked several times and found that the grouting layer you apply is less than one inch – in most cases we have found as little as a quarter of an inch.
Please correct the situation.

Sincerely,

Bill

Bill Schostak
Field Inspector

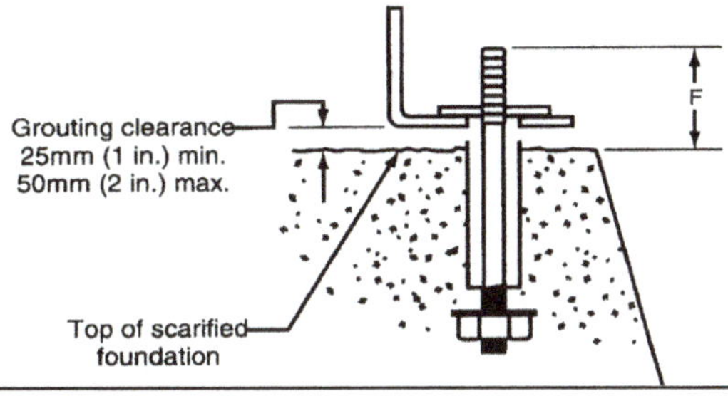

Grouting clearance
25mm (1 in.) min.
50mm (2 in.) max.

Top of scarified
foundation

F

Figure 3.5: Field inspector complaint.

3.4.2 Interface Problems and Their Solutions

Some Problems: In spite of formally accomplished coordination procedures and legally perfect contracts the owner–contractor interface can often be less than satisfactory if not adversarial. There may be some problems. For example:

- Basic personality differences – sometimes referred to as interpersonal *chemistry* incompatibility
- Interpersonal difficulties caused by language and culture. One of the authors had the opportunity to work on a C&S assignment in a European country where the engineering contractor was from the UK, the construction company English and French, and other staff members hailed from several other countries.
- Managerial errors – for example, putting the wrong person into the wrong position

Interpersonal problems can often be simply overcome by role statements and role delineation. A good project coordination procedure ought to help here. As we saw before, the coordination procedure is a project management document which supplements the contractual terms of the overall contractor agreement. It outlines detailed procedures to be followed in performing work and further defines the duties and responsibilities of the contractual parties. Frequently this procedure is not "customized," but a standard format which does not take the specifics of owner and contractor organizations into account. If they have not been able to influence the selection of effective owner–contractor team members before a project begins, knowledgeable project executives will see the signs of trouble and call on "transactional analysis" specialists for help. In team building exercises – they have to be timely – owner representatives and contractor personnel are interacting in off-work situations such as in a weekend camp. The outcome is that future interfacing will be on the basis of enhancing and maintaining self-esteem of others, of listening to each other with empathy and of asking for help in solving the problem at hand.

Managerial errors are frequently more difficult to overcome as the type of contractual arrangement will determine how a contractor is motivated. For example, as mentioned before at another place, on a lump sum project the contractor has a strong motivation to sacrifice the project reliability and asset availability goals to reduce cost and to improve his/her schedule. On a reimbursable cost project, the contractor has nothing to gain financially. Here the contractor might have the tendency to overreact to the project specifications and overdesign resulting in unnecessary high costs without any gain in ultimate asset functionality and availability.

Another obstacle to good owner–contractor relationships is the "cascading" of contracts. Here, one contractor has a subcontractor working for him/her, who in turn has another one working for him/her and so forth. The result of these arrangements are invariably communication problems. Whatever the conditions are, it is the responsibility of the owner's representatives such as field engineers, plant engineers and specialists to interface with the contractor in order to protect his or her company's investment, to get the best quality possible and to assure asset service factor goals are met.

Some solutions: The key to overcoming interface problems is the early assignment of experienced machinery technology and maintenance specialists to the project. They are best suited to interpret project operating standards and specifications by dealing with contractor personnel. Further, large projects have been successful by assigning early a coordinator between the future operating department, the owner project administration group and the contractor organization. This person could typically be the future startup leader, the senior equipment specialist or a former maintenance department head.

While this seems obvious, some of us have no doubt participated in projects where these assignments were not made at all; where and when they had been made, they were often either not timely or involved inexperienced personnel.

This is especially true for the assignment of maintenance representatives as alluded to earlier. Frequently, project professionals fail to understand the contribution maintenance specialists can make before they take over, because of a first cost philosophy widely shared by capital project specialists as opposed to the concept of total life cycle asset cost best understood by maintenance professionals. Additionally, there is an urgency to get facilities designed, constructed and on stream – without finding the time to invite maintenance. As very few EPCs are concerned with maintenance, having a maintenance specialist on board makes double sense. Field installation of the equipment must be followed on a day-to-day basis by the owner's representatives in close interfacing with the contractor's field supervisors. Not only is early ownership of the equipment by the people who will operate the plant desired, but reliability and maintainability of the plant will get a final boost.

Machinery startup engineers should assist the plant maintenance supervisor in identifying documentation to be collected and retained for future use.

3.4.3 Successful Interfacing Assures Equipment Goals Are Met

Owner–contractor (EPC) interfacing activities typically will occur during the following functional project steps connected to the C&S phases of the project:
- Inspection and test – FAT – at vendor's site
- OEM/EPC or vendor final drawing review
- Review of spare parts and documentation
- Equipment field handling, staging and potential storage protection
- Foundation design and execution – grouting
- Piping and connection to equipment
- Shaft alignment
- Lubrication systems/cleanliness
- Mechanical completion followed by equipment turnover/handover
- Pre-commissioning
- Commissioning
- Startup
- Initial operation

A number of key elements are indispensable to ensuring successful commissioning of rotating machinery in process plants. The mix of mandatory and contingency tasks to be accomplished is listed in Figure 3.6.

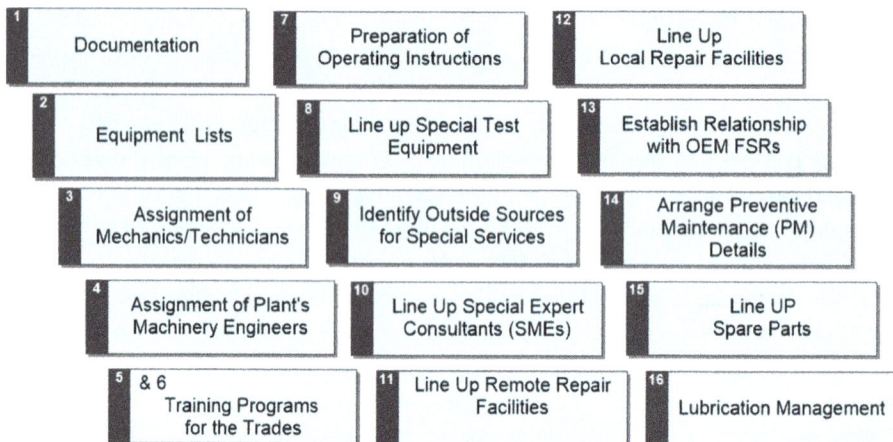

Figure 3.6: Overview of commissioning and startup review tasks.

3.5 Commissioning and Startup Review Tasks

3.5.1 Documentation Review

To assist in fulfilling this task, the reviewing owner's engineer generally instructs the contractor or equipment manufacturer to submit drawings and other data for his/her review. For example, many of the vendor data and drawing requirements (VDDR) are tabulated in the appendices of applicable API standards and can be adapted to serve as checklists for the task at hand. Other checklists may have to be derived from the reviewer's experience. Keeping track of the status of documentation reviews is best accomplished by first listing the VDDR for a given equipment category. Figure 3.7 shows an example of the VDDR listing for a special-purpose turbine purchase. Whenever possible, the listing should contain the data requirements of available industry standards.

Each "tracking sheet" is supplemented by three columns in which the reviewer can enter the review status, for example, "Preliminary Review Completed." Also, the "tracking sheet" shows seven columns which indicate a particular project phase during which the vendor is expected to submit certain drawings or analytical data for the owner's review. The decision as to when – at which project phase – data are to be submitted is best made by mutual agreement and interfacing of all parties involved.

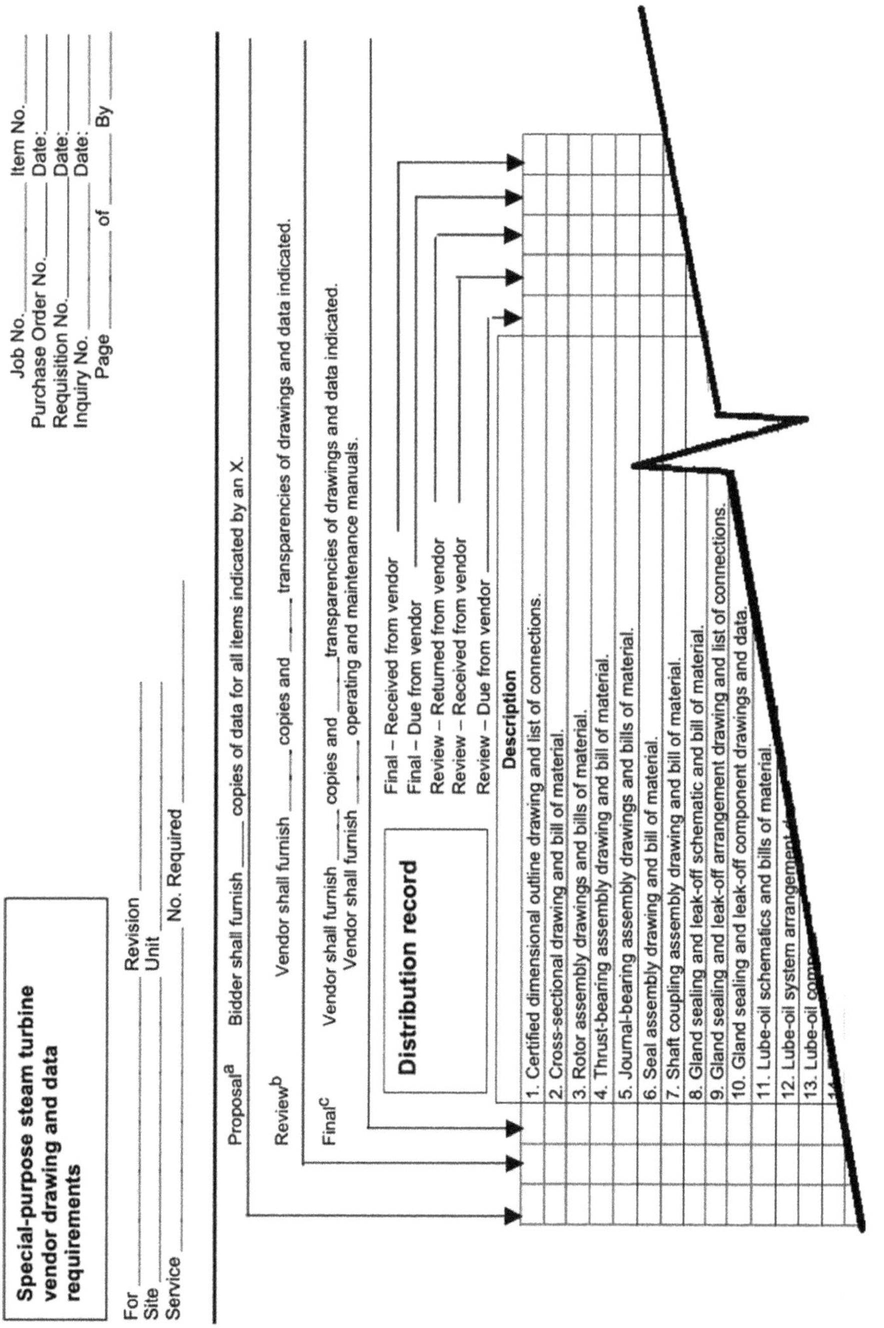

Figure 3.7: Example of a vendor drawing and data requirements (VDDR) listing (adapted from API 612).[4]

Problems in this area must always be anticipated. Since final documentation, as-built drawings and manual presentation by the contractor are frequently late in the project, it may happen that not enough resources are available in the vendor's organization to prepare the required documentation. Mediocre owner's equipment files are the result as well as the risk of lacking information when it is needed. There are cases where suppliers have forgone some 10% of their contractual fee for not having been able to supply the complete set of VDDR they owed. Further, as mundane and unimportant it may seem, missing data at C&S will impact future asset maintainability and ultimately availability. We know from experience that highly paid professionals have spent valuable time searching for missing important data.

3.5.2 Documentation for Effective Progress Tracking

We believe it to be important that owner records are established at startup. The equipment list would come first. In advance of major startup activity, a system of records keeping should be set up for the machinery of a project. In some cases, individual file folders have been used for each piece of machinery. In other cases, machinery records have been filed separately for special-purpose critical machine trains and combined for the general-purpose machines. In yet other cases, machinery files have been kept by combining the equipment of each plant in one folder. This has not changed with the advent of computer tablets. Usually, the latter choice is best since it conserves filing space while still performing the task. An effort must be exerted, however, to keep the file orderly by equipment identification number. This will avoid the accumulation of a disorganized collection of miscellaneous documents. To assist in maintaining an orderly file, it is advantageous to prepare individual but generalized data sheets for pre-startup checks and run-in information, data can then be entered in the file without the necessity of disturbing the file assembly. Data sheets should include complete equipment description, installation review checklists, preliminary evaluation, problem list and other progress tracking documents.

3.5.2.1 Start FRACAS Early

Think about setting up a failure reporting and corrective action system at this time. As an example, failures and repairs to "P-35B" should be electronically filed in the adjunct to the basic equipment file folder for "P-35B."

At the conclusion of the startup, the record files of the machinery should be maintained in a suitable location and updated as operating experience is accumulated. Such information becomes valuable during maintenance and future turnarounds.

3.5.2.2 Startup Plans Must Lead to Machine Commissioning

The preceding discussion of startup planning provides a good foundation for a successful startup. Objectives have been documented in the startup schedules, provision has been made for keeping up to date with the startup via the progress and status chart and a records system has been developed to assure that the condition of all machines at startup will prove satisfactory. This logically leads to the implementation of the planned activities. It also requires that specific instructions are available for individual machines – or machine type. Commissioning entails the development of equipment lists for general reference and progress tracking because machinery commissioning targets and accomplishments must be documented every step of the way if the startup is to progress in an orderly fashion.

Once a basic set of equipment listings has been prepared, it can be readily modified by replicating it in part and thus be adapted to a variety of documentation needs. Figure 3.8 will serve as an example. For clarification, Column 6, API 610 Type refers to standard pump type codes in reference;[5] Column 7, Driver type, M = Motor, ST = Steam turbine; Column 16, Type of mechanical seal: T = Tandem, P = Packing. S = Single, D = Double. Alternatively, users have referred to seal categories listed in reference[6] Column 18, number of pump mounted transducers for failure detection.

Figure 3.9 shows the previous equipment tabulation modified to track major checkout segments as the startup progresses. The form is now used for scheduling purposes. Subsequent columns identify anticipated and actual checkout dates, partial completion, etc.

Figure 3.10 represents another modification of the basic tabulation. Here, the emphasis is on important component data.

Pump outages experienced during pre-commissioning activities are logged to keep track of both duration and primary cause. Figure 3.11 represents a typical outage log sheet with entries following a time sequence of events.

A different set of startup documentation are the event logs for major machinery trains representing the trip/shutdowns for a given machine, respectively. It is, therefore, very important for the C&S effort to have DCS and SCADA data collection facilities available and ready. Note especially how Figure 3.12 gives an overview of several important entries: purpose of run, discipline responsible for shutdown, duration of shutdown, cumulative operating time and downtime.

Figure 3.13 shows a checklist used for centrifugal-pump field installation and initial operation. Used in conjunction with field-posted startup instructions this checklist will go a long way toward reducing installation and commissioning oversights.

Similar checklists should be developed for other machinery categories and should be used by mechanical work forces and operating technicians as mentioned earlier.

(1) Pump	(2) System	(3) OEM	(4) Model	(5) Serial #	(6) API 610 Type*	(7) Driver Type	(8) HP/kW	(9) Shaft speed rpm	(10) Service	(11) Pumpage	(12) SG	(13) Pumping Temp. °C/F	(14) Suction Press. psi/bar	(15) Disch.Press. psi/bar	(16) Mech.Seal Type**	(17) Flow gpm or m3/h	(18) # of Trasducers - P	(19) Asset Number
P-501 A/B	GC	UNION	3X4X10.5 VLA	7705JH115	OH2	M	45	3600	FG	SW	1.1	25	10	60	T	200	4	4501
P-502	GC	UNION	3X6X9 CW															
P-503 A/B	GC	KSB	3X4X10.5 VLA															
P-504 A/B	GC	KSB																
P-505 A/B	GC	FS																
P-510 A/B/C	GC	FS																
P-511 A/B	GC	UNION																
P-512 A/B	GC	UNION																
P-513 A/B	GC	KSB																
P-514 A/B	GC	KSB																
P-515 A/B	GC	FS																
P-516 A/B	GC	FS																
P-517 A/B	GC																	
P -518	GC																	
P-520 A/B	GC																	

Notes:
* Refers to 3-Letter Classification from API 610
** Category from API 682

Figure 3.8: Tabulation of important pump data.

(1) Pump	(2) System	(3) OEM	(4) Model	(5) API 610 Type*	(6) Train Type	(7) Grouting	(8) Strainer Unstalled	(9) Alignment	(10) Lubrication	(11) Driver Soloed	(12) Coupled up	(13) Final Piping	(14) Run in	(15) Target Completion	(16) Actual Completion
P-501 A/B	GC	UNION	3X4X10.5 VLA	OH2	M/M	✓	✓	✓	✓						
P-502	GC	UNION	3X6X9 CW	BB1	ST	✓	✓	✓							
P-503 A/B	GC	KSB	3X4X10.5 VLA	BB2	M/M	✓	✓								
P-504 A/B	GC	KSB	3X4X10.5 VLA	BB2	M/ST	✓									
P-505 A/B	GC	FS													
P-510 A/B/C	GC	FS													
P-511 A/B	GC	UNION													
P-512 A/B	GC	UNION													
P-513 A/B	GC	KSB													
P-514 A/B	GC	KSB													
P-515 A/B	GC	FS													
P-516 A/B	GC	FS													
P-517 A/B	GC														
P-518	GC														
P-520 A/B	GC														

Notes:
* Refers to 3-Letter Classification from API 610
** Category from API 682

Figure 3.9: Tabulation for major checkout steps prior to equipment startup.

(1) Pump	(2) System	(3) OEM	(4) Model	(5) API 610 Type*	(6) Train Type	(7) Service	(8) SG	(9) Seal Type/Arrangement**	(10) Seal Drwg.	(11) Coupling	(12) RPM	(13) Driver hp/kW	(14) Asset Number
P-501 A/B	GC	UNION	3X4X10.5 VLA	OH2	M/M	SW	1.1	2	C127X	M - DP	3600	45	4501
P-502	GC	UNION	3X6X9 CW	BB1	ST	SW	0.95	2	C128X	M - DP	3600	75	4502
P-503 A/B	GC	KSB	3X4X10.5 VLA	BB2	M/M	HC	0.91	2	C129X				
P-504 A/B	GC	KSB	3X4X10.5 VLA	BB2	M/ST	HC	0.85	2	C130X				
P-505 A/B	GC	FS											
P-510 A/B/C	GC	FS											
P-511 A/B	GC	UNION											
P-512 A/B	GC	UNION											
P-513 A/B	GC	KSB											
P-514 A/B	GC	KSB											
P-515 A/B	GC	FS											
P-516 A/B	GC	FS											
P-517 A/B	GC												
P -518	GC												
P-520 A/B	GC												

Figure 3.10: Equipment tabulation highlighting important component data.

Notes:
* Refers to 3-Letter Classification from API 610
** Category from API 682

AREA & ASSET	(1) Date Commissioned	(2) Date Recommissioned	(3) Days unavailable	(4) OEM DESIGN	(5) OEM ASSEMBLY	(6) EPC SPECS	(7) PRODUCT VISCOSITY	(8) IMPROPER SEAL FLUSH	(9) DEBRIS IN LINE	(10) LOSS OF SUCTION	(11) SEAL LEAKAGE	(12) BEARING FAILURE	(13) COUPLING FAILURE	(14) GASKET LEAKAGE	(15) OTHER	(15) DESIGN MODIFICATION
GCP-501 A	20-May	22-May	2		✓											
GCP-513 B	20-May	29-May	9	✓												
MP-02 B	20-May	24-May	4													
LP-01 A	20-May	21-May	1					✓								
LP-01 B	20-May	25-MY	5				✓									
GCP-518	22-May	23-May	1					✓								
GCP-512 A	22-May	24-May	2							✓						
GCP-513 B	22-May	24-May	2		✓											
GCP-514 A	23-May	26-May	3								✓					
GCP-515 A	23-May	24-May	1													
GCP-516 A	23-May	25-May	2					✓								
GCP-517 A	23-May	26-May	3		✓											
NP-08 B	24-May	25-May	1	✓												
NP-05 A	24-May	30-May	6										✓			
MEAP-180 A	25-May	27-May	2					✓							✓	
CP-511 A	26_MAY	31-May	5					✓								
CP-512 A	26-May	31-May	5					✓								
CP-515 B	26-May	1-Jun	6					✓					✓			
DEMP-524 A	27-May	28-May	1													
DEEP-525 A	1-Jun	2-Jun	1													
AMP-182 B	1-Jun	3-Jun	2													
SLEP-180 B	2-Jun	5-Jun	3													
RECP-182 A	5-Jun	6-Jun	1					✓						✓		
DEPP-521 A	5-Jun	9-Jun	4		✓											
DEBP-522 B	6-Jun	9-Jun	3													
DEBP-523 A	6-Jun	7-Jun	1					✓								

Figure 3.11: Equipment outage log sheet.

Event	Start	Trip	Total [Hrs.] Op. Time	Total [Hrs.] Down Time	Sum Op. Time	Sum Down Time	Cause
18-Jul-19	2:00 PM	2:45 PM	0.75	23.25	0.75	23.25	I & E
19-Jul-19	10:16 AM	5:11 PM	6.92	17.08	7.67	40.33	Mech.
19-Jul-19	10:00 PM		2.00	22.00	9.67	62.33	Mech.
20-Jul-19		6:16 PM	18.27	5.73	27.93	68.07	I & E
21-Jul-19	10:00 AM		14.00	10.00	41.93	78.07	I & E
5-Aug-19		5:38 PM	353.63	6.37	395.57	84.43	I & E
13-Aug-19	6:17 AM		17.88	174.12	413.45	258.55	I & E
26-Aug-19		1:59 PM	301.98	10.02	715.43	268.57	Operations
28-Aug-19	4:00 PM		8.00	16.00	723.43	284.57	Operations
22-Sep-19		5:45 PM	582.25	17.75	1305.68	302.32	Mech.
22-Sep-19	9:15 AM		14.75	9.25	1320.43	311.57	Mech.
3-Oct-19		2:15 PM	249.75	14.25	1570.18	325.82	I & E
3-Oct-19	4:18 PM		7.70	16.30	1577.88	342.12	I & E
19-Oct-19		2:31 PM	369.48	14.52	1947.37	356.63	Operations
22-Oct-19	4:45 PM		7.25	16.75	1954.62	373.38	Operations
1-Nov-19		10:15 AM	229.75	10.25	2184.37	383.63	I & E
2-Nov-19	6:17 AM		17.72	6.28	2202.08	389.92	I & E

Figure 3.12: Major process machine train trip and shutdown log.

They serve the purpose of verifying that the installation complies with the job specification and that it is ready for imminent startup. Having observed compliance or deviations, the inspecting engineer, technician or millwright should note his/her comments on the appropriate completeness summary. Figures 3.14 and 3.15 represent completeness summary forms for centrifugal pumps and general-purpose steam turbines.

Much the same summaries should be used for all other categories of rotating machinery. Again, there has to be leadership to assure that these tools are not just used in a perfunctory way. The summaries should be completed by the reviewer or inspector only after a careful on-site verification of compliance with a "detailed checklist for rotating machinery." The reviewer will normally compile such a detailed checklist from the owner's original specifications. For example, when the interconnecting field installed piping in a lubrication system for a compressor has been specified as being stainless, it would be imperative to make sure that stainless steel had been indeed furnished.

1.0	**Mechanical Preparation**
1.2	Preparing Pump and Baseplate for Installation
1.3	Pump Installation on Baseplate
1.4	Pump and Driver's coupling concentricity
1.5	Installation of Suction and Discharge Piping
1.6	Auxiliary Piping
1.7	Lubrication – Sleeve and Ball Bearings (AFBs)
1.8	Hot-Pump Piping Strain Check
1.9	Alignment of Driver to Pump
1.10	Coupling and Coupling Guard
2.0	**Operating Preparation**
2.1	Special Procedures
2.2	Machinery Electrical & Instrumentation Specialist Coverage
2.3	Preparation for Start-Up
2.4	Initial Start-Up Data to be taken
2.5	Test Run – Follow Startup Instructions

Figure 3.13: Centrifugal pump field installation and initial operation checklist.[7]

The work of the startup engineer can thus rapidly reach major proportions as the project size increases. Besides being responsible for the physical checkouts required for successful commissioning and the subsequent process operation of machinery, as shown earlier, the startup engineer must be in a position to discuss the progress being made in several areas, both in terms of startup progress and contractor progress. This added responsibility exists since the machinery startup engineer is often called upon to attend meetings with the contractor and project management or with the owner's project manager to discuss project status and identify critical path items. As an aid for keeping up to date on the continually changing status of a project, preparation of progress or status charts has proven very useful.

An example of a progress or status chart used during one of our startups is shown in Figure 3.16. While it is similar to Figure 3.9, it shows the major project management steps. On a project-wide basis, this chart traces the individual machine status from "acceptance from contractor" through "final acceptance" when the machine has demonstrated acceptable characteristics under normal operating

	Completeness Summary for Rotating Machinery	Rev./Date NOTE: This summary should be used with: **Detailed Checklist for Rotating Machinery**
Block # ____ System # ____	**Centrifugal Pumps**	Equipment # _____

Baseplate		Pipe Supports	
Case		Small-Bore Piping	
Packing or Seal Gland		Lubrication	
Packing or Seal Piping		Guards	
Suction Piping			
Strainer			
Discharge Piping			

Remarks, Incomplete Items, Deviations, etc:

Mechanical Inspector	(M)	
Process Inspector	(P)	Name _____
Mechanical Engineer	(T)	Date _____
Process Engineer	(T)	

Figure 3.14: Example of a completeness summary for centrifugal pumps.

conditions. The usefulness of the example chart is enhanced by the fact that it can be displayed on a bulletin board or featured on the computerized project or plant information system for all to see. Several individuals in the startup team will be responsible for keeping the information current on a daily basis. This eliminates unnecessary informal discussion of machine status for the various systems. Also, it invites early comment if scheduled progress is not being met. Of course, for the personal information of the startup engineer, informal charts conveying similar information can be used.

	Completeness Summary for Rotating Machinery	**Rev./Date** NOTE: This summary should be used with: **Detailed Checklist for Rotating Machinery**
Block # ____ System # ____	**G/P Steam Turbines**	Equipment # ____ ____

Baseplate		Small-Bore Piping	
Case		Lubrication	
Gland Condenser Piping		Guards	
Inlet Piping			
Strainer			
Exhaust Piping			
Pipe Supports			

Remarks, Incomplete Items, Deviations, etc:

Mechanical Inspector	(M)	
Process Inspector	(P)	Name _____
Mechanical Engineer	(T)	Date _____
Process Engineer	(T)	

Figure 3.15: Example of a completeness summary for general-purpose steam turbines.

3.5.3 Assignment of Mechanics

During the equipment erection period to become aware of the distinct project phases – we alluded to them before – namely:
– Receiving, handling and inspection of equipment at the construction site.
– Equipment staging and storage protection.
– Foundations, equipment installation and grouting

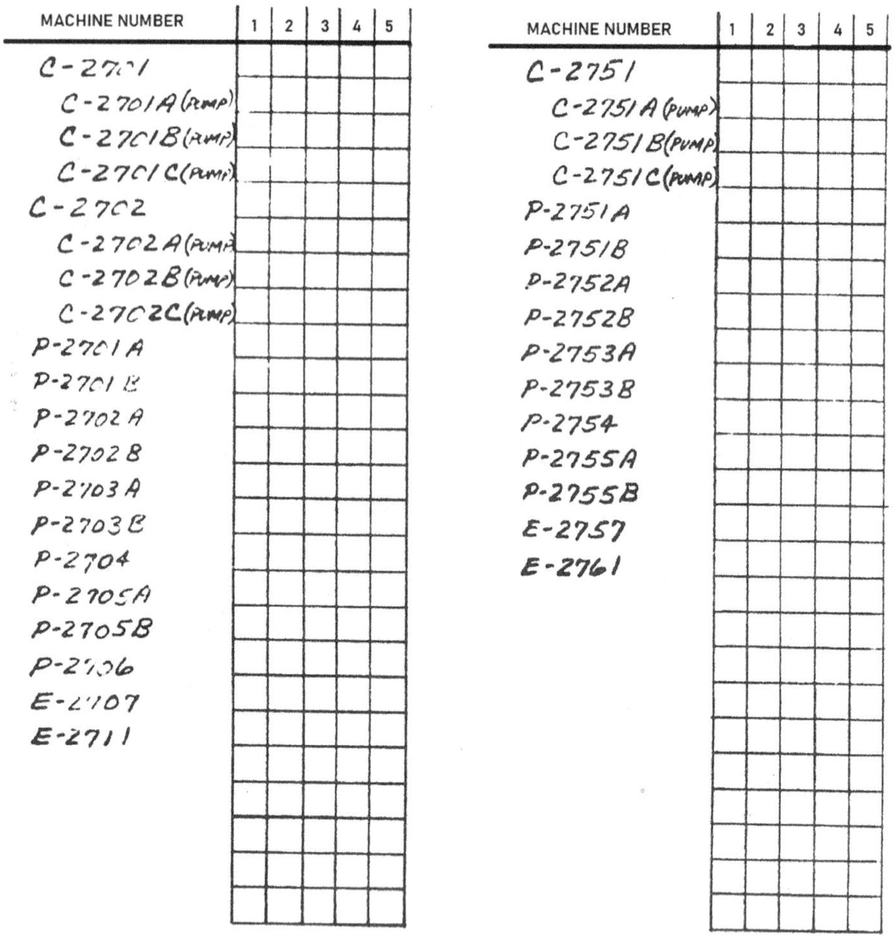

MACHINE NUMBER	1	2	3	4	5
C-2701					
C-2701A (PUMP)					
C-2701B (PUMP)					
C-2701C (PUMP)					
C-2702					
C-2702A (PUMP)					
C-2702B (PUMP)					
C-2702C (PUMP)					
P-2701A					
P-2701B					
P-2702A					
P-2702B					
P-2703A					
P-2703B					
P-2704					
P-2705A					
P-2705B					
P-2706					
E-2707					
E-2711					

MACHINE NUMBER	1	2	3	4	5
C-2751					
C-2751A (PUMP)					
C-2751B (PUMP)					
C-2751C (PUMP)					
P-2751A					
P-2751B					
P-2752A					
P-2752B					
P-2753A					
P-2753B					
P-2754					
P-2755A					
P-2755B					
E-2757					
E-2761					

1 Preliminary Acceptance   4 Machine Run-In

2 Driver Run-In   5 Final Acceptance

3 Coupling Installed

Figure 3.16: Example of a commissioning progress tracking list.

- Piping cleanliness for compressors, turbines, feed pumps and external flush systems of pumps. Best fitter practices such as parallelism of gaskets and flanges for major connections as well as proper piping supports.
- Correct machinery shaft alignment.
- Lubrication system material of construction and cleanliness.
- Millwrights (field installation specialists – also often mislabeled as machinists) should be assigned to witness and absorb machine assembly routines by participating in:

- Checking and recording critical clearances, and pre-operation settings and adjustments
- Cleaning, inspection and installation of bearings and seals
- Full inspection of the majority of centrifugal pumps
- Machinery-related instrument installation and adjustments

3.5.4 Assign the Plant's Machinery Engineers

To participate in startup activities full time. While it is not always possible to have a staff of machinery engineers or even one machinery SME (subject matter expert) on-site, large companies, especially in the hydrocarbon processing industry (HPI), can draw experienced personnel from their worldwide facilities and assign specialists on a temporary basis to a project wherever it may be. Another possibility, if we are dealing with a facility without any preceding operation – a grassroots facility – a new hire can be considered to become the resident machinery SME while getting familiar with his/her new plant during commissioning, startup and initial operation.

The purposes and benefits must be seen in the fact that machinery engineers have specialized training that enables them to respond swiftly to problems that arise during C&S. They frequently have sufficient distance from day-to-day issues allowing them to maintain overview. As part of the startup team they contribute to:
- Improved communication and implementation of advisor's recommendations
- Round-the-clock specialist manning
- Assurance of continuity upon departure of temporary startup advisors

Plant machinery engineers should join the startup team at about 50% – preferably less – completion of the project and continue their full-time startup assignment for one to two months after the plant has gone on stream.

3.5.5 Set Up Training Program for Special Machinery Operation and Repair

Ideally, the OEM FSR should be considered for this task. Frequently crafts personnel are sent to special off-site OEM training sessions. These sessions should not exceed three working days and should be subject to prior evaluation by the plant's training professional or a machinery SME.

On-site training should take place in a classroom environment and instruction should include practical demonstrations. Practical demonstrations should follow the principle of "show and tell" encouraging the OEM FSR to allow personnel assigned by the owner to do the work under his/her supervision.

If vendor personnel are not available, contract or on-loan startup advisors should be engaged as time permits. Subjects to be covered are to be:

- Verification of soundness of auxiliary systems, starting interlocks, shutdown protection – emergency shutdown functions (or ESD) – and alarms.
- Basic routines for checking and troubleshooting during machinery operations.
- Review OEM's instruction books, cross-sectional drawings and inspection procedures.
- Again, verification of completeness of records.

3.5.6 Train the Plant's Electricians, Instrument Mechanics and Operators

In machinery areas on machinery accessories such as:
- Machinery auxiliary systems, including starting interlocks, alarms, and shut-down features – ESDs
- Testing of auxiliaries during operation
- Machinery conditions requiring emergency shutdown
- Avoiding the kinds of operating errors that can damage machinery
- Lubrication systems such as major lube and seal oil supplies
- Self-acting dry gas seal support systems
- Hydraulic governors
- Prime mover management control systems
- DCS and SCADA facilities and their interfaces

3.5.7 Preparation of Specific Operating Instructions for Major Unspared Machinery

While operating manuals will not be finalized at this time, operating instructions must be prepared before run-in of the equipment. They remain "work-in-progress" until the plant is operating. They are the foundation for the final operating manuals that integrate vendors' instructions for prime movers and driven equipment – the train – into process startup instructions.

Specific startup and shutdown procedures must be prepared containing the elements of good operating manuals, such as:
- Monitoring
- Data on interlocks
- Alarms and shutdown features
- Process factors

3.5.8 Ascertain Availability of Test Equipment for Run-In and Operation

To avoid panic and consternation if and when the occasion arises, it would be well to have all necessary test equipment on hand. To mind come essential instruments which, of course, should be approved for the area where they are to be used:

- Stroboscope – high-speed coverage – to observe rotating parts such as drive couplings in motion
- Endoscope/borescope to inspect machinery internals such as combustors in gas turbines and otherwise inaccessible parts – see Figure 3.17
- Portable PMI – positive metal identification – instrument to ascertain material of construction of pipes and other components when in doubt – see Figure 3.18
- Handheld surface measuring gauge
- State-of-the-art handheld vibration-measuring equipment
- Handheld laser temperature measuring devices
- Thermal imaging devices such as thermal cameras. Thermal imaging is a mainstay of industrial monitoring and inspection for good reason. It allows the early detection of overheating, underheating, hot spots and other undesirable temperature-related conditions.
- Acoustic imaging camera
- Ultrasound leak testing device to check integrity of steam traps, compressed air, gas and vacuum equipment
- Portable real-time vibration analyzer for general-purpose equipment
- Oil condition monitor
- All conventional NDT tools and instruments

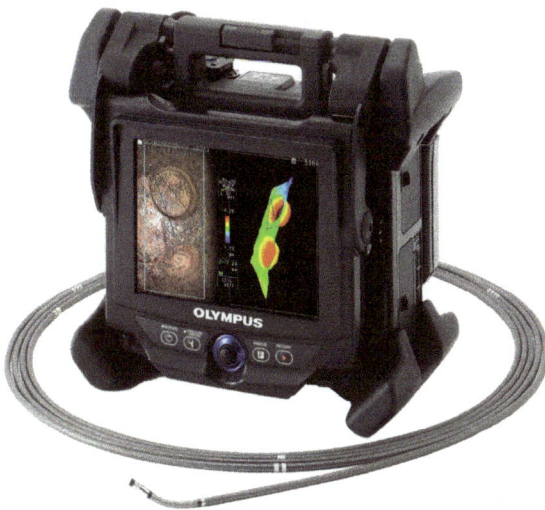

Figure 3.17: Modern borescope. (Courtesy of Olympus).

Figure 3.18: Handheld PMI instrument (courtesy of Olympus).

3.5.9 Identify Outside Sources for Special Testing or Balancing

The examination and definition of potential outside facilities is a typical contingency and risk mitigating step. These facilities may have to be relied on for rotor balancing, emergency repairs, or just plain routine work which cannot be reliably handled by the owner's or contractor's work forces. A typical question to be ask is: "What is our largest rotor size – can this particular balance machine handle this rotor?" Another example would be dismantling, cleaning and adjusting of mechanical, mechanical-hydraulic or electronic governors. A company specializing in this work may be interested in cataloging all governors installed at the facility about to be started up. Similarly, companies with expertise in shaft, bearing, labyrinth, or impeller manufacturing may want to perform cataloging and spare parts dimensional sketching services at no cost to the owner. In any event, their particular expertise needs to be identified just like that of the nearest repair facility which could rapidly handle critically important restoration of major machinery involved in a debilitating failure incident.

A list of potential outside services that could potentially be required would look like this:
- Vibration analysis – mechanical equipment
- Vibration analysis – electric motors and windings
- Dynamic balancing of rotors.
- Capacity to handle largest rotors.
- Balance quality achievable with available machines per recent experience
- Metallurgical testing laboratory

Dynamic balancing in place:[8]
- Computerized techniques available
- Special equipment required
- Skilled technicians required

3.5.10 Investigate the Availability of Expert Consultants and Contract Assistance

The best insurance against the risk of unexpected machinery problems lies in adequate pre-delivery audits and reliability reviews. They precede the C&S phase and are covered in detail in reference.[9] We alluded to this activity earlier. We see them as risk insurance measures that must be closely followed by conscientious supervision and completeness reviews of machinery installations during the time of field installation.

Commissioning instructions should be developed in cooperation with the owner's startup engineer or should, as a minimum, be submitted for his/her detailed review.

Ideally, then, the owner will draw on the expertise of his/her OEM senior professional personnel every step of the way. Unfortunately, these professionals are hardly ever available for the duration of a project – from its initial planning until the completion of the startup phase. Also, it would be rather presumptuous to assume that the owner's technical staff are thoroughly versed in all conceivable matters begging expeditious resolution during a major startup. Think, for instance, of difficulties with a sophisticated shutdown logic, electronic governors, high-speed gearing and bearings or other state-of-the-art components. Problems in any of these areas may be handled best by interfacing with capable consultants or SMEs.

Contract assistance may also prove helpful in machinery startup situations. Oftentimes, a plant can draw on the experience of capable and conscientious retirees whose background may allow them to serve as instructors, training coordinators, machinery repair supervisors, outside shop inspectors or expeditors.

One would, therefore, want to have a list of contacts available where potential availability would have been ascertained or where even a retainer would have been extended ahead of time. This would be to mitigate risks that could arise due to problems with:
– Electric motors
– Prime mover management controls
– Vibration
– Welding
– Metallurgy

3.5.11 Investigate Plan and Facilities for Repair of Remote Vendors' Equipment

– Identify possible locations
– Investigate qualifications by using a checklist – see Appendix 3.B
– Make advance contacts

3.5.12 Investigate Local Repair Facilities

- Larger machine tools than available in plant shop; special shop facilities or tools – investigate "swing" diameter on horizontal and vertical lathes
- Special casting repair techniques ("*Metalock*"®, metal stitching, and others)
- Field machining
- Welding and metallurgy

3.5.13 Vendor Assistance

The duty of the machinery startup team includes the determination of how many vendor representatives should assist in sharing machinery preparation tasks. This effort would pertain to not only their number but also their respective experience levels; even names should be identified for a given location.

Vendor assistance is required in cases where potential warranty disputes might arise, or where the owner's personnel simply do not have the expertise to assess or supervise the EPC or construction contractor. This is especially applicable to machinery packages that are put down on their foundations ready for the EPC or owner organization to provide the necessary connections. In past years, a single OEM FSR was able to commission and start up these packages. This person was trained in all necessary skills from millwrighting, for example, checking and correcting shaft alignment to instrumentation, to controls and PLC adjustment. Today, as equipment package content has become more sophisticated, several factory experts have to be scheduled and attend in order to get the equipment operational.

Vendor personnel could he given such additional assignments as spare parts review, possibly the expediting of missing parts and millwright training as alluded to above.

There should be clear procedures for obtaining services of vendor representatives for startup and run-in operations. There are definite steps:

- Determine official contact and responsible management
- Also make advance contacts for equipment where a service representative is to be on an "on call" only status, such as for governors on steam turbines, material-handling equipment, gearing, centrifuges and so on

The owner's engineer should assess the qualifications of assigned representatives quickly and obtain replacement if qualifications are unsatisfactory. If OEM's or vendor's erection advisor continues on as the startup advisor, verify that he is qualified in this area. Remember, utilize vendor representatives for training plant personnel as time permits.

3.5.14 Arrange Preventive Maintenance Details

In order to prepare the equipment for preventive maintenance interventions one should arrange for portable shelters, rotor lifting and supporting rigs and verify access path and lifting positions if mobile cranes. Do not accept statements from tradesmen to the effect that they have done it many times and no plan is required.

Further, pre-planned inspections of critical equipment for execution during unplanned, brief interruptions of plant operation have to be developed. Similarly, overall plans for preventive maintenance services on all equipment items have to be put in place.

Finally, establish computerized and easily retrievable records of inspections, repairs and replacements:

– Separate records for each major machinery item
– Use special forms, with sketches for recording vibration and other critical operating parameters
– Use vendors' instruction books and startup advisors for developing forms
– Plan overhaul technique of most critical machines in detail like:
  – Instrument air compressor
  – Fire water pumps
  – Generator sets

3.5.15 Understanding the Spare Parts Situation

3.5.15.1 Spare Parts Philosophies

This many-faceted topic is sometimes neglected because it seems to defy solution. After all, the determination of required spare parts seems to be an educated guess at best. Because of the randomness of rotating equipment failures in processing industries, it is necessary to have certain parts in stock. The focus is on *insurance* parts as opposed to operating time dependent *wear* parts. Insurance parts would be chosen based on probability of failure and the failure consequences. For example, a large electric motor – above 500 hp or 370 kW – would have a very low probability of failure or high reliability, but a winding failure could cause an outage of many weeks.

Experience has shown that some theoreticians extrapolate widely from data that apply to predictable wear-out failures. However, what transpires in HPI facilities, for example, is a function of numerous variables most of which have to do with human error. Therefore, spare parts predictions determined from mathematical models are generally too far off to merit intelligent discourse. Instead reasonable specifics are a function of equipment type, geographic location, skill levels of workforce members and other variables. For decades it has been understood that relevant answers require auditing or viewing a particular local situation.[10]

Additional risks have arisen by the emergence of legacy parts as a consequence of mergers and acquisition by engineering – OEM – companies. Reputable OEMs should have been in the position to discuss their obsolescence strategy for their products during the procurement process.[11] If this issue was not addressed, owner's engineers should be wary of any potential consequences of these market changes impacting spare parts availability.

Yet, unless a policy is established assuring spare parts have been procured and are timely on hand, a startup can be in serious difficulties before it even begins.

3.5.15.2 Spare Parts Decision Making

It is not prudent to leave spare parts decision making to mechanical FSLs and the OEM representatives alone because obviously both parties would have motives not necessarily compatible with the owner's business objectives: FSLs tend to cover all possible failure events regardless of their degree of probability of occurrence; vendor representatives are interested in selling parts. The selection of spare parts should, therefore, be made by experienced reliability professionals with input from OEM representatives and experienced millwrights or mechanical trade FLS. For instance, when it comes to a decision to stock a gear set for a critically important machinery train, one has to deliberate whether or not to stock a through-hardened gear at all. Many users never had to stock through-hardened gears because of their unique durability. Another typical example would be, again, a large electric motor drive where one would store winding material in hermetically sealed containers in order to expedite a rewind should the contingency occur. Similarly, one would consider creating both an actual and a virtual spare parts inventory by having the OEM manufacture "multipurpose" spare parts. Multipurpose parts are produced with oversize dimensions, which, upon finish-machining on site, will suit the specific requirements of a specific piece of equipment.[12]

A review of failure statistics can provide help with the justification for spare parts acquisition. Any recommendation should be based on a risk assessment.

Historically, problems with major unspared machinery have caused commissioning or startup delays in one out of two train installations without reliability reviews, and in one out of six or seven train installations employing pre-installation reliability reviews. In either case, centrifugal compressors and steam turbines have experienced the majority of the difficulties, followed by large gear speed-increaser units – see also Section 11.2 – electric motors in size categories above 1,000 HP or 750 kW, reciprocating compressors, gas turbines, and gas expanders.

Spare parts management has been continually evolving. The startup engineer must stay abreast of his/her company's spare parts policies and systems in order to potentially avoid losing his decision-making role to a newly introduced enterprise resource planning system (ERP).[13]

Essential machinery insurance spare parts are tabulated in Appendix 3.C. This table makes an attempt to list spare parts recommended to be on hand for startup and initial operations as well as spare parts to be stored for routine, post-startup operation. Risk-based, the relative frequency of replacement or repair is indicated by numbers ranging from 5 (high frequency) to 1 (low frequency).

A final advice that should not be overlooked: Whenever the spare part also fails, do not simply order another one. Spend time in thoroughly finding out the failure cause and proceed to design an engineered or suitably upgraded part and have it manufactured without delay.

3.5.15.3 Spare Parts Storage and Retrieval

The best source for information and sound recommendations on spare parts storage and retrieval is not necessarily an existing process plant in the same business. A reputable consulting company specializing in warehousing and inventory control is often better qualified to set up hardware and software systems to serve this function adequately and efficiently.

Inefficient storage and retrieval will result when major turbomachinery rotors are not hung vertically or when they have to be shipped to storage facilities away from the plant location.[14] Inefficient storage and retrieval will also result when spare parts for one compressor are stored at one end of the building and parts for another compressor are stored near the opposite end, and when parts are not properly cross-referenced, preserved or labeled with the vendor's parts and drawing numbers in addition to the owner's spare parts identification.

The advent of sophisticated electronic controls for machinery-related instrumentation – for example, electronic governors – makes it necessary to pay attention to their suitable preservation and storage. Dust, moisture and crush-proof packaging are indispensable. While spare parts for major machinery should be stored in designated areas clustered in close proximity, sensitive electronic components for the same machine are best not located in the same bin with heavy mechanical components.

The storage of large turbomachinery rotors presents special risks down the line as well as many opportunities for good or bad solutions. One preferred method is to preserve the rotors with an oil-derived coating and hung vertically from large cross beams. The building is humidity controlled, but does not require heating or cooling. However, we owe it to our readers to mention that many users have stored their rotors horizontally in well-designed shipping boxes in heated storage without any adverse effects and no requirement for balancing prior to deployment. These rotors were visited once a year and inspected.

Storage and retrieval methods must represent a logical compromise satisfying several critical requirements. For instance, the night-shift repair crew must have access to spare parts without the help of daytime personnel. At the same time, procedures should be in places which give assurance that the removal of parts from stores is recorded so as to have an up-to-date reading of true inventory levels and re-order requirements.

Remember that many assemblies or machines in storage could (and should) be stored while filled with oil mist. If you do not know about oil mist, you are 40 years behind in your reading and in your thinking.

3.5.15.4 Spare Parts Documentation

Money spent in documenting spare parts locations, inventory levels and reordering quantities, and in cross-referencing vendor designations, drawing numbers, bills of materials, owner's storage codes and so on is money well spent. Cross-sectional drawings of major machinery should be combined with component number designations and all other relevant cross-references to enable mechanical work forces to locate parts without wasted time or motion.

Spare parts identification sheets differ from conventional spare parts documentation or traditional storehouse information in a number of ways. They are primarily intended as an aid to mechanics, machinists and turnaround planners. These persons require that spare parts information be contained on a single sheet, not in separate catalogs or on computer printouts. In many cases, illustrations are required for positive identification of parts by personnel unfamiliar with either the machinery or the storehouse routine. An example is illustrated in Figure 3.19.

Major machinery spare parts documentation sheets must contain all the information needed by mechanical work forces to locate the parts in the storehouse. These documentation sheets must allow mechanics, turnaround planners and inspectors to verify stock levels, critical dimensions and suitability of parts. Cross-reference, design-change and inspection information complete the sheet and make it a stand-alone, highly useful document. The groundwork for the foregoing should be laid during pre-commissioning, commissioning and initial startup.

During the construction phase and before handover spare parts continue to be delivered. It is therefore necessary to have a good overview of the delivery status. Further, warehouses must be suitably organized for the pre-startup period by staging parts to be readily available to craftspeople. Finally, there ought to be a parts audit and a review of general house-keeping.

3.5.15.5 Machinery Maintenance Tools

These are special tools furnished by the equipment OEM. They need special attention as they can represent a risk for startup delays. They should be available, up-to-date and usable.[16] There have been numerous incidents in the past where important assembly and lifting tools furnished by the OEM had to be modified on site. All the while, the OEM's manufacturing department had been using changed-to-suit shop assembly tools without the necessary modifications to the tool kit furnished to the customer.

LIST OF SPARES

| WORKS ORDER | 352586 | PAGE | 5/ |

PROCESS GAS LOW STG   NUMBER 1 Set   MACHINE NO. LCT01

MANUFACTURED BY: HIROSHIMA WORKS

SERIAL NO.	PARTICULARS	MATERIAL	NO. OF SUPPLY		DRAW-ING NO.	PART NO.	SKETCH	REMARKS
			WORKING	SPARE				
24	5th Stage Diaphragm and Nozzle	SUS403 SC46	1 Set	1 Set	765 -81962 765 -81974	2515 -11-16 21-26 2715 -11-14	90 / 1205" SIN 66904-825	US $ 17,200
25	6th Stage Diaphragm and Nozzle	SUS403 SC46	1 Set	1 Set	765 -81963	2516 -11-16 21-26	90	
2~	7th St~							

SPARE PARTS CROSS-REFERENCE LISTING

```
                      .  6415-14          EA     18015811
                     MFG. CO TURBINE
                IND. 6415-14

         . CENTER FOR EXT CONTROL VALVE       EA
          PT.NO. 4814-13 ACME MFG. CO
          ACME MFG. CO              4814-13
          LCT01 LCT02

90432-728   SPINDLE, REGULATING VALVE, PT. 4424-11    ST    N0386
            DWG. 765-82245 FOR ACME MFG. CO TURBINE
            ACME MFG. CO HEAVY IND. 4424-11
            LCT01

90432-760   TUBE, COOLING FOR GLAND CONDENSERS PT.    EA    N036C
    6       DWG. 764-10391 FOR ACME MFG. CO
            TURBINE
            ACME MFG. CO HEAVY IND. 6
            LCT01

90432-763   TURN BUCKLE FOR GOVERNING LEVER           EA
            PT.NO. 5841-47 ACME MFG. CO TURBINE
            ACME MFG. CO              5841-47
            LCT01 VCT01

90432-764   TURN BUCKLE FOR GOVERNING LEVER PT.NO.    EA
            5843-32 ACME MFG. CO TURBINE
            ACME MFG. CO HEAVY IND. 4424-11
            LCT01
```

Figure 3.19: List of spares cross-referenced.[15]

3.5.16 Outline Lubrication Requirements

In order to escape confusion due to proliferation of lubricant types and future problems with housekeeping, man-hours must be dedicated to accomplishing the following tasks:

- Determine lubrication requirements for each machine – see Figures 3.20 and 3.21.
- Verify appropriate product equivalent of vendor-recommended lubricants in order to reduce lubricant inventory.
- Procure ample quantity for startup.
- Stock quantity to replenish possible seal leakage.
- Allow and plan for ways to discard initial charge.
- Plan for periodically sample test during operations.
- Determine oil sample laboratory.

3.6 Machinery Startup Review Tasks

The preceding section outlined startup preparation in broad terms. These preparatory tasks can be further broken down into completeness reviews, quality assurance tasks, cleanliness checks and other activities as shown in Figure 3.22.

3.6.1 Building Checklists

Executing these review tasks will be accomplished by a series of checklists. Generally, a commissioning checklist is used to easily perform comprehensive installation, functional, and operational inspections, resolve any detected failures and prevent costly project delays. These documents must be developed to help personnel conduct the reviews and execute the various tasks in reasonably uniform fashion. First and most important is a checklist restating every construction- and installation-related item contained in the job specification documents. Typical checklists, albeit not uniform in terms of format, are presented in the following chapters.

Machinery startup review personnel must use these checklists to verify that the installation complies with the job specification and that it is ready for imminent startup. Having observed compliance or deviations, the engineer, technician or millwright should note his or her comments on the appropriate completeness summary.

MANUFACTURER/SUPPLIER: BRAND NAME:	**LUBRICANT CLASSIFICATION FORM** OIL NO.: _____

| MINERAL OIL ☐
SYNTHETIC OIL ☐ |

| CLASSIFICATION BY TYPE OR USE: |

VISCOSITY:	Temperature (F)	Centistokes
	0°	
	100°	
	210°	

| VISCOSITY INDEX: |

| POUR POINT: |

| WATER COMPATABILITY (DEMULSABILITY): |

| NEUTRALIZATION NUMBER: |

| ADDITIVE TYPES: |

| COMPATABILITY WITH EQUIVALENT LUBRICANTS:

Can this oil be mixed with other oils of which the above characteristics are identical without any disadvantages?   Yes☐   No☐ |

| SPECIAL CHARACTERISTICS: |

| SPECIAL APPLICATIONS: |

| PRICE: |

| COMPANY CLASSIFICATION:

Universal Maintenance Catalogue Code _____

Company Product Number _____ |

Figure 3.20: Lube oil specification sheet.

MANUFACTURER/SUPPLIER:		LUBRICANT CLASSIFICATION FORM	
BRAND NAME:		GREASE NO. _____	
SOAP TYPE: calcium / lithium / sodium / lead / barium / aluminum / other			
TYPE OF SERVICE:			
WORK PENETRATION NUMBER (AT 77° F):			
DROPPING POINT:			
TEMPERATURE RANGE:			
RESISTANCE TO WATER:			
ADDITIVE TYPES:			
MINERAL OIL:		% of weight	Viscosity at 210° F
SPECIAL CHARACTERISTICS:			
SPECIAL APPLICATIONS:			
PRICE:			
COMPANY CLASSIFICATION: Universal Maintenance Catalogue Code: _____ Company Product Number _____			

Figure 3.21: Grease specification sheet.

3.6.2 The Mechanical Procedures Manual

Procedures are written routines or instructions that describe the logical sequence of activities required to perform a work process and the specific actions required to perform each activity. If there are no written procedures, there is no basis for monitoring performance, focus for improvement or mechanism by which to capture learning. The establishment of procedures and checklists allow more time and mental energy to deal with the unexpected, which always happen during commissioning. C&S procedures

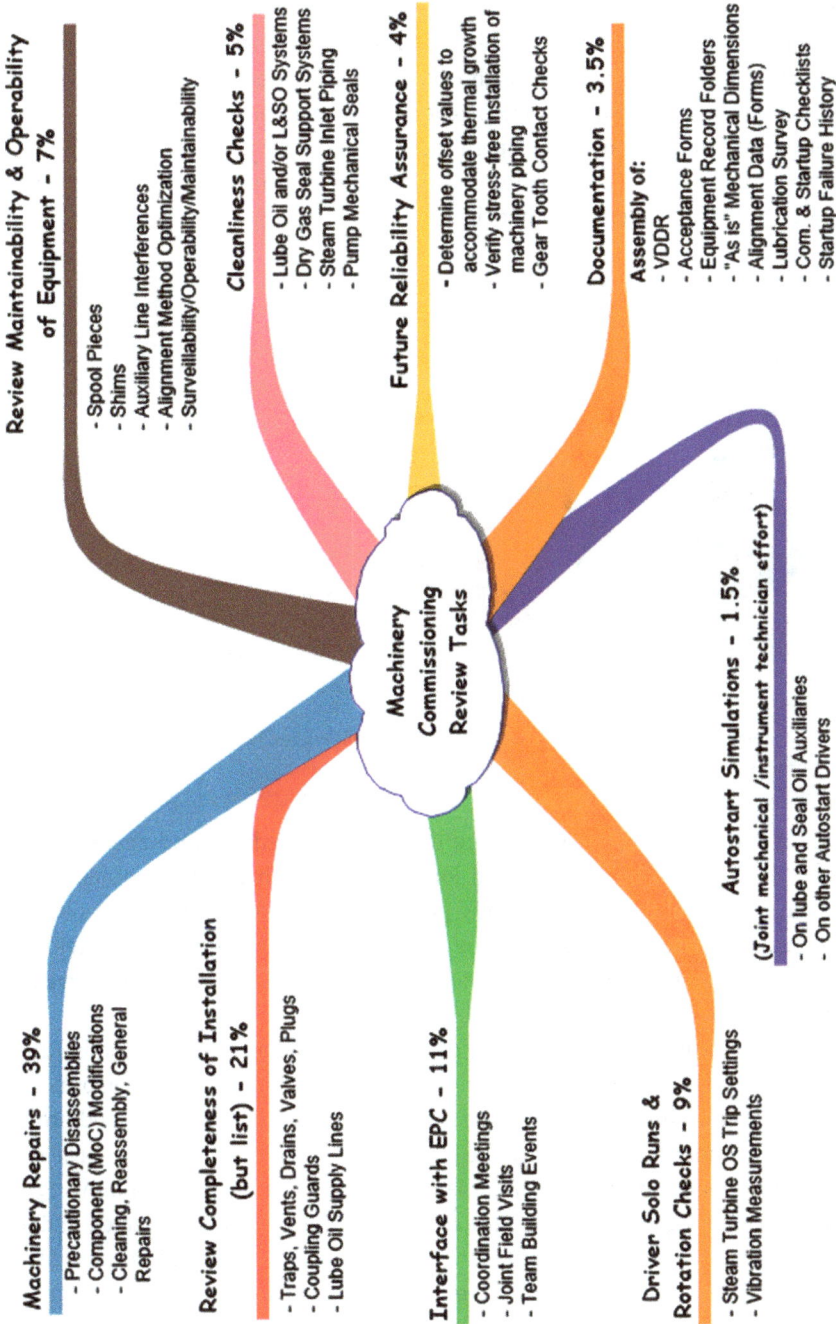

Figure 3.22: Startup team man-hour application percentage for field construction completeness review tasks associated with the startup of a large steam cracker (adapted from reference[17]).

The following content represents the mind-map diagram:

Machinery Commissioning Review Tasks (central node)

Review Maintainability & Operability of Equipment - 7%
- Spool Pieces
- Shims
- Auxiliary Line Interferences
- Alignment Method Optimization
- Surveillability/Operability/Maintainability

Cleanliness Checks - 5%
- Lube Oil and/or L&SO Systems
- Dry Gas Seal Support Systems
- Steam Turbine Inlet Piping
- Pump Mechanical Seals

Future Reliability Assurance - 4%
- Determine offset values to accommodate thermal growth
- Verify stress-free installation of machinery piping
- Gear Tooth Contact Checks

Documentation - 3.5%
Assembly of:
- VDDR
- Acceptance Forms
- Equipment Record Folders
- "As is" Mechanical Dimensions
- Alignment Data (Forms)
- Lubrication Survey
- Com. & Startup Checklists
- Startup Failure History

Machinery Repairs - 39%
- Precautionary Disassemblies
- Component (MoC) Modifications
- Cleaning, Reassembly, General Repairs

Review Completeness of Installation (but list) - 21%
- Traps, Vents, Drains, Valves, Plugs
- Coupling Guards
- Lube Oil Supply Lines

Interface with EPC - 11%
- Coordination Meetings
- Joint Field Visits
- Team Building Events

Driver Solo Runs & Rotation Checks - 9%
- Steam Turbine OS Trip Settings
- Vibration Measurements

Autostart Simulations - 1.5%
(Joint mechanical /instrument technician effort)
- On lube and Seal Oil Auxiliaries
- On other Autostart Drivers

and checklist also provide the basis for meaningful maintenance work instructions and checklists for critical assets when the plant is operating. "Then, assuring that people actually follow the work instructions is the most critical element of reliability. After all, equipment reliability is a result of what people do to the equipment."[18]

Root cause failure analysis in operating plants frequently shows that failures stem from nonexistent or inadequate procedures (see Figure 3.23). A similar justification applies to the existence of checklists.

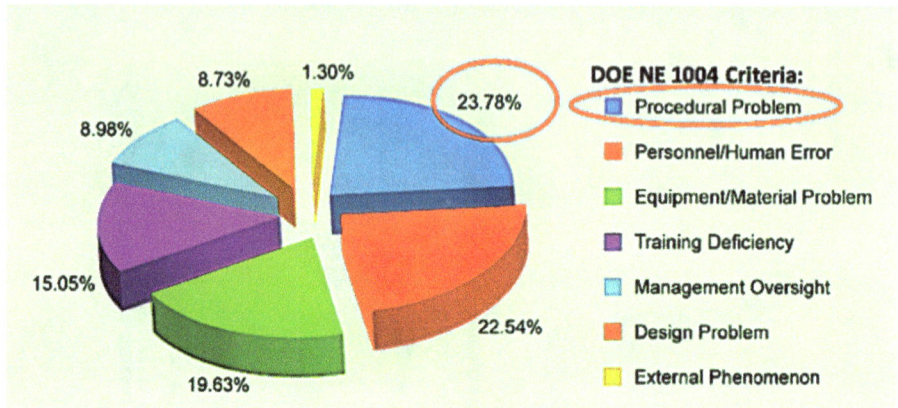

Figure 3.23: Experience suggests that lack of or ineffective procedures is the largest contributor to failures in process plants.[19]

The authors have had experiences where the use of procedures and checklists for equipment C&S had to be explained in long sessions with ensuing counter arguments from personnel that would greatly profit from the use of them. Sometimes it became a cultural issue, accompanied by personal pride, leading to the conviction that one would have no need for "crutches." Startup engineers and supervisors have to enforce the use of these very necessary tools.

Procedures have to be detailed and reflect the experience of the authors. It is often a good idea to consult or sequester experienced senior operating personnel in order to get the procedure written. A case in point stressing the fact that good procedures can have enormous paybacks: The cleaning of the inlet piping for a large mechanical drive steam turbine was imminent. Pre-commissioning steam-blowing to clean the pipe was indicated. The piping from the turbine was disconnected and an elbow installed to direct the steam flow to atmosphere. A polished aluminum plate was mounted a few inches or centimeters from the opening, serving as a target to monitor the depth and density of particles impacting during steam blowing to accomplish pipe cleanliness.[20] Refer also to Figure 5.5.

Velocity-based steam blowing procedures will mainly carry away loose debris. Conversely, weld splatter – often referred to as "icicles" – will break loose only when temperature cycling and the attendant expansion and contracting of piping is brought to bear. An effective steam blowing procedure will call for this temperature cycling. It would demand that a silencer be mounted at the upturned piping and low-velocity blowing continued until such time as the piping reached a temperature approaching that of the hot steam. At that time blowing would have to be suspended until the piping cooled off to near ambient temperature, and this heating and cooling cycle would be repeated several times. Undoubtedly, this steam blowing procedure will take longer than the often-unsatisfactory velocity-based method. It is known that in several cases the latter method was responsible for millions of dollars in damage, startup delays, outage time and litigation fees.

Cheap and fast procedures are seldom a best choice. Jobs hurried up may have to be repeated several times over. Every one of the best-in-class (BiC) performers arrived at the top by developing best practices using well defined, repeatable and recorded routines. They rarely rely on an employee's memory, placing, instead, confidence and expertise in the written word. Good procedures mitigate risk and create a safe work environment.

The goal of all planning and preparation is to place resources and create tools in order to follow and document the installation of plant equipment as part of the project's quality assurance effort. Ideally, the equipment is readied for turnover and subsequent commissioning as a joint effort by both owner's and contractor's personnel.

3.6.2.1 Standard Operating Procedures (SOP)

Most BiC companies have highly developed standard operating procedures (SOP) management systems. As an important part of the companies' technical and operational inventory, they mitigate risks arising from unplanned and rash activities in the workplace – see Figure 3.24. SOPs reflect the desire to safely and at the same time efficiently perform all operational sequences regardless how routine and commonplace they may be.

Figure 3.25 illustrates the administrative behavior around SOPs clearly reflecting the importance attached to the subject: The manufacturing manager sponsors the effort and signs off on completed forms.

A risk assessment is always a part of building a SOP. This extends to the modified version of any SOP. Figure 3.26a and b illustrates the approach.

In the following chapters, we do not intend to dwell on administrative matters and format, but shall try, instead, to convey technical details contained in procedures and checklists as we make the transition into the mechanical completion, turnover and pre-commissioning phase. Additionally, if owner–contractor interfaces will be harmonious, commissioning will be successful and asset service factors goals will be most likely met.

S/No.	Area	Equipment Type	Area	Procedures Description	Procedure Type	File No.
10	Site Wide	Blower	Utility	Centrifugal Blower Pre T/O Checklist	Commissioning	
11	Site Wide	Motor	All	Motor Solo Run (Log Sheet)	Commissioning	
12	Utility	Belt Press	Utility	Belt press commissioning Checklist	Commissioning	
13	Site Wide	Oil/Grease	All	Preserving Oil removal and preservation grease removal	Commissioning	
14	Site Wide		All	Annual GP Turbine Regulatory System Test Procedure	Commissioning	
15	Site Wide	Turbine	All	Turbine Checklist Operations	Commissioning	
16	Site Wide	Mixer	Utility	Machinery Mixer Checklist Operations	Commissioning	
17	Site Wide	Oil Mist	All	Oil Mist System commissioning & checklist	Commissioning	
18	Site Wide		All	Lube Oil System Flushing Procedure For Rotating Equipment With Skid.	Commissioning	
19	Site Wide	Oil Flushing	All	Manual Oil Flushing/Change Out Procedure For Between Bearing & Overhung Pump	Commissioning	
20	Olefin		Cracker	Mechanical Run Procedure	Commissioning	
21	Olefin		Cracker	Air Run Procedure	Commissioning	
22	Olefin		Hot End	ST-21A&B Electronic Over-Speed Trip System Commissioning & Turbine Solo Procedure	Safety Critical	
				ST-11A&B Electronic Over-Speed Trip System Commissioning &		

Figure 3.24: Example excerpt from a typical tabulation of standard operating procedures.

Commissioning and Startup Team
Chartered by
Manufacturing Manager

Facilitator
Appointed
(Process Engineer)

Machinery and I&EC
(SMEs & FLS)

SOP Development
(Team: PE, PFLS, ME, I&CE)

Operations
Input
(PE & PFLS)

Integrated
SOP
Narrative

Risk Assessment

SOP Narrative
Team Review

Process Simulation
Report*
(If Available)

Changes

Sign-Off
By
Manufacturing
Manager

Notes:
* Done by 3rd Party
PE   Process Engineer
SME Subject Matter Expert
FLS   First Line Supervisor
PFLS Process FLS
ME   Machinery Engineer
I&EC Instrument & Control Engr.

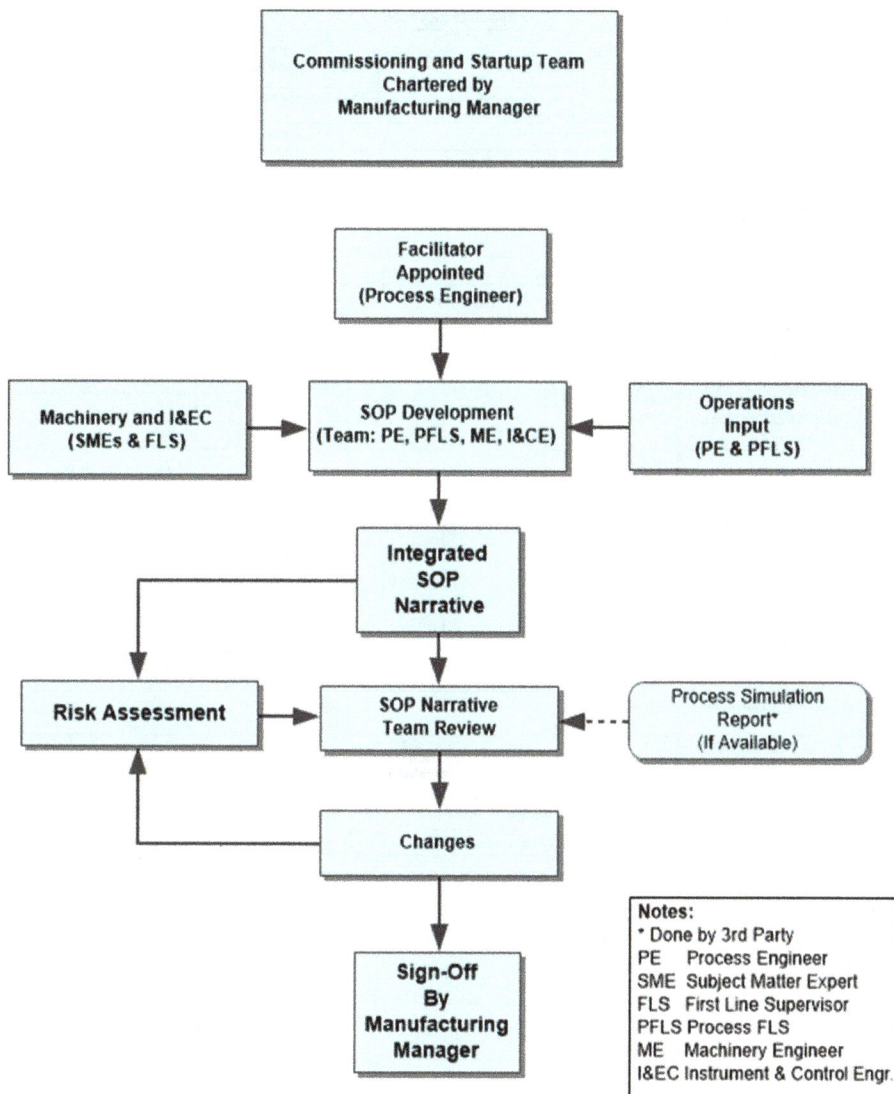

Figure 3.25: Standard operating procedure administration.

Chemical Plant & Refinery						Procedure Change Form					Rev. 3		
Name of Originator	Froelig, K.							Date Submitted					
Type of Change:	Operations			Maintenance				Change initiated by MOC?		Yes		No	x
	Modification/Others			New - (Requires IH & Environment Review)			x	If Yes Indicate MOC No.					
Procedure Title:	CT 1001 Cooling Tower Fan Commmissioning Checklist							New Revision Number			00		
Affected Area	SOP2	x	HE	x	RE	x	UT		M&E		Others		
	SPA		SPE		SPP		SAR		LAB		Analyzer		
Refinery BU	S1		S2		S3		S4 PAC		S4 Jurong		LAB		Others
	Machinery	x	Metals		IEA			Services					
SITE-WIDE	Safety Manual, Work Practices etc. Note: Approvals required for each BU, use multiple forms if necessary												
Change Assessment	Startup Team: Operations:				Maintenance:						/ Skills Planner		

Hazard Analysis	Training	Responsible Person	Initial Completion	Documents Updated	Responsible Person	Initial Completion
Previous Criticality Rating	FLS review with Technician prior to execution	FLS		MPSS	Jennie	
New Criticality Rating				Gems	Lor Shen Hong	
Priority 3				OIL		
Low Hazard	ROU / SSEP / MP Signoff			**Other Documents Affected by the Change**		
Not Operability Critical				Operating Manual	M&E Inspection	
Priority 2						
Significant Hazard	Conduct Classroom Training			Safety Manual	Environment Manual	
Operability Critical				Analytical / Laboratory	ISO / Business Controls	
Priority 1				Initiate an MOC to manage change of other documents affected by this procedure change		
(S) Safety or Health Critical	Conduct Certification Process:					
(E) Environmentally Critical	Completion Date			Document Name	Not applicable	
				MOC Number:	Not applicable	
(P) / (O) Critical Interface Outside Affected Area or Operability Critical	Responsible party to initial when their portion of the training and / or documentation follow-up is complete and return the original to the Procedure Administrator for closeout.					

		Check	Job Title	Initials	Signature	Date
Review	Required for all	X	Operations MC / Maintenance Supervisor	RGE		
			Contact Engineer			
	Minimum of one req'd for all procedures	X	Equipment Engineer	HPB		
			Industrial Hygienist			
			Other Engineer			
	All Priority 1		Senior Technical Engineer			
	Req'd for New procedures		IH Engineer			
			Environmental Engineer			
Validation	Minimum of one required for all procedures		FLS (Operations)			
			Technician (Operations)			
		X	FLS (Maintenance)	Al Walsh		
			Technician (Maintenance)			
Approval	Priority 1 (S) ↩		SOC or OSOC			
	Priority 1 (E) ↩		Environmental Engineer			
	Priority 1 (P)/ (O) ↩		Operations MLM or BTM of Affected Area			
	Required for all	X	Operations / Maintenance - MLM or BTM	Bill Schostak		

Figure 3.26: (a) Typical procedure change form. (b) Typical procedure risk assessment form.

Probability (X)

Evaluate the Probability of a Failure Event: Score = 1 for YES; Score = 0 for NO. Sum total for X.

	PROBABILITY SCENARIO	Score
1	Is the procedure complex? Does it involve many steps?	
2	Is the procedure infrequently used, i.e., used < 6 momths?	
3	Are more than three persons involved in executing the procedure?	
4	Is the activity NOT planned? Less than 15 min warning to execute.	
5	Are the persons involved in the work execution also handling other activities concurrently with this one?	
	Sum Total of X	

Consequence Risk Matrix (Consequences might have to be defined and explained for the site)

Circle consequence level for each category. The highest consequence level determines Y & Z score table below.

Consequence Category	Consequence Determinants			
	Health & Safety	Environmental Impact	Public Impact	Financial Impact
4	Fatality(ies) Serious Injury Requiring Medical of Members of the Public	Potential Widespread, Long Term, Significant Adverse Effects; Major Emergency Response; Long Term Cleanup	Significant Public Disruption; Extended National or International Media Coverage; Large Community Impact; Large Scale Evacuation; Major Road Closure > 24 Hours	On Corporate Level > $11M
3	Serious or Lost Time Injury or Illness	Potentially Localized, Medium Term, Significant Adverse Effects, Itermediate Emergency Response; Weeks/Months of Cleanup	Small Public Disruption; One Time National or Extended Local Media Coverage; Medium Community Impact; Small Scale Evacuation; Major Road Closure < 24 Hours	Business $1M - $11M
2	Restricted Work or Medical Treatment	Potential Short Term Minor Adverse Effects Local Emergency Response; Days/Weeks of Cleanup	Public Complaints; One Time Local Media Coverage Small Community Impact; Secondary Road Closure < 24 Hours	Site Operation $100k - $1M
1	First Aid or Minor Injury	Inconsequential or No Adverse Effects; Confined to Site or Close Proximity	Public Complaint; No Media Coverage; Temporary Closure of Side Road; Minor Inconvenience	Other < $100k

Check One Severity / Consequence Scenario.
Enter corresponding score for Severity (Y) and Consequence (Z) and Probability (X) in the last column "Risk Scoring Value"
Calculate Risk Score (R) with the formula Risk = Y (X+ Z)   Procedure with Risk Score >= 12 is a critical procedure

Check One	Severity / Consequence Scenario	SCORE		Risk Scoring Value	
		Y (Severity)	Z (Consequence)		
	4 · Risk Matrix Severity / Consequence	4	2	X	
	3 · Risk Matrix Severity / Consequence	3	1	Y	
	2 · Risk Matrix Severity / Consequence	2	1	Z	
	1 · Risk Matrix Severity / Consequence	1	1	**Risk =Y(X+Z)**	

Hazard Analysis critically rating indicate: S= Safety, E= environmentally, P= Critical Interface, O = Operability critical

Score R	Hazard Analysis	Criticality Rating	Additional Reasons for critical rating
>= 12	Priority 1		
> 8	Priority 2		
< 8	Priority 3	O& S	

Assessed By (Name)					
Signature					
Date:					

Figure 3.26 (continued)

Appendix 3.A: Project coordination procedure – example excerpt

KLM Technology Group Project Engineering Standard	PROJECT COORDINATION PROCEDURE (PROJECT STANDARDS AND SPECIFICATIONS)	Page 1 of 23
		Rev: 01
		Feb 2012

TABLE OF CONTENTS

Appendix 3.B: Checking out a mechanical contractor

#	Questions	Answers/comments
1	What kind of service does the company provide?	
2	Is it a full-service operation?	
3	What will it cost? (Price should not be the primary factor!)	
4	How long has the company been in business?	
5	Does it have assets?	
6	Does it carry insurance?	
7	What is its turnover rate?	
8	Does the potential contractor insist on surveying the client's premises?	
9	How dedicated to quality is the contractor?	
10	What kind of training does he/she provide?	
11	Does it have assets?	
12	Does it carry insurance?	
13	How are his/her employees hired and how well technically qualified are they?	
14	Are they permanent?	
15	Or are they "temporaries" hired for each job?	
16	What are the backgrounds, skill levels and experience of the contractor's supervisors or lead persons?	
17	It is important and necessary that the lead person have a background, for example, in compressor MR&O?	
18	What are the backgrounds of the owners or managers of the company?	
19	Will the contractor supply a complete report and record of all works performed?	
20	Are references, examples and so on available for review?	
21	What are the warrantee arrangements?	

Appendix 3.C: Major machinery recommended insurance spare parts – Based on component failure risk

Equipment and component	Frequency[1]	S/U spares	Post S/U spares
El motors (induction)			
Bearings	2	1 set	1 set
Oil seals	4 W	1 set	1 set
Stator	2	Windings[2]	Windings[2]
Rotor	1	–	–
Fan	1	1 unit	1 unit
Steam turbines			
Gland packing	4 W	1 full set	1 full set
Interstage packing	4 W	1 full set	1 full set
Bearings	2	1 full set	1 full set
Rotors	2	Full spare	Full spare
Nozzles	2	Partial set	Partial set
Diaphragms	2	–	–
Admission valves	2 W	Packing 2 sets	Packing 1 set
Governor	2	Full spare	Full spare
Hydraulic actuator	2	Rebuild kit	Rebuild kit
Governor drive gear	4	Full spare	Full spare
Turning gear	1	–	–
Nozzle ring fasteners	2 W	1 full set	1 full set
T&T	1	Internals	Internals
Gas turbines			
Combustors	4 W	1 full set	1 full set
Compressor rotor	2	Full spare	Full spare

---

1 W represents wear part.
2 Winding material has to reduce procurement time when needed.

(continued)

Equipment and component	Frequency[1]	S/U spares	Post S/U spares
Bearings	2 W	2 full sets	1 full set
Power turbine rotor	2	Full spare	full spare
Fuel controls	3	Partial set	Partial set
Speed controls	2	Partial set	Partial set
Gear – speed increasers			
Bearings	3	Full set	Full set
Bull gear	2[3]	Full spare[1]	Full spare[1]
Pinion	2[1]	Full spare[1]	Full spare[1]
Coupling	2	1 full spare	1 full spare
Oil seals	4 W	1 set	1 set
IC engines (gas fueled)[4]			
Main bearings	2	1 full set	1 full set
Valve gear	3 W	1 full set	1 full set
Piston rings	3 W	1 full set	1 full set
Piston	3	–	–
Lube oil filters	3 W	1 full set	1 full set
Spark plugs	3 W	1 full set	1 full set
Ignition system	3 W	Partial set	Partial set
Reciprocating compressors			
Crankshaft	1	–	–
Cylinder liners	2 W	Partial set	Partial set
Pistons	2	–	–
Piston rings/rider rings	3 W	1 full set	1 full set

---

3 Requires engineering study – gear designs differ. Some gear sets do not require spare parts.
4 With large fleets of gas engines throughout North America, just-in-time spare parts service providers abound. There is no requirement to stock expensive spare parts. IC engines require operating time-dependent IMO&R.

(continued)

Equipment and component	Frequency[1]	S/U spares	Post S/U spares
Piston rods	2	Partial set	Partial set
Valves	5 W	1 full set	1 full set
Crosshead slides	1 W	Partial set	Partial set
Crankshaft brg. inserts	2	1 full set	1 full set
Connecting rod bolts	1	1 full set	1 full set
Connecting rod bearings	1	1 full set	1 full set
Oil pump/lubricators	2 W	1 full set	1 full set
Helical screw compressors and PD blowers[5]			
Bearings	2	1 full set	1 full set
Seals	2 W	1 full set	1 full set
Thrust bearing	2	1 full set	1 full set
Rotors	2	1 full set[6]	1 full set[6]
Coupling	1	1 full spare	1 full spare
Timing gear[7]	1	1 full set	1 full set
Turbocompressors			
Bearings (journal)	3	1 full set	1 full set
Thrust bearing	2	1 disk and pads	1 disk and pads
Rotors (shaft/impellers)	1	Spare rotor	Spare rotor
Couplings	2	1 full spare	1 full spare
O-rings	3	Full set	full set
Labyrinth seal	2 W	1 full set	1 full set
Seals – DGS	2	1 full spare	1 full spare
Seals – oil film	4 W	1 full set	1 full set

---

5 Most helical screw compressor and blower vendors have replacement by core exchange policies.

6 Suggestion for a machinery population is >1.

7 Oil-free compressor.

(continued)

Equipment and component	Frequency[1]	S/U spares	Post S/U spares
Oil filter elements	4 W	1 full set	1 full set
RTDs for bearing temp.	3 W	1 full set	Full set
Turboexpanders			
See corresponding parts under steam turbines			
Frequency key			
No.	Event/train-year		
5	1 every 1–2 years		
4	1 every 2–3 years		
3	1 every 3–4 years		
2	1 every 4–5 years		
1	1 every 6–8 years		

References

1 The trade classification "machinist" varies regionally. Equivalent terms are *millwrights, artisans* – as used in South Africa – *mechanics, fitters and technicians.*
2 SCADA – Supervisory Control and Data Acquisition.
3 1. American Petroleum Standards API RP 686, 2nd Edition, 2009, *Recommended Practice for Machinery Installation and Installation.*
2. Design and/or API RP 1FSC, 1st Edition, 2013, *Facilities Systems Completion Planning and Execution.*
4 American Petroleum Standard API 612, Petroleum, Petrochemical and Natural Gas Industries – Steam Turbines – Special-Purpose Applications.
5 American Petroleum Standard API 610/ISO 13709, Centrifugal Pumps for Petroleum, Petrochemical and Natural Gas Industries.
6 American Petroleum Standard API 682, Pump-Shaft Sealing Systems for Centrifugal and Rotary Pumps, Fourth Edition, May 2014.
7 The actual detailed checklist can be found in Table 16.3.
8 Sulzer sales literature.
9 Heinz P. Bloch, *Improving Machinery Reliability*, Volume I, Series *Practical Machinery Management for Process Plants*, 3rd Edition, 1998, Gulf Publishing Company, Houston TX.
10 Bloch, H.P. & Geitner, F.K., *Machinery Component Maintenance and Repair*, Series *Practical Machinery Management for Process Plants*, Volume III, Fourth Edition, ELSEVIER, 2019, and preceding editions, 650 Pages, Page 500.
11 Klaus Brun and Rainer Kurz, *MYTH: MY GAS TURBINE OEM WILL BE AROUND FOREVER*, Turbomachinery International • May/June 2013, Page 40.
12 H. P. Bloch, *Consider both actual and virtual spare parts inventory*, Hydrocarbon Processing, May 2013, Pages 69 and 70.

13 K.R. Iyer, Manage project spares efficiently with these best practices, Hydrocarbon Processing, December 2020, Pages 49 to 53.
14 See reference 9, Page 364.
15 Ibid., page 19.
16 See Chapter 14, Figure 14.9. (Assembly Tool for Centrifugal Compressors)
17 Heinz P. Bloch, *Improving Machinery Reliability*, Volume 1, Practical Machinery Management for Process Plants, 1st Edition, 1982, Gulf Publishing Company, Houston TX.
18 Bob Williamson, The Challenges of Work Instructions, RAM REVIEW, Oct.2020, *bwilliamson@-theramreview.com*.
19 Drew Troyer, *Procedures + Precision = Reliability*, Asset Management, RAM's Human Factors, May 15, 2020.
20 See Reference 3.

Bibliography

[1] D.A. De Castro, Brasman Engenharia, Brazil, *Use the system approach to spare parts management*, Hydrocarbon Processing, February 2006.

Chapter 4
Preparation

4.1 Introduction

In order to establish acceptance by the contractor or EPC and arrive at startup baselines, commissioning activity on all equipment should begin during the contractor installation phase – if not earlier. The authors know of recent projects where assignment of leading SMEs for immediate and active involvement was made at the inception of the project to the effect that lines of communication to the contractor could be developed as early as possible. Close observation of the installation procedures should be routine. Questionable procedures which might lead to startup problems or shortened run lengths should be corrected as soon as possible. Since owner's acceptance of the machinery from the contractor will generally be subject to approval of the startup engineer, the contractor must be made aware of the intended inspections and the tolerances which will be applied. Following acceptance of the machinery, it is mandatory that process personnel are made aware of the requirements for commissioning. Finally, the rules for turning the equipment over to plant operating personnel should be established.

4.2 Installation Procedures Should Be Reviewed Critically

Startup delays can actually occur in the construction phase of a project if adequate care is not exercised by the contractor. Since the damage may not be discovered until startup begins, repair of any damage which occurs will detract from the scheduled startup time.

To orient the startup engineer toward reviewing installation procedures, some typical examples with the possible consequences are listed in Table 4.1.

Table 4.1: Potential problems and reactions during project installation phase.

Incorrect procedure	Consequences and remedy
– Lifting plan not prepared for large motor because of reliance on routine experience-based practices. – Machine parts (couplings, bolts, gaskets and hardware sundries) are not safely stored.	– Loss of critical motor resulting in project startup delay. Modification of lifting practices by addressing every lift associated with the project. – Missing parts lead to lost time looking for them and delays. EPC organization to store all machine parts in a safe, locked area to avoid loss or theft. Recommended are inventory checks by qualified millwrights who physically match parts to each machine.

https://doi.org/10.1515/9783110701074-004

Table 4.1 (continued)

Incorrect procedure	Consequences and remedy
– Blinds are not installed on machine flanges during construction – especially on equipment with vertical up nozzles.	– Casings with vertical up nozzles can collect trash such as welding rod, coupons, stones and other foreign objects stemming from erection work. Severe damage can occur if problem is not detected before startup. If it is detected, machine casings should be opened for cleaning. Equipment with other nozzle orientations is subject to internal corrosion if casings or not protecting by blinds. This can also cause startup damage or require opening casings for cleaning.
– Welding is being performed with grounds attached in locations that could enable electrical paths through machines.	– Electrical currents passing through machines can cause severe damage to bearings, gear teeth and couplings. Arc welding circuits should not include machinery as current carriers. When welding is required on machine base plates or other directly connected components all machine components must be electrically tied together. Damage from electrical arcing will typically be detected in the startup phase of a project and lead to costly delays.
– Protective coating –*Rust Ban*, for example – is being removed from bearing housings too soon.	– Sensitive parts such as bearings, shaft couplings and others should not have their protective coatings removed until installation is close to completion. This cleaning should preferably be deferred until a day or two before machinery is turned over to the owner. Corrosion damage especially in bearings can cause considerable startup delays.
– Shipping plugs and covers removed from equipment casing connections.	– Premature removal of protective coatings on machine casing connections can result in corrosion and foreign material entry. Covering should be maintained until permanent connections are to be installed. In most instances blinds with identifying tabs should accompany the permanent connections until flushing and cleaning is complete
– Machine casings used for piping support during construction.	– Machine casings should not be used as piping supports. Excessive loads imposed by pipe strains can cause permanent deformation of casing and/or baseplates. Significant delay in startup can be expected if this damage should occur on large critical important machinery.

This compilation is certainly not complete. However, it does present an example of the problems which have been encountered with contractor installation procedures. Considering the above examples should set the startup engineer on the correct path to review construction procedures. Early correction of an improper construction practice can avert unnecessary problems at startup.

4.3 Establish Machinery Acceptance Baselines with Contractor

Acceptance of a piece of machinery from the contractor means that the installation is acceptable to the owner represented by the startup engineer. Although considerable work is usually still required before the machine can be run, it is understood that the contractor has satisfactorily performed his work responsibility when the owner's representative officially signs the acceptance form. In some instances where installation work not essential to pre-startup activity or run-in has not been completed, the contractor's acceptance form will have a qualifying list – sometimes referred to as BUTT-LIST – attached.

4.3.1 What "Butt Lists" Are and Are Not

Nothing is ever perfect, but we need to make sure that we are on the same page, maximizing our common approach and understanding. To avoid acceptance problems, which can take the form of severe conflicts between owner and contractor, it is mandatory to establish the basis for acceptance in advance. This requires documentation of the agreements rather than verbal agreements which are easily misinterpreted or forgotten. This is not to suggest that an acceptable job is not intended, but rather that an acceptable job will not be achieved if the basis of acceptability is not established.

4.3.2 Development of the Acceptance Basis

Should begin with the previously established coordination procedure mentioned earlier, which, again, determines, the responsibilities of the contractor and the owner. The first task is usually to correct the coordination agreement to reflect realistic divisions of responsibility. For instance, contractors usually cannot be held responsible for hot alignment since this implies that normal process operations have been established. Contractor personnel are generally no longer at the plant when normal process operations commence. Further, contractors usually cannot have responsibility for steam line blowing since refinery practices usually require owner responsibility when significant utilities are involved. This is also a factor when large electric motors are run in.

Following review and revision of the coordination procedure, the contractor responsibilities must be defined. Clarify who installs, who checks, who calls the OEM's field service personnel. Detailed statements should be developed for the following construction phases:

– Preparation for shipment
– Handling, rigging and lifting of equipment
– Job site receiving and storage protection
– Foundations and baseplate grouting
– Shimming and doweling practices
– Piping cleanliness, supports and flange alignments
– Machine cold alignments and tolerances
– Coupling installation practices
– Bearing cleaning, inspection and lubrication
– Oil system flushing requirements/tests
– Dead air space venting
– Contractor representation for run-in

At this juncture, a meeting should be held with the owner, project management and contractor representatives attending. The discussions should deal with acceptance requirements, conflict resolution and establishment of procedures for implementing the agreements. Data sheets and check lists are appropriate methods of assuring the various agreements have been satisfied. It is also the point where the responsibility for establishment of commissioning and startup procedures and checklists should be determined. Owner or purchaser specifications should be congruent with documentation used by EPC or contractor as well as with external standards.[1,2] Memoranda from the startup engineer to the project manager will satisfactorily document the agreements.

4.4 Challenges

4.4.1 Process Personnel and Others Must Be Aware of Run-In Requirements

Once the acceptability criteria for machinery have been established with the contractor, commissioning requirements should be explained and agreed to by process, electrical, instrument engineers and other technical personnel. While schedules for startup were developed previously these do not include the details of cooperation which will be required when run-ins are occurring.

4.4.2 Process Engineers

Process engineers are required to identify the systems that will be available for run-in operations. This may require temporary piping connections, check valve removals, blinds, inventory capacity for run-in fluids and other details, thus showing the justification for the process engineer's involvement beyond the fact that he is responsible for the plant as a whole. As the time for run-in approaches, the process engineer will provide liaison with the operators who will actually perform the startup functions, that is, start the oil system, pressurize the process system, energize the machine, turn the valves and so on. Thus, it is the responsibility of the machinery engineer to present his machinery commissioning plan to the process engineer. For pumps, this plan should include the requirements for:

– Suction lines flushed
– Suction fluid level – water – available
– Fluid circulation system available
– Desired time for run-in
– Operator requirements

Explicit examples of the preparations for various types of equipment run-ins will be given in the next chapters.

4.4.3 Electrical and Instrument Engineers

Electrical and instrument engineers should also be consulted at this time. The efforts here should he to present electrical and instrument requirements including data transmission for the run-in and establish electrical/instrument availability in terms of facilities and manpower. Further, the machinery engineer's questions regarding the characteristics of his power and control systems should be answered at this time. Typically, the electrical engineer can be called upon to recommend motor starting limitations and cool-off requirements, identify the substations serving various sections of the project, and provide field manpower to assist during checkout and run-in work. Usually, it is necessary to discuss control circuitry with both electrical and instrument engineers since an interface exists when motor drivers are involved. The instrument engineer should review the interlock, start and auxiliary/emergency circuitry. The machinery engineer should then become completely familiar with the logic of this circuitry to assure that the design is proper. The machinery engineer's knowledge of these systems will also be valuable when run-in instructions are prepared since oftentimes certain control circuitry must be bypassed to permit run-in. A good example would be bypassing alarms and ESDs during startup and run-in of dry gas seal (DGS) auxiliary panels.

At this point, the machinery startup has been formulated to assure necessary support from the various parties involved. This leads directly to the task of detailing the machinery commissioning work by preparing instructions for the machinery checkout, run-in and subsequent process operation. The following chapters are devoted to this.

4.5 General-Purpose Machinery Commissioning

4.5.1 Attention to General-Purpose Machinery

General-purpose machinery, such as pumps, fans and small compressors – Categories I and II, Figure 2.8 – is often given little attention during startups. This is usually due to the considerable amount of time required to assure that large, un-spared special-purpose machinery – Category III – will startup without problems. However, if time is not devoted to the large quantity of general-purpose machines, for example in an oil refinery, problems during or immediately following startup can be anticipated. Difficulties can occur at startup when it is suddenly discovered that a pump scheduled for system flushing has insufficient driver power when flush water is being pumped. Bearing failures can occur during or shortly after startup from lack of proper preservation, improper removal of rust preventatives or dirt. Catastrophic failures can result from improper coupling installations. Future maintenance and troubleshooting will be difficult if startup data have not been kept. This list of potential problems can, of course, be extended considerably. Our message is that startup attention must be given to the smaller equipment.

4.5.2 General-Purpose Summary Sheets Provide Useful Reference

Data identifying the design requirements and the mechanical characteristics of all machinery in a project can be found in the mechanical catalogs. These catalogs, however, are quite voluminous since they contain details on all aspects of a project. Organization of the catalogs will typically have the machinery information dispersed throughout the many volumes, making easy reference difficult. This problem can be eliminated if time is taken initially to summarize the pertinent machinery information in tabular form. An example of this is shown in Figure 4.1. This summary sheet, similar to Figure 3.8 shown earlier, is, again, useful as a field reference during equipment checkout and as a reference during process discussions. It is also useful for run-in performance calculations.

	P-501A	P-501B	P-501C	P-502A	P-502B	P-503A	P-503B	P-504A	P-504B	P-505A	P-505B	P-506A	P-50B	P-501A	P-501A	P-501A	P-501A	P-501A	P-501A
OEM	FS																		
Type	HCRA-1																		
Service	Feed																		
Temp. - °F	400																		
Capacity - GPM	1920																		
MCSF - GPM	300																		
Suction Pressure - psig	25																		
Discharge Pressure - psig	347																		
Δ Head - ft.	855																		
SO Head - ft.	1000																		
S.G.	0.868																		
NPSH available/required	25/17																		
RPM	3580																		
Efficiency - %	71.5																		
Brake HP	505																		
Driver	M																		
Driver Rating - HP	700																		
Impeller Dia. - in.	14.25																		
Running Clearance - in.	0.02																		
Thrust Bearing	3313M																		
Radial Bearing	3313PM																		
Seal	BW UR-3250																		
Flush - API Plan	52																		
Cooling	None																		
Lubrication	T-48																		

Figure 4.1: Typical comprehensive record sheet for centrifugal pumps.

4.5.3 Cold Alignment Estimates Are Based on Geometry and Operating Temperature

Equipment summary sheets are useful references when determining the cold alignment estimates for general-purpose machines since the operating temperature data is given. However, before the calculations can be performed, the support geometry and dimensions for the equipment must be examined. These data can be collected from the mechanical catalogs, but it is often better to physically collect the data in the field. This only requires a few minutes of time per machine and provides additional exposure time to the installation activity.

Collection of the geometric alignment data requires categorizing the machinery according to the following support descriptions:
- centerline casing support,
- centerline casing support and bearing housing support,
- bearing housing support and
- base support.

For each driver and driven piece of equipment, the distance between base and machine centerline must be recorded. Systems with casing centerline and bearing housing supports also require the distance between the supports be measured and recorded – or provide adjustable bearing housing supports as discussed later.

With the above data, cold alignments are calculated as shown in Figure 4.2, where T is the operating temperature, °F; L is the base to shaft centerline length, ft.; ΔL is the thermal growth, inches; A is ambient temperature, °F.

Cold alignment figures must be specified for the difference between expected thermal growth of the driver and the driven equipment. This also pertains to, for example, reciprocating compressor alignment. Here, thermal growth is often easy to assess; however, when there is a stack-up of support components as often found in offshore facilities, thermal growth considerations can become quite complex.

Sample contractor alignment tabulations are shown in Appendix 4.A. Simplified alignment guidelines may be used if there is no time for data collection and calculation. Details on machinery alignment may be found in reference,[3] but keep in mind that simplified guidelines may occasionally lack accuracy.

4.5.4 Run-In Performance of Pumps Should Be Estimated

4.5.4.1 Centrifugal Pump
Performance information is usually available in terms of differential head[4] versus capacity or in terms of pressure rise for a given specific gravity (SG) fluid versus capacity. Normal startup practice, however, usually permits pump run-ins on water at ambient temperature rather than on the design fluid. Since the characteristics of water

Centerline Casing Support :

$$\Delta L = 0.008 \times \frac{T - A}{100} \times \frac{L}{3}$$

Centerline Casing Support and Bearing Housing Support :

$$\Delta L_{Casing} = 0.008 \times \frac{T - A}{100} \times \frac{L}{3}$$

$$\Delta L_{Brg.\,Housing} = 0.008 \times \frac{1}{2} \times \frac{L}{3}$$

Bearing Housing Support :

$$\Delta L = 0$$

Base Support :

$$\Delta L = 0.008 \times \frac{T - A}{100} \times L$$

Figure 4.2: Calculating cold offset.

will rarely be the same as the design fluid, it is necessary to determine the expected water performance prior to pump run-ins. Two data points are of particular importance for water run-in. These include the expected shutoff pressure and the required run-in horsepower. It is recommended that these data be tabulated with appropriate caution-ary notes and be made available for the run-in tests. The comprehensive record sheet shown earlier in Figure 4.1 contains sufficient information to prepare this supplemen-tary information by calculating water test data such as suction pressures and expected discharge pressures. Further, water run-in tabulation sheets should record the shut-off pressure rise of the pumps as well as the recommended run-in pressure rise considering possible power limits. The shut-off pressure rise is a useful and easy performance check to record during water run-in. Since the flow is zero, it is only necessary to record the suction and discharge pressures during shut-off conditions. The resultant pressure rise can then be readily compared to the calculated pressure rise. Significant differences (>5%) between the calculated and measured pressure rise may indicate improper im-peller diameters, improper pump speed, gross wear ring clearances – even lack of wear rings – or other deficiencies. Elimination of these types of problems during run-in operations will save time during plant startup. It is further useful to check the power consumption at shut-off during run-in and compare this to the calculated power requirement. This check, however, is directly applicable only to motor drive pumps where electrical power is easily measured. Turbine drive pumps cannot usually have their power measured due to the lack of instrumentation. Turbine governor valve stem position, however, is a suitable check to detect marginal power situations.

4.5.4.2 Positive Displacement Pumps

Positive displacement pumps must be reviewed to determine if no-load run-in is required and if water run-in is possible. In general, pumps requiring lubrication from the pumped fluid such as rotary screw- or gear-type pumps should not be run-in on water. Considerable damage resulting from inadequate lubrication can occur on these types of pumps if they are run on water. On the other hand, plunger- and diaphragm-type positive displacement pumps can be run-in on water without concern assuming water is compatible with the materials of pump construction. Water compatibility should be specifically checked against pump packing and gasket materials. With respect to performance, a positive displacement pump will produce capacity and pressure rise on water which duplicates the design conditions. Capacity can be conveniently checked by operating the pump at maximum stroke length – or at a known speed, when dealing with a constant displacement type. Maximum pressure reached is, of course, a function of the safety valve setting.

Any pumps, whether centrifugal or positive displacement, which cannot be run in on water should be brought to the attention of the process engineers as soon as possible. This will permit alternate plans to be developed for run-in scheduling and plant flushing operations. Of course, the mechanical staff should be aware of water run-in limitations also, to avoid mistaken operation of pumps on water and thus avoid potential machine damage.

Finally, it is important to include bladder-type pulsation dampeners in this review. As a general rule, these dampeners should not be part of the flushing circuit because debris may become lodged in them and cause premature failure. This is also advisable for sensitive valves such as anti-surge or recycle valves. They should be removed during the flushing and cleaning activities and replaced by a spool piece.

4.5.4.3 Installation Checklist Helpful for Checking EPC

The summary lists of general-purpose machinery are a useful field reference for checking the completeness of equipment installations. Such documents will provide the operating pressure conditions, flushing requirements, horsepower requirements, machine rotation, cooling requirements and other parameters that can be checked easily on the field installations. See Appendices 4.A to 4.C.

4.6 How to Deal with Deficient Installations

When deficient items in design or installation procedure are found to exist over an entire project, it is preferred to advise the project manager by a separate written document – see also Figure 3.5 earlier. As things go, deficient items of this type usually require the attention and concurrence of the contractor's engineering department and management. Typically, a meeting to discuss the details of the deficiency will be conducted between the involved parties, that is, the owner's representative, project management, contractor and startup team.

An example of a design or installation error is the installation of Y-strainers on turbine inlet piping downstream of warm-up bypasses. The error existed across the entire project and significant cost was incurred to correct the error. Agreement to correct the error was not difficult to achieve once the deficiency had been discussed in terms of the damage potential to the machinery. Of course, the original design violated the appropriate practice on Y-strainers for general-purpose turbines.

Other examples have been the installation of eccentric reducers where frequently throughout an entire system somebody chose to install them with their "baggy" part up. However, it is advantageous to achieve contractor concurrence on engineering design deficiencies through logical engineering argument without relying on any rules. The respect which can be achieved in this manner will be useful throughout the project.

4.7 Make Final Acceptance a Formal Transaction

It is usually the intention of the EPC or contractor to have machinery accepted as part of the plant's "mechanical completion" acceptance by the owner. This arrangement is perfectly acceptable in terms of marking the time at which the newly constructed plant becomes the property of the owner. This final acceptance, however, should only be a formality with specific acceptances by the individual specialists – machinery, electrical, process and other disciplines – taking place in advance of "mechanical completion." To be satisfied that machinery acceptance requirements have been met – or will be met – at the time of mechanical completion, it is necessary to inspect and accept each piece of machinery prior to mechanical completion.

Inspection and accepting are tasks that deserve explanation. Because they are interwoven, these tasks are best accomplished by preparing standard inspection procedures for the various types of general-purpose machinery of a project. Standardization of inspections assures uniformity in the degree of inspection and the methods employed. Standardization avoids unanticipated acceptance requirements and thus further assures that the inspection will be complete while maintaining a cooperative relationship with the contractor. Finally, inspection standardization permits the inspection work to be delegated to several individuals who may be equally capable to perform final inspection tasks.

The final acceptance of each piece of machinery should be conducted on an official basis. Since the inspections have been standardized, all of the necessary inspection instruments such as dial indicators, scales and feeler gages should be available when an inspection is requested. In addition, alignment brackets should be installed and coupling run-in plates with attaching bolts should be immediately available for installation. It is generally the responsibility of the contractor to have the above items available at the time of inspection.

Following completion of a satisfactory inspection, an inspection form should be signed by the owner's inspector and the contractor's representative, usually a millwright, a machinist foreman or a first-line supervisor. Copies of the signed inspection form should be retained in the files of both the owner and the contractor. Thus, when the time of mechanical completion approaches, the exact status of each piece of machinery in a given plant can be quickly established. Incomplete contractor work can then be readily attached as exceptions to the mechanical completion acceptance should this become necessary.

Machinery acceptance forms for various types of general-purpose machines should be developed. An example is presented in Figure 4.3. It is evident that creating forms of this nature requires establishing criteria for acceptable installations which are directly the responsibility of the contractor or which can be negotiated as the responsibility of the contractor in advance of acceptance inspections. Work which is not established as the contractor's responsibility automatically becomes a pre-startup activity and a startup responsibility.

Pump Number	Foundation Grout	Soft Foot Check		Pipe Flanges		Driver to Pump Alignment		Check Valve	Temporary Strainer	Bearing And Lubrication	Witness And Date
		Pump	Driver	Bolts Loose	Face Separation	Pipes Loose	Pipes Tight				
						◯	◯				
						◯	◯				
						◯	◯				
						◯	◯				
						◯	◯				
						◯	◯				
						◯	◯				
						◯	◯				
						◯	◯				
						◯	◯				
						◯	◯				

Figure 4.3: Pump acceptance from contractor.

If acceptance criteria and work responsibilities are not established, the result may be additional unanticipated manpower requirements during the pre-startup activity phase. Items of this category obviously include doweling, coupling spacer greasing and installation, packing or seal replacement after water run-in, for example.

4.8 Pre-startup Checks Represent Last Chance to Avert Problems

In preparation for actually running a machine, it is necessary to once again review the installation. This typically requires confirmation that such items as motor megger tests have been performed and found acceptable, as, for example, steam lines have been blown, fluid lines have been flushed, strainers have been reinstalled, run-in circuits are prepared, fluid inventory is established, instruments have been calibrated. The machine should similarly be checked to assure proper lubrication, rotation is free and run-in plates are tight. These checks are tabulated in run-in log sheets similar to the one shown in Appendices 4.B and 4.C. Following satisfactory completion of the pre-startup checks and with previously agreed-to representation in attendance, the run-in can begin.

4.9 Equipment Run-In Should Be Uneventful

No problems will occur if the work of all involved parties has been performed satisfactorily to the point of actual run-in. The absence of problems will, in turn, create a definite feeling of accomplishment and satisfaction. The usual feeling of anxiety preceding each initial run-in will naturally vanish without satisfaction, resulting in what may be termed an anti-climactic situation. This, however, is the desirable conclusion of all of the preceding preparations. At the same time, it presents a situation requiring intentional caution. We must not conclude that problems will not occur. Further, it cannot be assumed that problems will occur immediately, thus making it unnecessary to maintain close surveillance of a machine which is seemingly trouble-free after a few minutes of operation. On the contrary, it is conventional practice to run-in machines for a period of four hours. Earlier we dealt with the phenomenon of infant mortality. One real example are mechanical shaft seals that require a certain amount of time before they can be considered fully effective and reliable after startup. Machinists or millwrights are quite familiar with this circumstance, and they wait a while after a seal repair and the initial startup before leaving the site. Expert observation should be available at all times during the run-in. Typical problems such as loose internal parts, unbalance or mechanical interference created by hydraulic or centrifugal forces will become evident immediately after starting the machine. Other problems such as casing distortion, bearing overloading or mechanical interference resulting from thermal changes may not appear for a considerable period of time. In fact, many thermally related problems will require design process operation before malfunctions are detectable. This further suggests that expert machinery observation be arranged for initial process runs especially when process operation is significantly different in temperature, pressure or load when compared to the run-in condition.

4.9.1 Special Run-In Considerations

It would be impractical to review each type of run-in which may be encountered in a startup. However, certain types of machines are normally found in large numbers on practically all projects. These machinery types which include motors, turbines, pumps and fans justify some special discussion in this context. Special step-by-step checkout and run-in instructions for these machinery types are included in the following chapters.

Appendix 4.A Pump pre-grout acceptance record

TOP VIEW OF THE INSTALLATION

Pump	Pipe Flanges Loose	Pump Pedestal Level		Driver Pedestal Level		Shim Stock		Coupling Spacer Length (Check)	With Machine Bolts Tight Driver-to-Pump Alignment
#	Flg. #	(L)	(R)	(L)	(R)	Pump	Driver	in./mm	Four Clock Readings

Appendix 4.B Typical pump alignment correction record sheet

$L = \ell/3$ [ft.]

TIR = Total Indicator Reading
PAD = Pump above Driver
PBD = Pump below Driver

| Pump Number | PUMPS | | | Thermal Growth | MOTORS | | Thermal G. | TURBINES | | | Vertical | TIR | Remarks |
| | ℓ | L | T | $0.008 \times (T-100)/100 \times L$ | ℓ' | $0.004 \times \ell'$ | | Type | Exhaust Temp. | Growth from Chart | Offset | | |
	ft.	ft.	°F	in.	ft.	in.	in.		°F	in.	in.	in.	
P-501 A/B	1.75	0.583	130	0.001	1.31	0.005					0.004	0.008	PAD
P-502	1.75	0.583	130	0.001	1.31	0.005					0.004	0.008	PAD
P-503 A/B	1.09	0.363	110	0.000							0.004	0.008	PAD
P-504 A/B	1.09	0.363	110	0.000							0.004	0.008	PAD
P-505 A/B													
P-510 A/B/C													
P-511 A/B													
P-512 A/B													
P-513 A/B													
P-514 A/B													
P-515 A/B													
P-516 A/B													
P-517 A/B													
P-518													
P-520 A/B													

Appendix 4.C Typical pump installation review log

UNIT: _____

Pump Number	API 610 Type	Eccentric Reducer	Strainer Installed	Check V. Installed	Check V. Drilled	Warm Up Line	Vent	Drain	Flush	Cooling	Lube	Rotation	Auto Start	Turbine Traps	Safety Valves	Checker
P-501 A/B	OH2															
P-502																
P-503 A/B																
P-504 A/B																
P-505 A/B																
P-510 A/B/C																
P-511 A/B																
P-512 A/B																
P-513 A/B																
P-514 A/B																
P-515 A/B																
P-516 A/B																
P-517 A/B																
P-518																
P-520 A/B																

References

1 API RP 686, *Recommended Practice for Machinery Installation and Installation Design*, 2nd Ed., December 2009.

2 API RECOMMENDED PRACTICE 1FSC, *Facilities Systems Completion, Planning and Execution*, FIRST EDITION, JULY 2013.

3 Geitner, F.K. and Heinz P. Bloch, Series *Practical Machinery Management for Process Plants*, Volume 3, *Machinery Component Maintenance and Repair*, Fourth Edition, ELSEVIER, 2019, and preceding editions, 650 Pages, Pages 205–263.

4 Differential head in ft. or m. In order to convert to pressure engineering units (psi, kPa or bar), the head value has to be multiplied by SG and divided by a constant.

Chapter 5
Reaching Consensus

5.1 Special-Purpose Machinery Commissioning and Startup

5.1.1 Introduction

Most probably, any given project will include a few special-purpose machines – Category III. Even though only a few are typically involved, the commissioning time requirement can be considerable. This is due to the complexity of special-purpose systems as well as the fact that commissioning delays will often result in plant startup delays. Thus, the responsible machinery engineer has a strong incentive to spend considerable time on the special-purpose machines. If the commissioning requirements have been properly planned and the work requirements and assignments for the general-purpose equipment have been adequately resolved as discussed in the previous chapters, the special-purpose machines will receive the proper attention.

5.1.2 Machinery Requiring Personal Attention of the Startup Engineer

Machines that satisfy any of the following categories should typically be given the personal attention of the startup engineer:
- Cat. III – high speed > 5,000 RPM
- High horsepower > 1,000 HP or 750 kW
- Cat. II

They are critically important machines. The degree of personal attention which must be expended will vary from actually performing the individual inspection steps to accepting the assurance that each inspection step has been satisfactorily performed. The startup engineer must determine the required degree of personal involvement. He does this by assessing risk-based priorities – see Chapter 2 – and matching the available machinery talent to the specific type of machinery under consideration. In addition, the experience of other individuals who might assume the inspection tasks must be judged. The startup engineer must generally be prepared to accept the personal inspection responsibility for initial startup type items, that is, line cleaning procedures and inspections, instrumentation checkout and simulation, uncoupled driver run-in, coupled runs and other tasks. The startup engineer should also be prepared to accept the personal inspection responsibility for activities such as line fit-up, clearance measurements and coupling alignment.

Whenever the startup engineer is required to personally conduct the inspection tasks, it is suggested that an owner's representative be simultaneously trained in

https://doi.org/10.1515/9783110701074-005

the procedure. This way one can develop an alternative inspector for the same task during the remainder of the commissioning as well as train somebody to perform similar inspection during future turnarounds.

5.1.3 Special-Purpose Commissioning Requirements – General

Special-purpose equipment, regardless of the type, will generally have some commissioning requirements in common. For this reason, we have elected to cover the general items first, while the items specific to several different machinery types will be covered separately. Common commissioning requirements are identified as:
- Installation review
- Lube and seal oil systems cleaning
- Instrument checkout and simulation
- Pipe fit-up review
- Alignment determination and checks

Seal oil system cleaning is obviously not common to all special-purpose machines. However, due to its similarity to the lube oil system, we will consider them together.

5.2 Installation Review Should Begin Early

5.2.1 Handling, Rigging and Lifting

Preceding the installation review, equipment handling, rigging and lifting practices should be scrutinized. They are important project activities that are often habitually relegated to an area left to the "experts" or entities that are "doing it all the time." A leading standard[1] devotes two very informative pages to this topic, and yet we believe it deserves additional consideration in our context.

Our experience has shown that handling, rigging and lifting accidents do happen.[2] Interviews with longtime crane operators[3] convey that many government rules and regulations concerning their trade have emerged during their lifetime. This attests to the fact that the risk of oversights and accidents continue to exist in need to be eliminated or mitigated. Rules and regulations issued by OSHA[4] and OHSA[5] as well as by ASME/ANSI B30[6] that apply to lifting and rigging of machinery. API RP 686 should be consulted as it is intended to supplement the rules just mentioned.

Our recommendation is therefore, to make lifting plans for all equipment categories II, III and large multistage pumps mandatory. Figure 5.1 shows an example of a risk-based lifting plan.

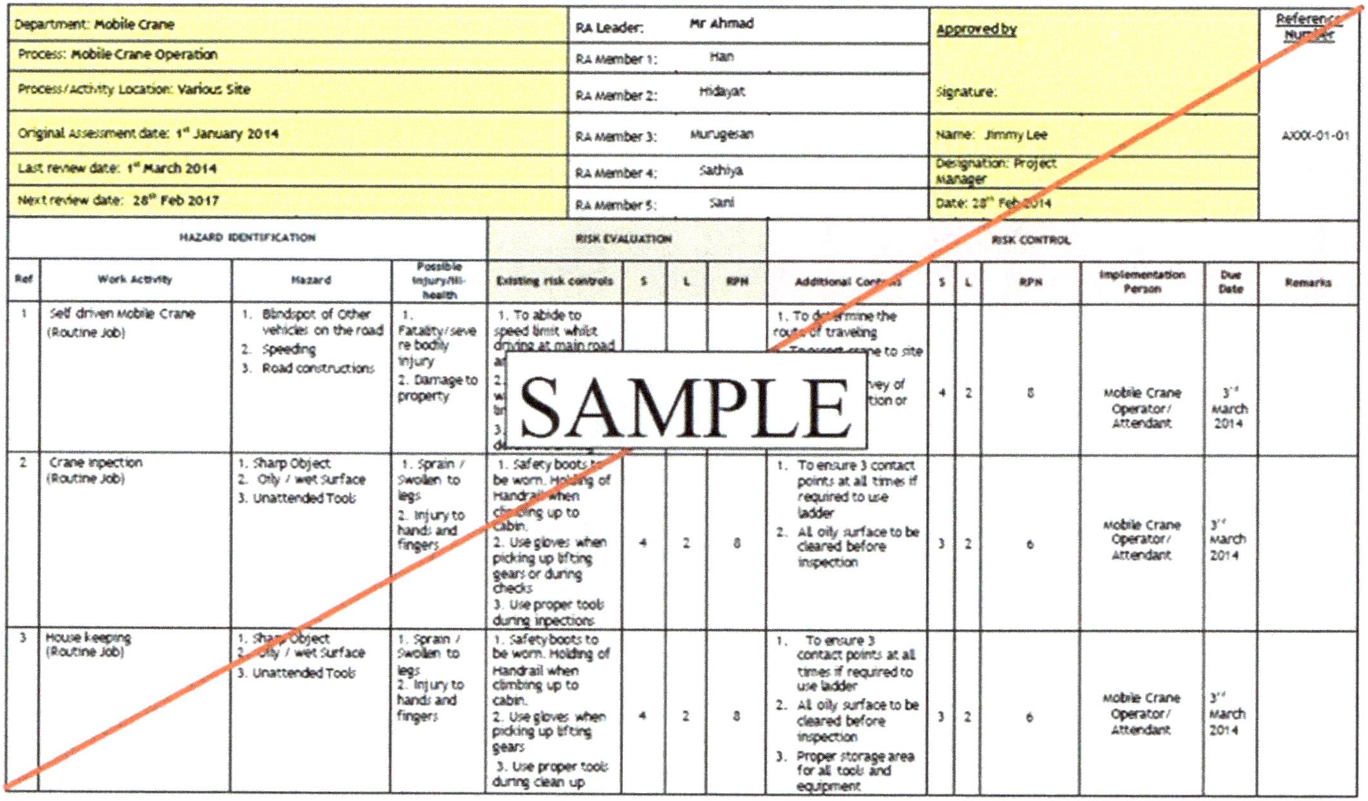

Department: Mobile Crane		RA Leader:	Mr Ahmad	Approved by		Reference Number
Process: Mobile Crane Operation		RA Member 1:	Han	Signature:		AXXX-01-01
Process/Activity Location: Various Site		RA Member 2:	Hidayat			
Original Assessment date: 1st January 2014		RA Member 3:	Murugesan	Name: Jimmy Lee		
Last review date: 1st March 2014		RA Member 4:	Sathiya	Designation: Project Manager		
Next review date: 28th Feb 2017		RA Member 5:	Sani	Date: 28th Feb 2014		

		HAZARD IDENTIFICATION		RISK EVALUATION					RISK CONTROL					
Ref	Work Activity	Hazard	Possible Injury/Ill-health	Existing risk controls	S	L	RPN	Additional Controls	S	L	RPN	Implementation Person	Due Date	Remarks
1	Self driven Mobile Crane (Routine Job)	1. Blindspot of Other vehicles on the road 2. Speeding 3. Road constructions	1. Fatality/severe bodily injury 2. Damage to property	1. To abide to speed limit whilst driving at main road ...				1. To determine the route of traveling ... ...e to site ...ney of ...tion or	4	2	8	Mobile Crane Operator/ Attendant	3rd March 2014	
2	Crane inspection (Routine Job)	1. Sharp Object 2. Oily / wet Surface 3. Unattended Tools	1. Sprain / Swollen to legs 2. Injury to hands and fingers	1. Safety boots to be worn. Holding of Handrail when climbing up to cabin. 2. Use gloves when picking up lifting gears or during checks 3. Use proper tools during inspections	4	2	8	1. To ensure 3 contact points at all times if required to use ladder 2. All oily surface to be cleaned before inspection	3	2	6	Mobile Crane Operator/ Attendant	3rd March 2014	
3	House keeping (Routine Job)	1. Sharp Object 2. Oily / wet Surface 3. Unattended Tools	1. Sprain / Swollen to legs 2. Injury to hands and fingers	1. Safety boots to be worn. Holding of Handrail when climbing up to cabin. 2. Use gloves when picking up lifting gears 3. Use proper tools during clean up	4	2	8	1. To ensure 3 contact points at all times if required to use ladder 2. All oily surface to be cleaned before inspection 3. Proper storage area for all tools and equipment	3	2	6	Mobile Crane Operator/ Attendant	3rd March 2014	

SAMPLE

Figure 5.1: Example of a lifting plan.

Along with the lifting activity, it would be well to maintain and complete a pre-lift checklist in order to reduce the risk of an oversight. A typical pre-lift checklist is shown in Figure 5.2.

	Yes	No
Is the crane configured in accordance with the lift plan?		
Has the crane been inspected and the condition acceptable?		
Has the rigging equipment been inspected, secured, and in acceptable condition?		
Is the supporting surface stable?		
Are proper crane mats placed under outrigger floats and at a 90-degree angle to the outrigger cylinders? Are crawler cranes on proper crane mats?		
Are outriggers (if applicable) fully extended with tires off the ground?		
Is the crane within 1 degree of level? Has the levelness of the crane been checked with a four-foot carpenter's level or other acceptable method? The "target" level in the crane cab can be used for initial leveling but should not be considered reliable for critical lifts.		
Is the exact load weight known?		
Is the location of the center of gravity of the load known and the crane hook positioned directly above it?		
Was the load radius measured exactly? For heavy lifts, has the potential increasing load radius due to deflections in the boom, tire, and/or carrier been considered?		
Was the boom length determined exactly?		
Was the boom angle determined exactly?		
Are wind conditions acceptable? If wind speeds are in excess of 30 mph, the lift should not be made; if wind speeds are more than 20 mph, consider postponing the lift.		
Is the rope reeving balance to prevent boom twist?		
Is the rigging capacity acceptable?		
Is the weight of the rigging known?		
Has the clearance between the boom and the load been considered and is it sufficient?		
Has the clearance between the boom tip and block been considered and is it sufficient?		
Is the crane operator experienced and qualified?		
Has a qualified crane signalperson been assigned and method of communication between the crane operator and signalperson established.		
Is a person assigned to control the load with the use of a tag line?		
Is the area clear of obstacles (including power lines, pipelines, and unnecessary personnel)?		
Has a pre-lift meeting between the crane operator, signalperson, supervisor, and other affected persons been conducted?		

Figure 5.2: Pre-lift checklist.

Checklists reflect good practices. They are not to be confused with mandatory requirements. Of course, even the most comprehensive review checklist will be of limited usefulness if its desirable features have not previously been part of the owner's equipment purchase and installation specifications. Note that the checklists offered in this text will enable you to critically examine if your existing specifications reflect the procurement and implementation guidelines applied by the reliability-minded competition.

Often critical equipment like a turbocompressor and its prime mover arrive late at the job site, and there will be pressure to put it in place. In fact, now the risk of procedural shortcuts has to be addressed and mitigated by the commissioning team.

There is always the risk of events in this phase taking the wrong turn as illustrated by these examples:

- When dry gas seals for critical turbocompressors and helical screw machines are shipped lose, the box should be taken out from the common compressor package, clearly identified and preserved in an air-conditioned room in the warehouse.

– On receipt of the material, equipment boxes should be marked with equipment tag, serial number and unit of installation to avoid any unintended exchange between same makers and models.

5.2.2 Job Site Receiving and Storage Protection

Another project phase preceding the installation review are job site receiving and equipment storage protection practices. The conscientious C&S practitioner would be well advised to delve into the intricacies of optimized lubrication, oil mist technology and storage preservation which are thoroughly covered by two sources.[7,8] These texts discuss, for example, the much-favored vapor phase or vapor space inhibitors and their limitations as opposed to modern storage protection methods.

The concern about this project phase ranges from the original equipment manufacturers (OEMs) preparation for shipment to protection during short-term field staging at the installation site and during the period prior to commissioning. In between are just-in-time shelter requirements in connection with off-shore applications and long-term preservation needs. In off-shore projects, some of the most significant efforts are made to protect equipment against damage and degradation prior to being put into final operation. Here, preservation, connection covers, purging and other means may need to be employed due to significant construction durations for shipboard applications. Receiving and storage protection measures in an off-shore environment are frequently taken as far as providing an equipment mockup during shipment and skid installation. This approach permits the critical machinery to be stored in a clean, preserved state for as long as possible before the required installation date, which minimizes the risk of degradation.[9] Further, and finally, we must not forget that now and then the requirement for longer-term equipment storage and protection emerges. There should be a contingency plan.

After discussions among the owner's engineer, the representatives of the engineering, procurement and construction (EPC) company and the OEM, suitable policies for the project's receiving, storage and staging protection practices should ensue. Reference 1 presents a typical example of recommended practices for equipment job site receiving and storage protection that could be used in the absence of specific pre-existing owner's standards. However, suitable policies only become useful when they reappear in form of precise procedures to fulfill the requirement of the job at hand. Table 5.1 is a good example of a straightforward checklist that has served us well in the past.

The authors feel they cannot warn their readers enough about the grave consequences of inadequate equipment storage protection. Reputable equipment vendors are trying to address the problem by providing engineering advice for staging and field storage often going beyond the limits of their responsibility. A good example is presented in Appendix 14.A. However, OEMs naturally lose their influence when it comes to long-term storage. In the following chapters, we shall revisit this subject

Table 5.1: Checklist – machinery field handling and storage protection.

Designation: _____
Location: _____
Service: _____
Note: Cross out items that are not applicable

1. Receiving Inspection: Have qualified machinists been assigned to assist in checking for damage and other inconsistencies?	_____ (Yes or no)
2. Has the machine – unit – been checked for transit damage?	_____ (Yes or no)
3. Are blinds on flanged openings still tight? (If not, retighten or renew.)	_____ (Yes or no)
4. Are all other openings plugged or blinded?	_____ (Yes or no)
5. Is the paint covering on machine – unit – still good? No signs of rust? (Rust should be removed and area repainted.)	_____ (Yes or no)
6. Check all items against packing list. Anything short?	_____ (Yes or no)
7. Have specific instructions regarding rotation of rotors, crankshafts and other parts been included in the vendor's service manual?	_____ (Yes or no)
8. Have oil reservoirs been checked for presence of water and drained if necessary?	_____ (Yes or no)
9. How much time is expected between receipt of equipment and start of installation? Give strong consideration to using oil mist as a "preservative blanket" for all machine internals.	_____ (Yes or no)
10. Has a program been established for regular rotation of shafts – two turns at 2-week intervals? Include draining water from oil.	_____ (Yes or no)
11. If time in (9) is over 1 month, have blinds been removed and the machine inspected?	_____ (Yes or no)
12. Are all exposed machined surfaces coated with rust preventive?	_____ (Yes or no)
13. Have reciprocating compressor valves been stored in a container of light oil?	_____ (Yes or no)
14. Has it been arranged that lube and seal oil units will be installed as soon as possible in order to put them into operation? (They can be flushed by discharging directly back to the reservoir waiting for hook-up to machinery piping – see flushing and cleaning procedure Chapter 5.)	_____ (Yes or no)
15. Have all major equipment not stored within a warehouse been stored in a place where damage risk from construction activities and traffic is least likely? Again, has oil mist preservation been considered?	_____ (Yes or no)
16. Gear units – preserved in vendor's shop for extended storage should be stored such that unit will not be turned. No oil to be added until finally installed. Consider use of appropriate diester or polyalphaolefln synthetic lubricants.	_____ (Yes or no)

Table 5.1 (continued)

17. Have all loose items been restored in closed boxes?	_____
(a) Have these been stored in limited access areas?	(Yes or no)
(b) Has a record been made of where these items are stored?	
18. Following (9), has the rotor or crankshaft been turned two complete	_____
revolutions? (This includes small pumps as part of a package.) On	(Yes or no)
reciprocating compressors, operate the hand pump if available, and crank	
the cylinder lubricator if the machine has the cylinders installed.	

by highlighting the handling and storage requirements for specific critical equipment. We believe we can limit the use of machinery storage preservatives to the three products as shown in Appendix 5.A. Appendix 5.A1 also shows the example of how to manage a preservation and storage operation.

5.2.3 Foundations and Mounting Plate Grouting

The installation of all special-purpose machines should be reviewed at the earliest opportunity, preferably when the foundations come into being. Applicable standards such as API RP 686[10] and GMRC[11] would serve as guidance. This will permit rapid familiarization with the basic machine support structure. Also, the top foundation surface can be inspected to ensure a rough texture has been prepared for adhesion of the grout. Baseplate hold-down bolt installation can also be inspected, if caught early enough, to assure they are appropriately fastened to the foundation reinforcing bars. Costly mistakes can, for example, be avoided by having profound agreement about installation techniques for grouting large skid-mounted equipment.[12]

5.2.3.1 Why Do Machinery Foundations Deserve Our Attention?

Many foundation flaws and shortcomings will appear years after startup and initial operation of the equipment. For example, in the mid-1900s, reciprocating compressor foundations were expected to last 20 to 30 years. However, many of the machines installed in the 1960s or 70s are still running on their original foundations which are now in need of repair or replacement because of poor design, construction or, sometimes, inadequate maintenance.[13]

Particular problems may signal that there is something wrong with the foundation. These problems could be a broken anchor bolt, twisting movements of the machine frame or excessive vibration indicating looseness with oil oozing from under the machine base.

Often, the machine has to be stopped in order to make repairs – or it may stop by itself. In many cases, unplanned shutdown could be avoided by reviewing:
- maintenance records,
- the original construction and
- the as-built situation.

Moreover, modern machinery's performance – horsepower and flow capacity – accuracy of alignment and environmental requirements with respect to sound and vibration due to stricter government regulations have been increased radically over the past decades. This has made the need for increased attention to the foundation, its anchorage and upkeep an important issue.

Typical foundation faults are:
- *Loose or broken hold-down bolts* – they are easy to spot. Frequently, the compressor trips on a broken anchor bolt triggered by high vibrations. Machine operators judge that there is something wrong with the foundation if one of the anchor bolts fails. While this is indeed a logical conclusion, it is also possible that the cause of a bolt failure is to be found somewhere else. For example, sometimes it is easy to over-torque bolts with hydraulic tools. It is, therefore, advisable to look at the contractor's tool management practices.
- *Excessive use of shims and shim plates* – often indicates that there is a need to "fill up" space. This condition can be an indication that there is something "moving" or a mistake has been made during the design phase. The maximum number and thickness of shim plates differ depending on the guideline, region and local engineering standards. Different guidelines agree on not using more than three shims in a pack, as more shims lead to spongy – soft – pads. Unfortunately, the use of adjustable jacks, steel shim blocks or wedges is still frequently advised for alignment of pumps skids, base frames or even compressors while they remain in place. These steel jacks make the equipment stand on "high heels," forming a direct steel contact between the machine frame and concrete foundation. Instead of having the necessary constant compression on the ground layer, the machinery stands on non-compressive steel blocks which allows oil and water to penetrate between the steel and the grout. Even worse, neither the anchors nor the grout can function according to their design objective as pre-tension is lost in the steel jacks and grout becomes merely an aesthetic cover.
- Another problem with steel shim blocks in the ground is that they tend to corrode. This is a very common condition that causes serious problems such as cracked grout and a machine tilted out of alignment.
- *Edge lifting and delamination of grout* – where edge lifting is caused by the difference in the rate of thermal contraction between epoxy grout and concrete. Generally, the main reason for edge lifting is poor or inappropriate application and the use of bad quality concrete.

- *Delamination* – often occurs between cementitious grout and concrete. It is mostly caused by the poor adhesive property of the grout or it is due to poor preparation of the concrete block. Edge lifting and delamination do not always become an immediate threat to the machine foundation since the part under the machine – if well grouted – is under constant pressure and therefore in better condition. However, oil and water can intrude into the foundation causing more problems. It can also be a sign of poor application and therefore a reason to suspect more problems.
- *Cracks in the grout layer* – may have different causes such as sharp corners, fast curing or thermal expansion. In many cases, these cracks do not represent an immediate threat. However, it is important to look for their root cause and to seal them to prevent further damage.
- *Deterioration of the grout layer* – Penetration of oil and other fluids into cementitious grout will over time weaken its compressive strength and the adhesive capacity of cementitious grout in the anchor pockets. This process will continue up to a point where the grout will crumble between your fingers. Oil penetrating along the anchors will eventually jeopardize the fixation of the machinery.
- *Cracks and cold joints in the concrete foundation* – can be due to various factors such as weather conditions and thermal or aggregate expansion. They can be found at re-entrant angles such as the corners where the foundation of the crosshead supports of reciprocating compressors is connected to the main block but also in the sump area, and, of course, cracks running from the anchors toward the outside of the foundation.

When observing cracks in the concrete foundation, one has to realize that vertical cracks are less "dramatic" than horizontal cracks. Horizontal cracks can cause alignment disorders, while vertical cracks do not.

The compressor on its foundation must form a tightly integrated structure. Vibration energy travels in the form of waves down and out through the foundation where the soil can absorb it. Breaks, cracks or separation in the integrated compressor-foundation structure will prevent the vibration waves from traveling downward. Horizontal cracks create "separated" parts in the foundation and therefore, it would become unable to transfer vibration into the soil. The monolithic structure – as illustrated in Figure 5.3 – is disconnected. The same problem can be caused by construction joints, also called "cold joints." Because concrete does not bond very well to itself, separation can occur if different parts of the foundation are not poured continuously.

- *Concrete carbonation and spalling* – if carbon dioxide from the air reacts with calcium hydroxide in concrete, it forms calcium carbonate. This reaction is called carbonation. It is a slow and continuous process progressing from the outer surface inward and has two effects: It decreases mechanical strength of concrete and decreases alkalinity essential for corrosion prevention of the reinforcement steel.

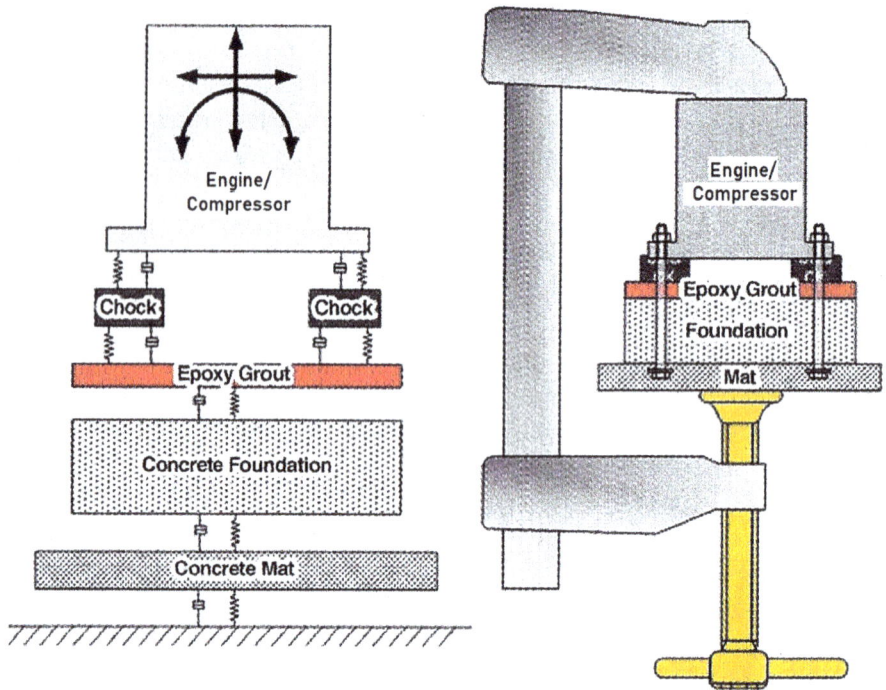

Figure 5.3: Foundation parts forming a monolithic structure (from reference 13 by permission).

Carbonation and rebars located too close to the concrete surface can cause rebar corrosion. The expansion of iron oxides induces mechanical stress that can cause the formation of cracks, disrupt the concrete structure or make outer parts of the foundation fall off, a failure mode called spalling. These phenomena are easy to recognize during visual inspection of the foundation.

– *Foundation displacement* – caused by soil instability as shown in Figure 5.4. Therefore, displacement of the foundation can indicate that the soil is not adequately supporting the foundation. This condition can be checked with tools such as digital water level and tapeline, or more precisely by laser scanning. Trending measurements over time can provide more suitable information.

The foundation design should be reviewed regarding:

– *Concrete mat and block design* – The information on required dimensions and minimum weight of the concrete block can be found in the aforementioned guidelines.
– *Rebar details* – In general, there should be a minimum bar size of 0.75 in or 19.05 mm – GMRC guidelines for high-speed compressor packages. There are

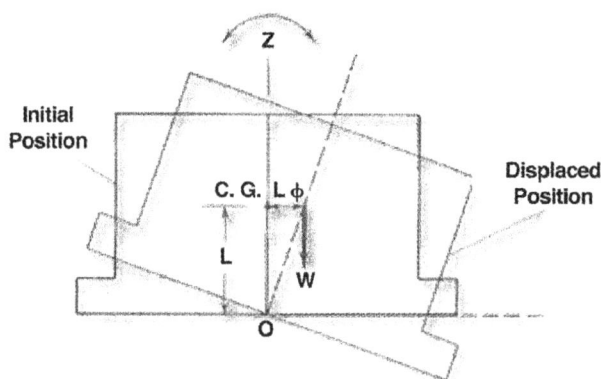

Figure 5.4: Tilting foundation (from reference 13 by permission).

also rules and practices that can be found in GMRC and API RP 686 to check rebar spacing – or density – and material.

- *Concrete properties* – The concrete should have a minimum compressive strength of 4,060 psi or 28 MPa according to GMRC and API RP 686, although the design value in many cases does not reflect reality since concrete can be weakened by oil or just because the actual poured concrete was not at an optimum quality. If the properties are in doubt, it is advised to have them checked in a laboratory.
- *Soil and pile setup* – A foundation should never be installed without thorough analysis and characterization of the underlying soil. First of all, one should find out if a study has been done before designing the foundation. Further, look for any changes in the soil properties, for example, due to rising water levels. To assess this type of situation, the advice of a geotechnical consultant is required who would determine important elements like mass density layer variation and stiffness of the soil. Based on this data and on the type and mass of the compressor system, a determination can be made as to whether pilings need to be used as well as the proper type of foundation for the project. A tool that could be helpful in checking whether the design of a compressor foundation might be susceptible to any problems is the compressor foundation analysis tool. This program is available to R&D members of the European forum for reciprocating compressors.
- *Anchor type and dimensions should be reviewed.* It is of major importance to gather all information on the anchor bolts. Some installations incorporate outdated anchor bolt designs that can eventually lead to problems. At the very least, the following information should be obtained on the anchorage:
  - Bolt preload and design clamping force
  - Bolt material, size and type
  - Design stretch length or elongation

- Coating or surface treatment
- Pre-installed versus post-installed
- Pocket size and type information
- Edge distance of the anchor pockets
- Type of grout to be used for bonding

Most standards and guidelines such as reference[1] provide helpful guidance on the subjects mentioned above.

Baseplate and soleplate installation will follow foundation preparations. Again, prior to grouting, bases must be adequately supported to prevent twists and bows and to ensure the machine feet will be mounted on a level surface. Grouting preparations, as with general-purpose machines, should provide for removal of supporting shims and wedges once the grout has set. It is also advisable to inspect during the grout pour to assure measures are taken to prevent formation of air pockets. As a general rule, an adequate quantity of grout should be poured with frequent vigorous agitation.

5.2.3.2 Pump Baseplate Issues

Do not allow a contractor to say "we have always done it that way." There are instances where the installers have always done it wrong and where the resulting repair frequency over the lifetime of the equipment has kept an entire plant from ever becoming a best-in-class performer.

A case in point is the acceptable methods of baseplate leveling for centrifugal process pumps. Frequently, there is an issue with the pump unit not having adequate access on the mounting pads. Therefore, the machined mounting surfaces could not be used for leveling without removing the pump. The OEM had suggested using the machined surface of the discharge nozzle instead. However, Reference[1] states that nozzles should never be used for that purpose. Clearly, experience-based texts[14] recommend that the pump and driver be removed from the common baseplate prior to leveling. True, the pump and driver had been pre-mounted by the vendor on the baseplate to ascertain bolt locations and fit and also to facilitate transporting the pre-mounted pump-driver set as a single unit. However, it should not be considered the best approach to a long-term, reliable field installation. Full access to machined surfaces will be needed for proper baseplate leveling. Does the contractor need a technology update?

Are there other choices for pump baseplates? The answer is yes. In the above case, for unexplained reasons, the EPC provider had chosen to specify conventional baseplates for the process pumps on this project. Baseplates prefilled with epoxy – as demonstrated in reference[15] – would have been viable contenders here. Ideally, the owner's representative involved in the machinery quality assessment may have looked into the matter and could have asked to examine cost justification, as well as long-term reliability issues. All parties may have been surprised by the findings.

5.2.4 Shaft Alignment Practices

In Chapter 4, we talked about alignment routines. What could go wrong is the installation of the machine casings on their respective bases. Here, for the purpose of marking foot bolt locations for drilling, we should insist on shaft-to-shaft sidewise – lateral – alignment within three times the tolerances which will be required for final acceptance. Similarly, proper shaft end-to-end dimensions should be ascertained. For instance, 2 mils or 50 μm is usually an acceptable hub misalignment while 1 mil or 25 μm is an acceptable face misalignment. Thus, without regard for the vertical alignment at this time, the side-to-side alignment should be adjusted to within 6 mils or 152 μm hub and 3 mils or 76 μm face before the foot bolts are being marked. Multiple case trains should be handled similarly over the entire train before the first bolthole is marked. This procedure assures proper coupling spacing and clearance for sidewise adjustments in machine case positions without the necessity of considering such unsatisfactory devices as undercut bolts – commonly known as "cheater-bolts."

5.2.4.1 Best Shimming Practices

When the casings are next installed on their bases, it is necessary to consider the shimming practice which will be followed for all future alignment adjustments. This requires that one casing be selected as a reference with all other casings being shimmed to permit adequate lowering adjustments relative to the reference casing. To achieve this flexibility in vertical adjustment, it may be necessary to shim the reference casing. Should this be required, the reference case shims should be solid with minimum ¼-in thickness, surface ground to assure face parallelism. It is further suggested that these shims be fit by doweling them to the base, thus making them part of the base. Note, however, that doweling should be done after hot alignment checks are complete.

Once permanent shims – if required – have been installed, the reference casing should be mounted on the base. Shims and mating surfaces must always be free of dirt or corrosion and oil. Areas which have new shims installed should be loaded – bolts tightened – to crush the shim and eliminate false clearance and alignment readings. Levels should be checked as prescribed by the manufacturer. Any leveling corrections which are necessary should follow normal shimming practice, that is, use full size single shims where possible. This procedure should be followed by "soft-foot" or unequal foot loading checks and corrections.

Soft foot is the condition of a machine casing, where a foot and/or several feet are not within the same plane supporting the weight equally. This seemingly unimportant error has detrimental consequences. It can significantly reduce service life of the machine. For example, electric motors can suffer frame distortions which in turn affects the winding ultimately leading to electrical imbalance. Electrical imbalance tends to elevated winding temperatures and ultimately leads to a reduction of motor service life.

With the casing positioned properly relative to the train centerline, a minimum thickness feeler gage – from 0.0015 to 0.002 in or from 38 to 51 μm – should be used to determine if any appreciable gaps exist under any foot. If the feeler gage passes under any foot, stainless steel or brass shims should be added to reduce the clearance to within the measuring capability of the feeler gage. A more accurate method of measuring "soft feet" employs dial indicators mounted on the base and indicating the machine feet. With all foot bolts tightened and all dial indicators zeroed, loosen the bolts of one foot and note any deflection on the indicator. Deflections over 2 mils or 50 μm will typically require a shim addition. When the deflection of one foot has been reduced to within tolerance, tighten that foot and similarly test the remaining feet. Once all feet have been checked and corrected, it is good practice to recheck all feet by once again loosening and retightening each foot in sequence. Upon completion of the shimming work of the reference casing, it is good practice to block the machine in position using the lateral jackscrews if they have been provided. Doweling of all casings should be performed after hot alignment checks are complete.

Following the manufacturer recommended alignment figures or alignment figures calculated according to thermal expansion theory – carbon steel has a coefficient of thermal expansion of approximately 8 mils/ft/100 °F or 12.0 mm/10^6/°K temperature change, the remaining casings of the train should he aligned to the reference casing. It is common practice to initially establish any shaft angularity requirements in the vertical plane. Secondly, the vertical elevation is established. Finally, the side-to-side alignment is adjusted. Jack bolts are indispensable for alignment work both in the vertical and in the horizontal planes. It should be possible to produce any required shim thickness using three individual shims, providing a complete inventory of shim sizes is maintained.

Following completion of the cold alignment, provision should be made to protect all exposed components of the machine casings since it will probably be a minimum of several weeks before the machines can be completely closed and oil circulation established. All openings including oil supplies, oil drains, vents, gas connections, instrument connections and other openings should be plugged to prevent foreign matter from entering and to prevent atmospheric corrosion. If a long-time elapse is anticipated before oil circulation can be established – approximately in excess of 30 days – N_2 purge should be considered for all internals or for areas where Rust Ban® or similar preservatives have been removed, that is, bearing, seal, coupling and governor housings.

5.2.5 Piping

5.2.5.1 General

Continuing the installation review, we consider process piping erection the next phase of construction. We trust that the contractor will follow best practices documented in form of procedures incorporating the standard referred to earlier[16] with site and project specific additions or modifications. Piping and piping supports should be designed with respect to operation and maintenance to facilitate piping spool removal and avoid support removal. Access should be checked for ease of maintenance. In the case of reciprocating process compressors, for example, as a minimum requirement, access is required for the non-drive end of the compressor. Proper spaces and areas should be respected for withdrawal of piston, cooler bundles removal, laydown area and other ancillaries. But clean the pipe first. Do not depend on a temporary line filter. If the gas or air being compressed may, at times, contain dust, sand or other abrasive particles, a gas scrubber or air cleaner must be installed permanently and serviced regularly. Some process machines are extremely sensitive to the ingestion of dirt, rust, welding beads or scale. In a reciprocating compressor, such foreign objects will cause scored packing rings, piston rods, cylinder bores and pitted, leaky valves. These problems translate into extensive delays.

When piping problems occur during commissioning, we look back and ask questions such as:

- Were the dead weight of the piping and the insulation included in the design parameter?
- Was the live weight of the test and service fluid, ice and snow in the design calculations?
- Did calculations take into consideration dynamic loads due to wind, earthquakes, impact and surge of fluid as well as high vibration?
- Did the designer consider thermal expansion as well as contraction effects and effects of settlement of equipment foundations and or piping support caused by natural forces such as winter-to-summer thaw or natural soil compaction?
- Why was the pipe interior not clean?

A lot of our concerns about pipe cleanliness could, of course, be allayed if we conveyed the need for cleanliness to the pipe fabricator. Frequently, we "forget to tell the welder," because we fail to deliver the message through appropriate instructions. When fabricators know that they are dealing with "clean" pipe it will be most often free of welding beads, slag and spent welding rod. For instance, chill rings could be used for butt welds in piping as a recommended practice. This prevents welding beads from getting into the pipe to carry through, not only during the first startup of a compressor, but later on during its operation. Further, it is important that the piping be fabricated with sufficient flange joints so that it can be dismantled

easily for cleaning and testing. It is far better to clean and test piping in sections before actual erection than after it is in place.

If it is necessary to conduct the final test when the piping is in position, care
should be taken to provide vents at the high spots so that air or gas will not be
trapped in the piping. Make provision for complete drainage after the test is completed by providing low-point drains if necessary. These connections should, of
course, be planned in advance.

5.2.5.2 Pipe Cleaning

When piping is cleaned in sections before erection, it is possible to do a thorough
job of eliminating all acid or residue stemming from the fabrication process. This is
difficult to do with piping erected and in position, because carry-over of acid into
cylinders or casings is almost certain to occur when the machine is started. This can
cause extensive damage.

After hydrostatic tests have been performed and the pipe sections have been
cleaned as thoroughly as possible on the inside, the piping should be pickled using
one of the procedures outlined in Appendix 5.B.

On large piping – where a man can work inside – pickling procedures can be
omitted if the piping is cleaned mechanically with a wire brush, vacuumed and
then thoroughly inspected for cleanliness. Time and trouble taken in the very beginning to ensure that the piping is clean will shorten the break-in period, and may
save a number of expensive shutdowns.

5.2.5.3 Piping Installation

Since the machine is cold aligned, a fixed reference has been established for the
piping installation. Thus, the piping should be installed to accurately match the machine flanges. However, care must be exercised during the piping installation to
avoid excessive loading on the machine nozzles. Compressors, and for that matter
all other pieces of machinery, are not pipe anchors.[17] An acceptable construction
procedure for piping installation requires the piping to be completed from the process side toward the machine. The final welded joint in the pipe should be between
the pipe support closest to the machine and the pipe-flanged section connected to
the machine. Prior to performing this final weld, however, certain procedural steps
should be followed to eliminate difficulties. These are best presented in a summary
form as follows:

- Fabricate pipe section including flange at machine and adjacent pipe length.
- Remove machine nozzle cover and install the fabricated section with blind flange
  observing correct flange thickness.
- Provide auxiliary support for above pipe section.
- Fabricate piping from process plant to pipe section flanged to machine.
- Install and adjust all pipe supports for piping system.

- Align piping at juncture with section attached to machine.
- Weld the juncture observing specified weld procedures.
- Loosen flanged joint at machine nozzle and check flange alignment. Faces should be parallel within 1/64 in or 0.4 mm. Flange bolts should pass from flange to flange without interference.

Also, the pipe support system should be sufficient to hold the pipe in position. Pipe support flexibility should be evident by deflections created with small applied loads such as pushing with a small crowbar.

If necessary, pipe flange alignment should be corrected by the contractor using the localized heat yield method – "rosebud heating" in pipefitter's jargon – cutting and rewelding.

When the pipe alignment is finalized, the pipe flange at the machine should be blinded until run-in approaches.[18]

Appendices 5.C and 5.D incorporate suitable procedures and checklists that will help in making the correct moves to connect piping to machinery.

5.2.5.4 Small Bore Piping

A vexing problem can be small bore piping connections (SBC) of diameter size 2 in. or 50 mm and below on machinery. A standard must be agreed upon in order to avoid confusion and mistakes. In initial piping designs, these small diameter connections are often not considered or fully defined.[19] Some of the important rules involving small-bore piping are:

- Wall thickness should be schedule 80 or higher.
- Fittings with higher stress concentration – such as threaded connections – should be avoided in higher risk areas such as on or near the compression equipment. Therefore, weldolets should be used over threadolets.
- Branch-line lengths should be kept as short as possible with minimal attached weight to increase their structural natural frequencies.
- If possible, locate the branch line away from valves, reducers, bends and tees in the main line where flow turbulence and acoustic induced vibration may cause problems.
- Threaded fittings should be tight and back welded such that there are no exposed threads. Threadolets are generally not recommended.
- Standard or higher schedules are recommended for main line piping to reduce the stress concentration at the connection to the SBC.

5.2.5.5 Small Bore Piping Restraints

The installation of an external restraint – such as supports and bracing – will add stiffness to the branch line, raising its minimum mechanical natural frequency and reducing the risk of excessive vibration. Because there are an almost unlimited number of potential branch line configurations, it is not possible to recommend specific restraint configurations that will cover all feasible layouts. However, the following general guidelines might be helpful:

–   The stiffer the restraint, the more effective it will be. A minimum restraint stiffness of 10,000 lbf/in or 1,751 N/mm in all three directions is recommended.
–   A good restraint should be triangulated in multiple planes to provide stiffness in multiple directions.
–   Simple weight supports and springs typically provide very little vibration control.
–   It is typically preferred to brace the branch line back to the mainline piping. Bracing the branch line to a very stiff external support can increase bending stresses due to relative displacement from vibration and thermal expansion. A relatively stiff external support should not be installed too close to the branch connection.
–   Strap-type clamps are typically more effective than U-bolt-type restraints. If U-bolts are used, they should be used in pairs to prevent rotation about the U-bolt.
–   Gusset plates can add stiffness but also add high stress intensification factors and should be used with caution. If used, gussets should attach to a reinforcing pad and not directly to the mainline.
–   To prevent fretting, all clamps should be lined with a resilient liner material.[20]

Finally, there should be agreement as to flange torquing and tagging responsibility after the pneumatic pressure test and cleaning of the piping system connected to machinery. A *flange map* based on the piping and instrumentation diagrams (P&IDs) could assist in reducing the risk of flanges being missed by the torquing team.

5.2.5.6 Cleaning of Piping Systems for Steam Turbines

The following steps should be considered for pre-commissioning and cleaning of feedwater, heat transfer and steam piping. Startup team personnel must assure the contractor has procedures in place that can be mutually agreed upon in order to eliminate the risk of startup delays. Pertinent cleanup and commissioning procedures are aimed at the removal of scale and dirt in the feed-water system which could later foul boiler tubes. Boiler cleanup is to provide a clean metal surface for most efficient heat transfer and avoidance of long-range problems, including loss of boiler service factor due to boiler tube corrosion. In higher pressure units, this is a stepwise procedure to remove trash, oil film and, finally, scale and oxides.

The clean-up of steam piping to the steam turbines is directed at the elimination of particles in the size range of 1/16" or 1.6 mm and smaller entering the turbine in high-pressure units. A number of high-pressure steam turbines – 600 psi or 41 bar and above – have shown a marked susceptibility to damage by particles of this size.

We must note that most large companies in the hydrocarbon processing industries, for instance, have their own steam cleaning practices developed. These in-house rules should govern in case of conflicts.

Pre-commissioning cleaning procedures should be reviewed by the startup advisor typically responsible for an area that is customarily referred to as "Offsites."

A. *Steam Systems Operating Under 600 psi or 41 bar Nominal Rating*

The minimum precautions for low-pressure steam systems are as follows:

- Flush entire feedwater system with clean water to remove trash and loose scale.
- Flush steam-generating section of boiler with clean water to remove trash.
- Perform alkaline boil-out of steam generating section for removal of oil and loose material.
- Flush boiler with treated feedwater and drain for removal of alkaline solution and sludge.
- Open lower drum and headers and remove accumulated debris.
- Mechanically clean steam lines to turbine to the extent practical and necessary after erection.
- Blowdown superheater and steam piping to turbine with repeated blasts of steam or air from steam drum. See later notes on blowdown techniques.
- Blowdown should be accompanied by mechanical rapping of steam line and weld seams, when practical.

The objective of any cleanup procedure is to maintain a consistency of approach and cleaning objectives throughout the system. Chemical cleaning of boilers is desirable even in low-pressure systems for maximum assurance of a high availability operation. When boilers receive a chemical cleaning in addition to alkaline boil-out, other portions of the system should be similarly prepared by acid cleaning.

B. *Steam Systems Operating at 600 psi or 41 bar and Higher Nominal Rating*

- Feedwater system
    - Rubber lined or stainless steel piping in high-purity systems should be cleaned with high-velocity clean water flushing.
    - Carbon steel pipe should be first cleaned by mechanical or chemical means and followed with high-velocity clean water flushing.
    - The interior of storage tanks, deaerator and other low-velocity facilities should be opened for mechanical cleaning followed by hosing with a high-velocity clean water stream.
    - After cleaning, vessels should be opened for manual removal of loose debris.

- Steam generating section of boiler
    - Flush with clean water for trash removal.
    - Alkaline boil-out for oil removal.
    - Flush with clean water for alkaline and sludge removal.
    - Acid clean with maximum circulation velocity of 8 fps for oxide removal.
    - Thoroughly neutralize and flush utilizing minimum circulation velocities of 10 fps or 3 m/s.

- Open lower drum and headers and remove accumulated debris. Commence air- or steam blow.

- Superheater
  - Acid clean as above if flow can be properly directed. For superheaters, which are not completely drainable, acid cleaning is not recommended unless flow through individual tubes can be verified.
  - If acid cleaning is not performed, use mechanical cleaning (grit blast, grind or equivalent) prior to tube erection and use a clean weld process (TIG, MIG or others) for all closing welds, followed by an air blow at high velocity prior to connection of external piping to remove all residual material. Avoid moisture prior to cleaning since this may increase the difficulty of removing trapped material.

- Piping
  - Mechanically clean weld joints to remove weld bead and slag. Remove line scale and oxides by chemically cleaning in place after erection if flow and velocity criteria can be observed and system is free of dead areas.

For large diameter pipe (8″ or 20 cm and larger), chemical cleaning may be ineffective, impractical or too costly. This decision requires knowledge of the availability and capability of local chemical cleaning equipment or contractors. When these limitations exist, use mechanical cleaning of entire line interior prior to erection and close piping with a clean weld technique.
- Provide a vertical dead leg or other disengaging device in the steam line close to the turbine inlet to throw out solid material from the steam path.
- Provide a Y-type strainer with 1/8" or 3 mm maximum opening.
- Provide temporary piping at the turbine inlet for diverting blowdown flow to atmosphere.
- Provide quick-opening valve – motor operated, butterfly valve or valve with equivalent quick opening ability – to assure high-velocity line flow prior to dissipation of blowdown pressure. As an alternative, leave the turbine end of the line open and provide a conventional or quick opening valve at the steam drum outlet. Some utility companies blow down their main steam line by removing the bonnet from the turbine stop valve to make necessary diversion pipe connection.
- Provide soft metal targets in front of blowdown pipe to gage the effectiveness and progress of blowdown – refer to Figure 5.5 showing blowdown target arrangements.
- Repeat blowdown with new target until target is essentially free of particle impact evidence. Use 400–600 psi or 28–41 bar steam or air to pressurize the steam drum and all downstream piping as a minimum.

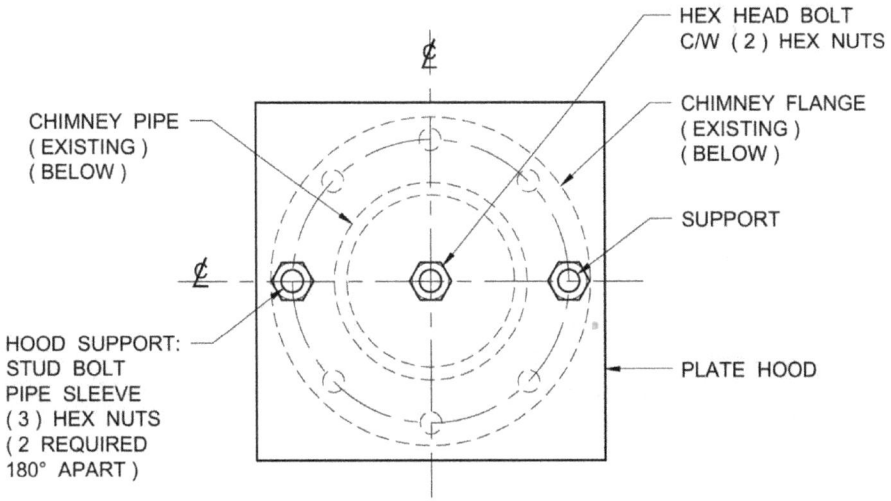

HEX HEAD BOLT
C/W (2) HEX NUTS

CHIMNEY FLANGE
(EXISTING)
(BELOW)

CHIMNEY PIPE
(EXISTING)
(BELOW)

SUPPORT

HOOD SUPPORT:
STUD BOLT
PIPE SLEEVE
(3) HEX NUTS
(2 REQUIRED
180° APART)

PLATE HOOD

PLAN VIEW

HEX HEAD BOLT
C/W (2) HEX NUTS

(2) HEX NUTS
(TYP)

PLATE HOOD

STUD BOLT
(TYP)

PIPE SLEEVE (TYP)

TARGET

HEX NUT
(TYP)

EXISTING
CHIMNEY FLANGE

EXISTING
CHIMNEY PIPE

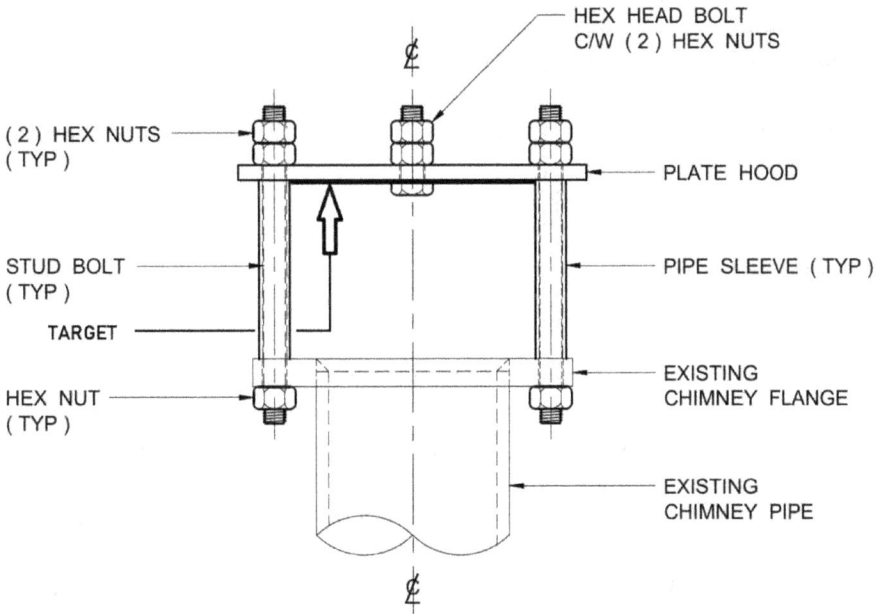

Figure 5.5: Turbine steam piping blowdown targets.

- When practical, the entire steam line should receive a mechanical rapping or vibration during the blowdown procedure to free tenacious material from the line interior. This may not be possible if insulation is in place.

- Steam turbine
  - The integral steam strainer in the turbine steam chest or trip and throttle valve (TTV) should be wrapped with a fine mesh material with approximately 1/16" or 1.6 mm openings for the initial 1 or 2 weeks of operation. This liner is inserted to check steam line cleanliness and gage the effectiveness of the Y-type line strainer. If left too long, it can fail and cause damage to the turbine.
  - A 20-mesh strainer of 26-gage-type 304 stainless steel has been suggested as a liner material by at least one turbine manufacturer. The strainer vendor is normally familiar with the design and fitting of fine mesh liners.

C. *Additional Precautions Pertaining to Chemical and Mechanical Cleaning Procedures*
- When acid cleaning is used, we recommend contracting to a specialist in the acid cleaning business. In addition, we recommend that the procedure to be used by the chosen contractor be reviewed by the startup advisor for guidance.
- Proper circulation of acid and chemicals must be maintained. A special circulating pump is required downstream of the deaerator during the alkaline flushing, acid cleaning, neutralization and deionized water washing steps. If the boiler feedwater pump is in place, a suitable bypass should be provided around the pump during the cleaning. In addition, temporary pipework is required to complete the cleaning circuit. The mechanical or machinery startup engineer must check if the chemical cleaning procedure was carried out, particularly for all the lines between the steam drum and the turbine stop valve. He or she should also make sure that the pH and the hydrazine content of the boiler water is constant at the proper value before starting final steam blowdown operations. If not, the circuit should be water flushed until the neutralizing fluid has been removed from all tubes.
- The alloy steels used for turbine blading are susceptible to attack from even minute traces of the acids and some neutralizing solutions used for chemical cleaning. Trace amounts of acid from the cleaning operation can cause blade failures by stress corrosion. Therefore, neutralization and water flushing steps must continue for sufficient duration to ensure a neutral pH in the system before final blowdown.
- Chemical cleaning will not remove weld "icicles" hanging in the pipe. The main steam line should be checked internally. If icicles are evident, the line should be grit blasted or mechanically routed to remove them. These icicles erode away in tiny beads which pass through the strainer and damage the turbine nozzles and buckets, lines should be mechanically cleaned prior to chemical cleaning or flushing.

- Mechanical cleaning of steam lines with erosive materials must be done with a very extensive follow-up effort to remove all cleaning material from the line interior. Silica-containing materials will vaporize in steam systems which operate above 600 psig or 41 bar and will redeposit in the steam turbine causing a loss of efficiency and premature forced shutdowns when the steam rate becomes intolerable. Alternately, any fine erosive material left in the line will eventually erode the turbine blading and also cause a shutdown.
- It is recommended that any erosive mechanical cleaning be done with steel grit or other silica-free material. Chilled iron grit is a less desirable alternate since it tends to fragment and form a dust.
- If the unit is not to be commissioned immediately following mechanical of chemical cleaning, the clean surfaces of all components must be protected from oxidation by filling with a non-aggressive atmosphere such as deionized water or nitrogen. Deionized water should be treated with 100 ppm hydrazine for oxygen scavenging and suitable dissolved ammonia gas to raise pH to 9 (see Figure 5.6).

D. Steam System Cleaning Recommendations
Detailed procedures can be found in Appendix 5.E.

5.2.6 Lubrication and Seal Oil System Cleaning and Flushing

5.2.6.1 Introduction

Process machinery lubrication is essential. It is accomplished by means of grease or oil application. Cleanliness of the lubricant, in both cases, is of utmost importance for obvious reasons. While grease lubrication systems require somewhat less attention due to their smaller containments and conveyance lines, oil systems can range in volume contents from a few gallons or liters to containments of some 700 gal or 2,650 L with extensive supply and drain piping systems.

Self-contained lubrication systems are primarily used to lubricate anti-friction bearings (AFBs). The bearing housing is the lubricant reservoir. There is no filtration for these self-contained systems. There is no temperature control for these self-contained systems. Self-contained systems depend on radiant cooling or sometimes they may be fitted with a liquid cooling coil located inside the housing. It is strongly discouraged using liquid cooling coils immersed in the oil because when they leak the cooling fluid into the lubricating oil, it is not readily detectable. Self-contained systems circulate oil within the bearing housing by oil rings – sometimes called slinger rings – attached to the rotating shaft (see Figure 5.7). Oil rings have been used for many years and are effective. The oil rings generate an oil mist when rotating through the oil sump.

Oil mist lubrication is a very effective method of lubricating AFBs and ensuring the bearing housing of hydrodynamic bearings are purged.[21] Major hydrocarbon

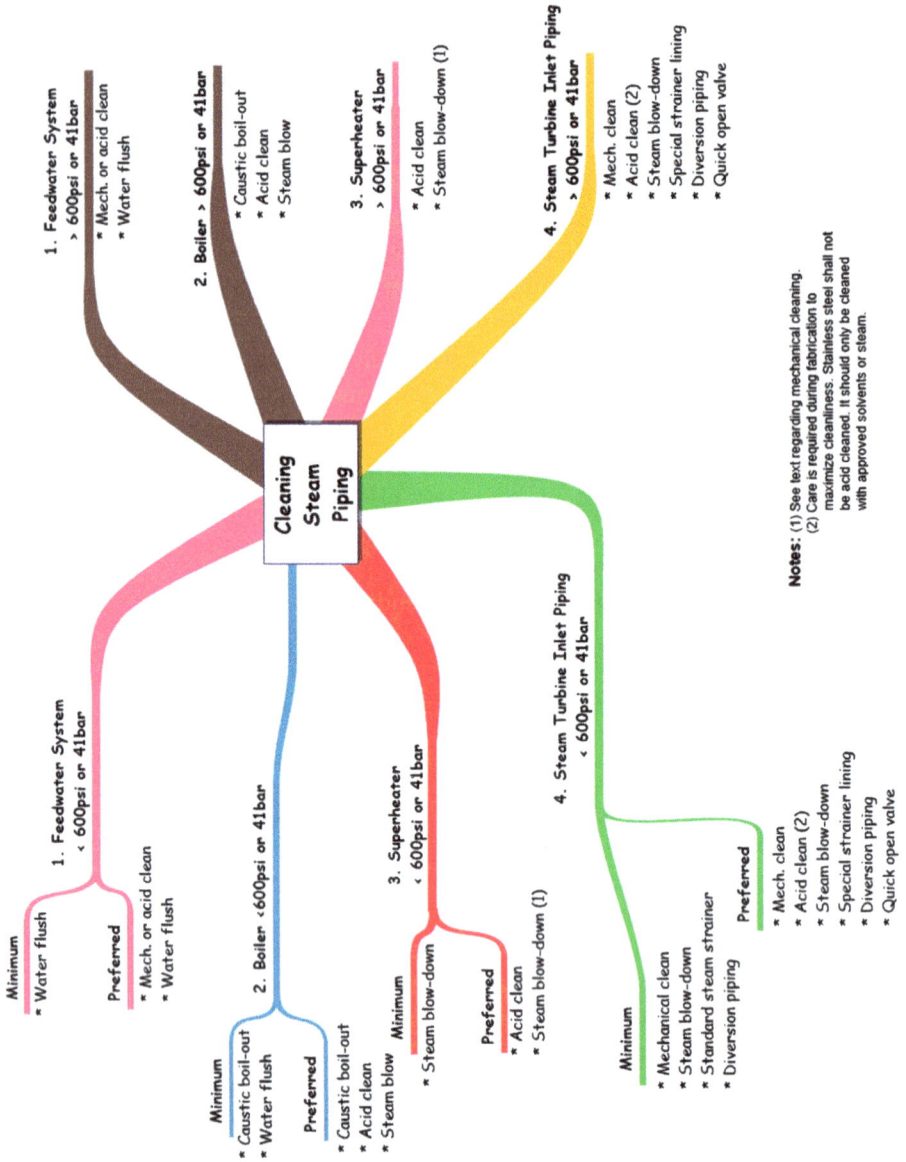

Figure 5.6: A guide to cleaning of steam piping.

Cleaning Steam Piping

1. Feedwater System > 600psi or 41bar
* Mech. or acid clean
* Water flush

2. Boiler > 600psi or 41bar
* Caustic boil-out
* Acid clean
* Steam blow

3. Superheater > 600psi or 41bar
* Acid clean
* Steam blow-down (1)

4. Steam Turbine Inlet Piping > 600psi or 41bar
* Mech. clean
* Acid clean (2)
* Steam blow-down
* Special strainer lining
* Diversion piping
* Quick open valve

1. Feedwater System < 600psi or 41bar
Minimum
* Water flush
Preferred
* Mech. or acid clean
* Water flush

2. Boiler <600psi or 41bar
Minimum
* Caustic boil-out
* Water flush
Preferred
* Caustic boil-out
* Acid clean
* Steam blow

3. Superheater < 600psi or 41bar
Minimum
* Steam blow-down
Preferred
* Acid clean
* Steam blow-down (1)

4. Steam Turbine Inlet Piping < 600psi or 41bar
Minimum
* Mechanical clean
* Steam blow-down
* Standard steam strainer
* Diversion piping
Preferred
* Mech. clean
* Acid clean (2)
* Steam blow-down
* Special strainer lining
* Diversion piping
* Quick open valve

Notes: (1) See text regarding mechanical cleaning.
(2) Care is required during fabrication to maximize cleanliness. Stainless steel shall not be acid cleaned. It should only be cleaned with approved solvents or steam.

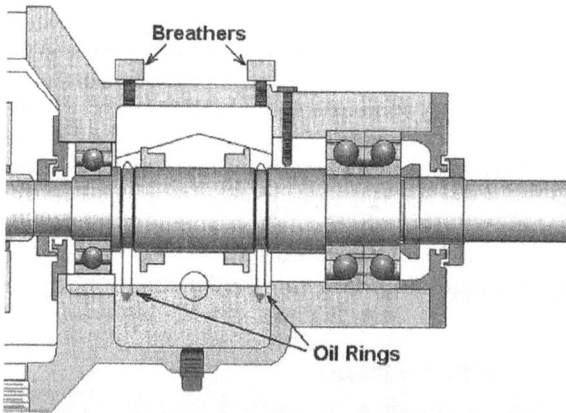

Figure 5.7: Pump bearing housing with self-contained oil sump with oil rings..

processing favorites, for example, have demonstrated significant increase in the mean time between failures of AFBs after applying oil mist lubrication to their equipment. Figure 5.8 illustrates the concept.

Machinery equipped with hydrodynamic bearings, such as most large prime movers, compressors, gearboxes and large pumps require forced circulating lubrication systems. Such a lube oil system for a special-purpose machinery train, for example, typically includes an oil pumps and its auxiliary, two coolers, two filters, frequently an accumulator and pressure regulating control valves as shown schematically in Figure 5.9. These items are assembled on a steel platform or skid usually referred to as console.

In a combined lube and seal oil system – see oil seal in Chapter 6 – a very similar system is used, except that high-pressure pumps and additional control valves will be required. Separate lube and seal oil systems essentially require two complete systems, one system for lubrication and the other for sealing oil. These systems supply filtered, temperature-controlled oil at a controlled flowrate and pressure to the many close clearance spaces that separate the stationary parts from the moving parts of machines. At the machine, lube oil will be supplied from a header through laterals to the bearings and couplings of the machinery train. Seal oil supply pressure will typically be maintained in an overhead tank and fed to the seals by laterals from the main header. Cleaning of these oil systems during pre-commissioning requires removal of debris from the internals of the system to avoid foreign particle damage to the bearings, seals and couplings – if the latter are lubricated.

Field installation of an oil system will usually have the contractor assemble the interconnecting lines between the oil reservoirs and the headers at the machines. Cleanliness of these interconnecting lines should be assured by inert environment welding of the – preferably – stainless steel lines followed by steam blowing before

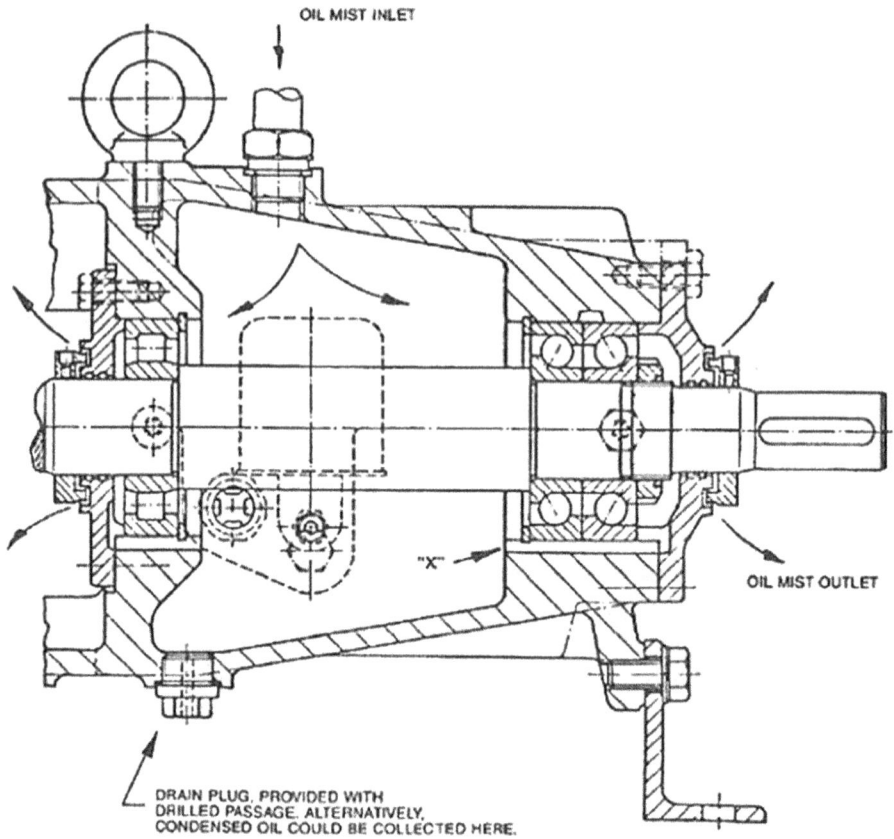

Figure 5.8: Pump bearing housing with oil mist lubrication..

installation. Carbon steel lines should be chemically cleaned with the cleaned surfaces protected by an oil coating prior to installation.

Early during the front-end engineering and design stage of capital projects for process industries and specifically, as an example, for the hydrocarbon processing industry, a subtle yet critical decision is taken that will impact the life cycle costs of the project. This decision pertains to the choice of material of construction for the lube and seal oil system.

Historically, selection of the material of construction rests with the project team and is based upon the difference between the first capital cost between stainless steel versus carbon steel. This is a simple decision based on the cost difference between these two materials. The first capital cost difference between carbon steel – the base cost – versus stainless steel which is in the range of two to four times the material cost of carbon steel.

Further, stainless steel piping systems differ from carbon steel systems as follows:

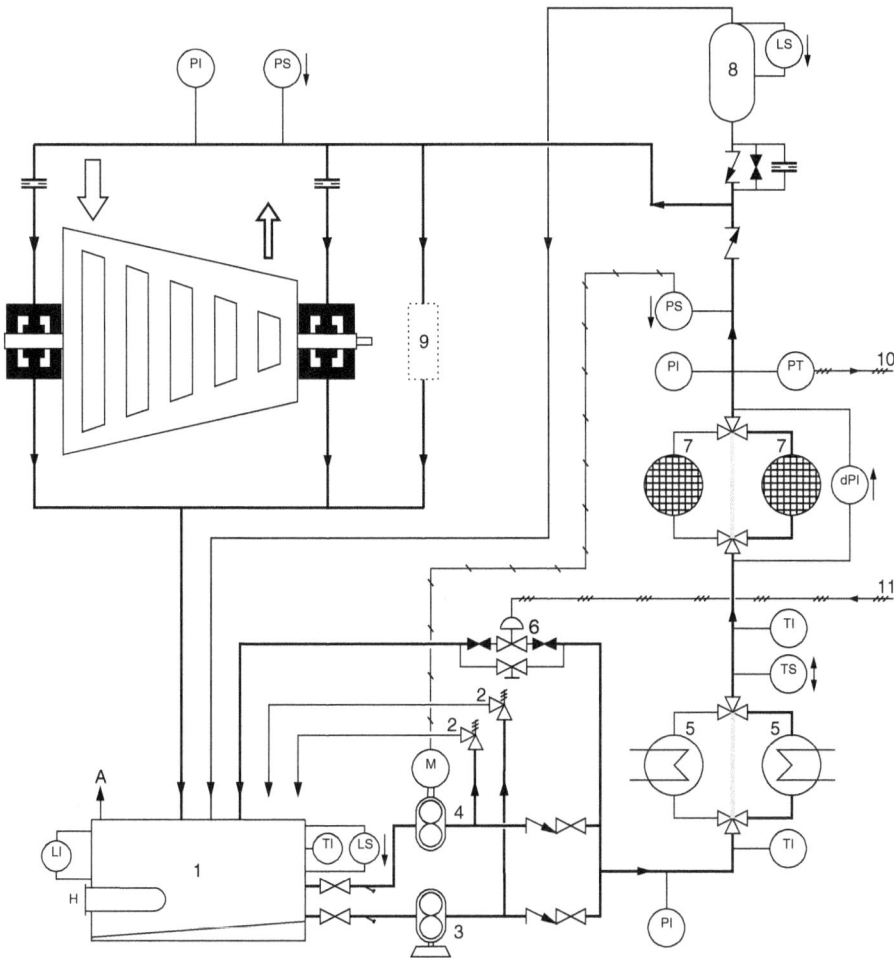

Figure 5.9: Lube oil schematic for a turbocompressor. 1 – reservoir; 2 – safety valve; 3 – main oil pump; 4 – auxiliary oil pump; 5 – cooler; 6 – pressure regulating valve; 7 – filter; 8 – overhead run-down tank; 9 – for prime mover and other users; 10 – pressure input signal to DCS pressure controller; 11 – valve output signal from DCS pressure controller; PI – pressure indicator; DPI – differential pressure indicator; PS – pressure switch; TS – temperature switch; LI – level indicator; LS – level switch; H – oil heater; A – reservoir vent. (Arrow direction indicates device activation on increasing or decreasing parameter). (Modified from original *Mannesmann Demag* source.)

1. The cost to fabricate oil supply systems is marginally higher for stainless steel because of reduced welding requirements, fewer qualified welders needed, more rigorous and associated inspection required.
2. It is widely recognized that a stainless steel system will result in a higher fabricated quality and fewer defects during fabrication.

3. A carbon steel lube oil and seal system will impact the profitability of the enterprise over its entire life in that
    a) cleanliness of the oil system over equipment lifetime is a critical parameter. Carbon steel systems are difficult to clean during the commissioning phase of a project,
    b) during system down times the carbon steel systems are prone to develop rust and subsequently contaminate an otherwise clean lubricant.

All lubrication systems have one thing in common, namely, they have to be meticulously cleaned before commissioning and startup.

5.2.6.2 Lubrication Systems Commissioning and Startup

The startup team charged with following lube and or seal oil systems would be well advised to look at what happened to the console before it arrived at the construction site. Lube and seal oil consoles are usually inspected during fabrication and assembly at the OEM shop. A flushing operation is carried out in the fabricating shop as part of the factory acceptance testing activities.

On consoles and seal oil drainer packages, the procedure outlined in Table 5.2 should be followed by the inspector.

Table 5.2: Lube and/or seal oil console – mechanical run test record.

Designation: _____
Location: _____
Service: _____
Note: Cross out items that are not applicable

1. Are high-point vents and low-point drains provided?	_____ (Yes or no)
2. Are adequate drains provided?	_____ (Yes or no)
3. Is there easy access to fill and drain connections for operating and maintenance personnel?	_____ (Yes or no)
4. Are chemical cleaning vets and drains provided?	_____ (Yes or no)
5. Are fill and drain connections of sufficient size and orientation?	_____ (Yes or no)
6. Does the design eliminate any air pockets?	_____ (Yes or no)
7. Is the oil piping at the pumps properly supported? Follow small bore piping standard.	_____ (Yes or no)
8. With pump piping flanges unbolted, is the piping alignment satisfactory?	_____ (Yes or no)

Table 5.2 (continued)

9.	Are all valves and strainers accessible and correctly manufactured and installed?	_____ (Yes or no)
10.	Have all piping and valves been inspected internally, probing with a magnet? Finally, no machining chips and burrs, weld spatter and dross, burn through, flux and other contaminants?	_____ (Yes or no)
11.	Has all lacquer been removed from bends and fittings?	_____ (Yes or no)
12.	Have cooler bundles been pulled and checked for cleanliness?	_____ (Yes or no)
13.	Reservoir (a) Reservoir interior clean? (b) Paint work satisfactory?	_____ (Yes or no)
14.	Pumps (a) Alignment to drivers satisfactory? (b) With dial indicators on pumps, is alignment satisfactory while jumping on console base? (If not, check whether console is to be grouted, or is supported as level in shop as expected in the field, and correct pump installation as required.)	_____ (Yes or no)
15.	Whole system successfully hydrostatically tested?	_____ (Yes or no)
16.	Flushing (a) Are all console discharges connected to the tank by temporary bypasses? ☐ (b) Is the flushing oil temperature at 180 °F or 82 °C? (c) Are vibrators being used to shake pipework during the flushing process? ☐ (d) Are all control valve bypasses, four-way valves on filters and coolers, being swung periodically? ☐ (e) Is this initial flushing carried out for 8 h uninterrupted? (f) Control function tested successfully? ☐ (g) After initial flush, system checked by installing felt pads, backed up with SS mesh in the temporary bypasses and filter outlets. System re-flushed for 2 h, oil at 180 °F or 82 °C, flow as high as possible, pipework vibrated with temporary vibration devices? ☐ (h) Bypass pads clean? ☐ If the answer is no, flushing must continue. (i) Filter outlet pads clean? ☐ If no, filters to be overhauled to determine the cause of the filter leakage.	_____ (Yes or no)
17.	Will the console be shipped with the temporary bypasses installed? (Required, so that flushing can start in the field as soon as the console is set on its foundation.)	_____ (Yes or no)
18.	Drainer packages (a) Are all valves accessible? ☐ (b) Have all volume chambers, sight glasses, traps, pipework and valves been inspected internally, probing with a magnet? ☐ (c) Check made for no evidence of lacquer? ☐	_____ (Yes or no)

Flushing and readying a lubrication and/or seal oil system is usually considered a pre-commissioning effort, and a critical part of preparing for the complete commissioning of the associated machinery. With shipping and possibly long-term storage the record reflected by the completed checklist in Table 5.2 somehow seems to fade away and it is like the proverbial box of chocolates where one does not know what one will get, until the first cleaning test is made. This can be disappointing and will impact on the project schedule, and, as a result, the commissioning team must allow sufficient time to deal with any surprises.

There are number of stages to prepare for the flushing process:

1. Prepare the system for flushing. This requires the removal of critical components and fabrication of pipe spools to bypass critical equipment items during flushing. Think of components that can become dead ended and therefore, unable to pass a flushing stream. There are great opportunities for shortcuts if and when there is time pressure – with dire consequences.

2. Use both lube oil pumps simultaneously to achieve the highest flowrate possible. Sometimes, based upon the system hydraulic design both lube oil pumps cannot operate simultaneously. The suction conditions – manifested by net positive suction head available and/or suction side turbulence – will not allow for the operation of both pumps at the same time. If this occurs, it is an unfortunate situation. The flowrate during flushing will be reduced and the time required to flush and clean the system will be extended thereby delaying commissioning of the associated machinery.

3. During the flushing process, the temperature of the oil must be cycled between lowest temperature and highest temperature achievable with the lube oil heaters and coolers. The heating process can be further accelerated by admitting wet steam into the waterside of the coolers. Cycling the temperature will accelerate the remove of contaminating particles from the system.

4. The welded areas of the piping system should be impacted by rapping or vibrating during the flushing process to break free any trapped foreign objects and dirt such as debris like weld slag or contaminants entrapped during assembly, that is, tape, lost washers and cigarette butts.

5. Today's best practices suggest that the flushing process should continue nonstop until the system is judged to be clean and fit for purpose.

6. When is the oil system clean? It is strongly recommended that quantitative criteria be used and agreed to before the flushing process begins. A well-regarded standard[22] defines fluid cleanliness based upon particle size as measured by 100 mesh and cloth temporary screens installed between flanges of the system during the flushing process. Details for the placement of these 100 mesh stainless steel and cloth screens must be shown on a marked-up copy of the P&ID or a simplified flow plan that must accompany the written procedure (SOP) and checklist. An arrangement to check oil cleanliness is illustrated in Figure 5.10.

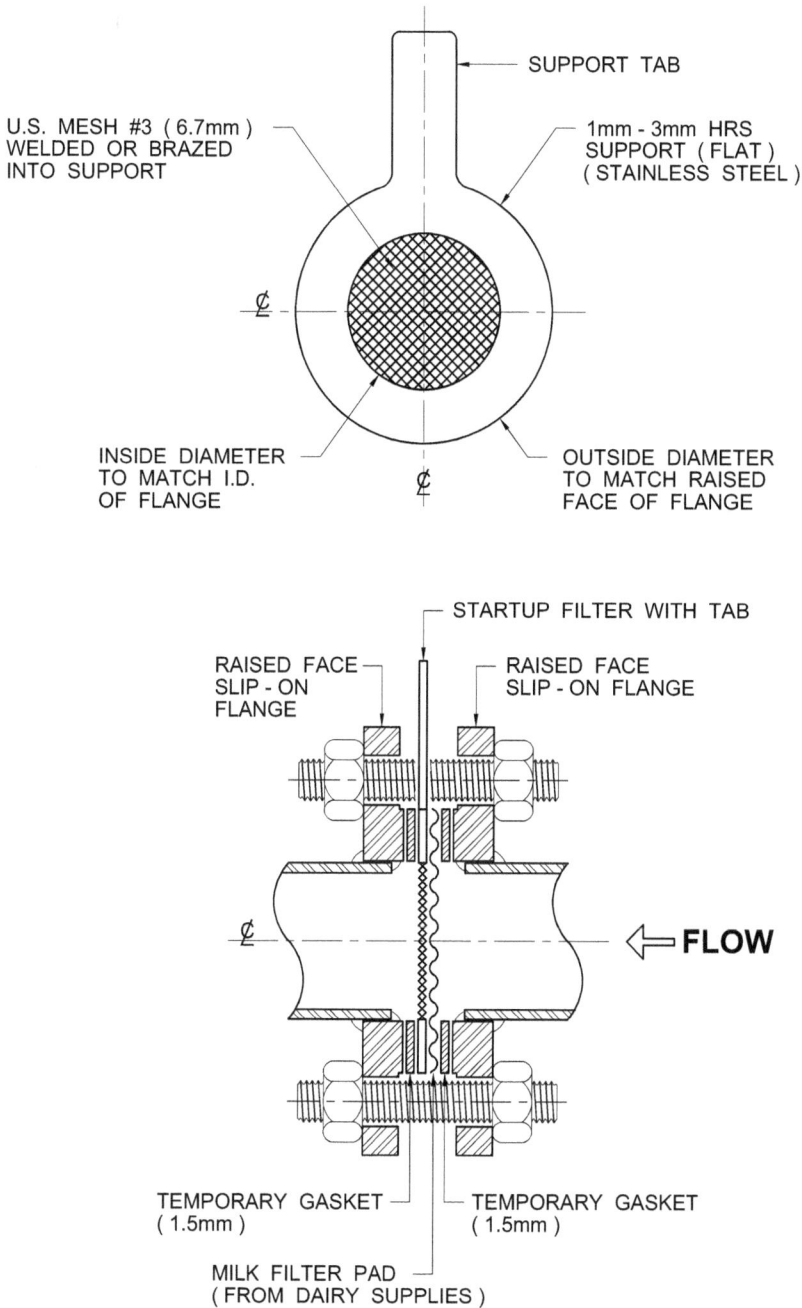

SUPPORT TAB

U.S. MESH #3 (6.7mm)
WELDED OR BRAZED
INTO SUPPORT

1mm - 3mm HRS
SUPPORT (FLAT)
(STAINLESS STEEL)

INSIDE DIAMETER
TO MATCH I.D.
OF FLANGE

OUTSIDE DIAMETER
TO MATCH RAISED
FACE OF FLANGE

STARTUP FILTER WITH TAB

RAISED FACE
SLIP - ON
FLANGE

RAISED FACE
SLIP - ON FLANGE

⇐ FLOW

TEMPORARY GASKET
(1.5mm)

TEMPORARY GASKET
(1.5mm)

MILK FILTER PAD
(FROM DAIRY SUPPLIES)

Figure 5.10: Screen insert for oil cleanliness check.

7. After the first 24 h of flushing, turn off the system. Remove all stainless steel and cloth screens. Label the screens with the locations they were removed from to help identify the parts of the system they have monitored during the flushing process. This helps identify the areas of the system that have not achieved the required level of cleanliness.
8. Install clean 100 mesh stainless steel and cloth screens at all locations previously used as noted on marked-up P&IDs.
9. Continue flushing as described above. The duration of the flushing can now be reduced to 8 h.
10. After 8 h of flushing, turn off the system. Remove all 100 mesh stainless steel and cloth screens. Label the screens with the locations they were removed from to help identify the parts of the system they have monitored during the flushing process. This helps identify the areas of the system that have not achieved the required level of cleanliness.
11. When the agreed level of cleanliness is achieved, remove all pipe spool bypasses and reconnect the system as designed.

Our attention should always be directed to the following general principles:
- Control valves in lube and seal oil piping must be stroked and the valve body disassembled before a system can be certified clean. Although undesirable downstream of filters, block valves are sometimes found between filters and bearings or seals. They must be thoroughly cleaned and valve stems cut – to prevent tampering or inadvertent block-in – before the system can be considered ready for startup.
- Final inspection by borescope of piping and suction nozzles of multistage pumps and centrifugal compressors before tightening suction flanges.

5.2.6.3 System Ready for Commissioning
The details of commissioning the complete oil system are defined in Table 5.3, a lube oil system commissioning checklist.

Additional procedures and checklists for the commissioning of lube oil supply systems are featured in Appendices 5.F and 5.G.

5.2.6.4 General Cleanliness
There can never be enough emphasis put on the issue of general site and specific equipment exterior. Sometimes these admonitions sound routine and jaded as it is often referred to as housekeeping. Around machinery, and especially before an initial startup, it is essential that things are clean and uncluttered. Startup personnel have to be able to detect liquid and gas leaks as well as be sure that no external foreign object impedes the required free movement of installed equipment. We are reminded of an incident that involved a large axial compressor – refer to Figure 14.1 – that experienced high vibrations due to misalignment after its initial startup. The cause was

Table 5.3: Oil systems commissioning checklist.

MANUFACTURER: _____			
MODEL NO.: _____			
SERIAL NO.: _____			
ASSET Number: _____			
	EPC witness	OEM witness	Owner witness
1. Review sizing of lube and seal oil system components.			
2. Review instrumentation list.			
3. Review sizing of lube and seal oil systems components.			
4. Review P&IDs.			
5. Review electrical device list.			
6. Review lube oil console. Check all pipe connections, pumps, drivers, sight glasses and so on.			
7. Review lube oil control scheme.			
8. Review seal oil console – if applicable. Check all pipe connections, pumps, drivers sight glasses and so on.			
9. Check all interconnecting piping. Do NOT use Teflon™ tape. Use *Never-Seez*™ thread compound or *Loctite*™ Teflon thread compound.			
10. Check interconnecting lube/seal oil piping. See piping specs.			
11. Do NOT use Teflon™ tape on small bore piping, that is, 3/4"or tubing. Use *Never-Seez*™ thread compound or *Loctite*™ Teflon thread compound.			
12. If equipped with liquid film seals, review sour seal oil traps, sight glasses, vents and drains. Disassemble and check operability of floats.			
13. If equipped with gas seals, review and check gas seal panel.			
14. Review lube, seal and control oil buffer gas systems against flow diagrams. Check for proper 1/2"/ft or ~ 42 mm/m slope on drain lines. Ensure there are no liquid trapping.			
15. Check all small-bore piping for rigidity and support. Check for compliance with seal welding and gusseting requirements (small-bore piping policy).			
16. Clean all lube/seal oil and buffer gas piping not furnished by lube oil system manufacturer. This auxiliary piping is stainless steel. Test auxiliary piping for leaks.			
17. If item 16 is completed, have Owner's inspector review contractor documents for compliance.			

Table 5.3 (continued)

18.	Make up filter cloth and screens for flushing. These should be 100–150 mesh stainless steel. Install in flanges marked in white.
19.	Open all reservoirs for cleaning and inspection. Clean all lube/seal oil and buffer gas piping not furnished by the lube oil system manufacturer. Make sure this is stainless steel. Test auxiliary piping for leaks. When completed, have owner's inspector review contractor documents for compliance.
20.	Ensure that all piping is installed correctly. Do not overlook drums, balance lines, buffer gas for startup and for operation and so on.
21.	Temporarily remove orifices. Tie each orifice to pipe from which it was removed.
22.	Calibrate and install temperature sensing bulbs, TI dials and pressure gauge
23.	Install flexible hose temporary bypasses around the following: a. All compressor bearing and seals b. Both driver bearings c. Driver controls d. Control switches e. Dead-end lines and non-flowing branch piping f. Lube/seal overhead oil tank gas reference lines g. Close all control valves and open bypasses in the oil systems. If bypasses are not furnished, remove valves and install spool pieces. h. Any other consumer
24.	Lubricate oil pump and driver bearings. Rotate both shafts and check freedom.
25.	Check run-out on both shafts. TIR is not to exceed 0.002" or 50 µm max.
26.	Connect electrical supply to motor pumps.
27.	Bump check motors for rotation and run-in for 4 h.
28.	Reinstall the coupling between the pumps and their drivers.
29.	Fill oil reservoir with _____ oil up to maximum level indicated on the reservoir. Close valves marked in red and all other drains and vents.
30.	Open valves and plugs marked in green.

Table 5.3 (continued)

31.	Heat oil to 150–160 °F or 65–70 °C with reservoir steam coil and additional temporary coils if needed. Block off water to coolers.
32.	Start main oil pump to fill the system and vent high points. Switch flow valves to fill both coolers and filters. Bleed air from both sets. Check operability of three-way valve.
33.	Confirm reservoir is full to minimum operating level after system is full after the downstream equipment has been filled by the pumps.
34.	Check for leaks in the system. Ensure that oil flows through every lube point. Check return line sight glasses for flow through screens. Circulate for 1 h. Turn off heaters; open water lines on cooler and cool oil to 90 °F or ~ 30 °C.
35.	After 12 h shut down main oil pump, start auxiliary oil pump and switch to other cooler and filter. Continue circulating for the last 12 h.
36.	During the last hour of circulation, open all control valves that will normally be in flowing system, for example, PCVs.
37.	Shut down system and clean all screens. Reinstall screens.
38.	During the last hour of circulation, open all control valves that will normally be in flowing system, for example, PCVs.
39.	Turn on heaters and raise oil temperature to 160 °F or 71 °C.
40.	Start main oil pump to circulate the maximum flow through system. Use two pumps if possible.
41.	Circulate oil for 2 h. Hammer full length of all piping at the end of each hour (especially those around welds) using a piece of wood or a soft hammer. After completion of each hammering, close and immediately open pump discharge valve to "shock" the system. Repeat twice after a five-minute interval. **Note:** Oil should be heated and cooled alternately during this 2-h process. One hour for each setting is sufficient.
42.	If the pressure drop between the upstream and downstream side of the filter exceeds 20 psi or 138 kPa (~1.4 bar) switch filter and replace the elements in the one just used.
43.	Shut system down and inspect temporary screens. If material is found, repeat steps 38–43.

Table 5.3 (continued)

44.	Cleanliness of the system must be approved by OEM representative, EPC representative and owner's engineer.
45.	Reconnect oil piping wherever temporary bypasses were installed. Reinstall 100 mesh stainless steel screens. Reinstall all orifices.
46.	Fill reservoir with _____ oil to the maximum operating level indicated on the lube and seal oil reservoir.
47.	Set all pressure switches at their prescribed setting.
48.	Check setting of cooling water control valves.
49.	Check setting of oil pressure control valves.
50.	Check safety and bypass relief valve settings.
51.	Start main oil pump and fill system. Be sure to vent air out of both the filters, coolers and other high points.
52.	Check proper operation of low reservoir level alarm when level in tank falls.
53.	Activate pressure switches and bleed out air.
54.	Activate pressure regulators.
55.	Slowly reduce pressure to each pressure switch and check proper operation of alarm and trip circuits.
56.	Repeat the above for level and temperature switches.
57.	Shut off main oil pump and simulate a low-pressure condition.
58.	When stand-by pump starts, check for continued stable condition in lube oil and seal oil supply headers (no pressure excursion/bump).
59.	Shut down oil system and check out permissive start for compressor driver (motor or steam turbine). Be prepared for the driver to start up during this check.
60.	Inspect temporary screens. If material is found clean the screens and continue to circulate oil until no material is found. Cleanliness of the system must be approved by OEM representative, EPC representative and owner's engineer. Remove screens after approval.
61.	Fill oil reservoir to normal operating level and make final check to ensure that all previously blanked-off components are recommissioned.

Table 5.3 (continued)

62.	Check that all valves are in their normal operating positions (note all valve positions).
63.	Label all alarm and trips with their set points
64.	Replace both filter elements.

found in the inability of the machine to move and accommodate thermal expansion because construction debris had plugged the sliding key arrangement similar to the one indicated in Figure 5.11.

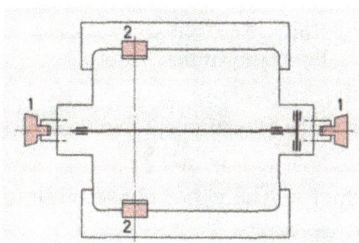

Figure 5.11: Why good housekeeping is important (source: DEMAG).

It is therefore important to include statements pertaining to the responsibility for housekeeping and degree of cleanliness in the project co-ordination document or any other suitable written form.

Appendix 5.A: Machinery storage protection products.

Properties of Product "A"

The folowing text is based on references[23] and[24]. Accordingly, Product A is a rust inhibitor circulation oil with an ISO viscosity grade (VG) of 32. In essence, Product A is a turbine oil to which a special formulation or percentage of anti-foaming agents, oxidation inhibitors and three-phase rust inhibitors have been added. A machine requiring turbine oil can probably be operated at its normal rating when using Product A. However, contacting the equipment manufacturer would be recommended if Product A substantially deviates from the manufacturer's stipulated oil type and performance parameters.

Product A and similar oils are made by major lubricant marketers whose respective formulations remain closely guarded secrets. The three-phase additive package is really (a) a rust inhibitor intended to protect surfaces that are lightly coated with oil, (b) a liquid phase inhibitor intended to protect surfaces submerged in oil and (c) certain vapor phase inhibitors intended to protect surfaces that are exposed to oil vapors.

Oil companies are usually careful not to expose themselves to lawsuits. That is why one can be reasonably certain that the advertised claims of at least one major producer of storage protection fluids are correct. This producer states that Product A, an enhanced or fortified turbine oil with ISO VG 32:

 - protects steel surfaces in the vapor (air) spaces that do not come into contact with the oil;
 - gives rust protection for metal surfaces from which the oil has been drained and where the product leaves a thin residual coating or film;
 - incorporates demulsibility properties similar to those of a turbine oil;
 - protects surfaces that are submerged in the oil;
 - protects other metals occasionally found in machinery;
 - has superior oxidation stability;
 - is suitable for protecting a wide range of equipment in a wide range of temperatures.

There is usually a caveat for Product A: The vapor phase inhibitors will rapidly deplete at temperatures above 120 °F or 49 °C.

Product A and its nearest competitors have the following principal properties:

- Specific gravity	0.86–0.88
- Pour point	–7 °C
- Flash point (ASTM D92)	120–240 °C
- Viscosity index	95
- TAN (ASTM D664)	0.4
- Demulsibility (ASTM D1401)	0
- Rust protection (ASTM D665)	Pass, distilled and/or seawater

Properties of Product "B"

Product B is an oil and is more viscous than Product A. Operating certain machines with Product B in the lubrication system or oil sump is possible for short periods of time, but oil rings may no longer qualify as the lube application method because of high viscous drag. Most oil rings are designed for ISO VG 32 and will not turn properly in the higher viscosity Product B. However, Product B provides effective rust preventive films on the internal surfaces of machinery and it is this capability that makes it a highly suitable storage protection fluid.

Generally speaking, Product B oils displace water from metal surfaces and form relatively strong, water-resistant films on metal surfaces. Corrosion and rust cannot form while the film is intact. If water is already in the system, Product B absorbs the water, thereby turning it into water in oil emulsion. Therefore, the surfaces remain protected.

In most applications, the residual oil does not need to be removed by flushing when re- commissioning or commissioning the machinery. There are exceptions to this rule if:

- The manufacturer or vendor of Product B finds it incompatible with the lubricant recommended by the equipment manufacturer. In this case, flushing with the recommended lubricant would be advised.
- Residual quantities of Product B, regardless of the lubricant application method, would significantly alter the viscosity and temperature performance of the regular or recommended lubricant. An example is a high-pressure hydraulic system where water in oil emulsions could reduce anti-wear properties of the system's oil.
- Draining of the machine's casing or sump is difficult due to its internal geometry or construction features, for instance the presence of pockets.
- Presence of the rust preventive agents would reduce the ability of the new oil to shed water. This is important because the lube oil also serves as the motive fluid in hydraulic stroking mechanisms and actuators.

Product B and its nearest competitors have, approximately, the following principal properties:

- Specific gravity	0.92
- Pour point	−20 °C
- Flash point (ASTM D92)	~ 190–210 °C
- Viscosity index	95
- TAN (ASTM D664)	0.4
- Demulsibility (ASTM D1401)	0
- Rust protection (ASTM D665)	Pass, distilled and/or seawater

Overall, Product B encompasses lubricants that also function as rust preventives. They are especially useful if machines are subjected to repeat cycles of operation under moderate load, then followed by a shutdown. Except for the aforementioned instances, the use of Product B can save the costs of cleanup or flushing prior to again placing the equipment in service.

Properties of Product "C"

Product C is a rust preventive, highly viscous, grease-like substance that is intended for applications with external components, such as exposed shafts, couplings and wire ropes. It is a premium rust prevention grease that can be sprayed or brushed onto a variety of surfaces. Product C exhibits excellent water displacing properties and forms thin, tenacious films. These films protect surfaces, even under high moisture conditions. They are pliable at −35 °C and will not drip at temperatures up to +60 °C.

When formulated to provide wet, oily and extremely thin barrier films that are not tacky, Product C will not unduly attract dirt or dust. Product C also survives light acid fumes and serious exposure to salt laden moisture environments, such as those encountered on a ship.

A slightly different formulation of Product C forms a grease-like film that protects wire rope from corrosion. It stands to reason that this particular formulation should be sought out regardless of whether the wire rope is in operation or dormant, such as laid up and not moving for any length of time.

Product C and its nearest competitors have the following principal properties:

– Specific gravity	0.86–0.88
– Pour point	−7 °C
– Flash point (ASTM D92)	120–240 °C
– Dropping point, ASTM D2265	Up to 63 °C

Best practice companies and/or professionals advocate vapor-related and old-style conventional storage protection methods for short-term preservation of assets only.

Grease-like conventional products can be obtained with viscosities that allow pouring the product into cavities or brushing it on external parts. In general, these products are applied while equipment is in transit from the manufacturer to the purchaser. Working with experienced product manufacturers will prove advantageous.

When trying to extend the period of protection to reliably last longer, vapor-related and old-style conventional storage protection methods will require potentially costly reapplication and maintenance. Their removal and cleanup will incur added expense later. Once these facts are considered, it will often make them unattractive for long-term preservation tasks.

The use of vapor-related and old-style conventional storage protection methods will be risky propositions once risk-taking decision makers get involved. The reliability professional may go along with these methods under the assumption that storage preservation is needed for 3 to at most 6 months only. However, the pro may be without recourse when the risk-taking managers claim that there is no budget item for maintenance, product replenishment, shaft rotation or re-caulking.

Appendix 5.A1

Nos.	Equipment Tag Nos.	Discription	QTY	Frequency	Preservation Requirements
				AREA 18	
1	UWP-01-A	Cooling Water Supply Pump	1	24 weeks	1. Inspect Flange Covers, Caps etc.
2	UWP-01-B	Cooling Water Supply Pump	1	24 weeks	2. Inspect service coatings and repair as necessary. Use Cortec VpCI368 for machined bare surfaces.
3	UWP-01-C	Cooling Water Supply Pump	1		
4	UWP-01-D	Cooling Water Supply Pump	1	12 weeks	3. Connect pump to Oil Mist. (Check & refill oil mixture of 5 parts VG46 to 1 part of Cortec 329)
5	UWP-01-E	Cooling Water Supply Pump	1		
				12 weeks	4. Connect turbine to Oil Mist. (Check & refill oil mixture of 5 parts VG46 to 1 part of Cortec 329)
				12 weeks	5. Connect gearbox to Oil Mist. (Spray with oil only, mixture of 5 parts VG46 to 1 part of Cortec 329)
				10 weeks	6. Rotate Pump Shaft 10-1/4 turns.
				10 weeks	7. Rotate Lube Motor Shaft 10-1/4 turns.
				weekly	8. Rotate Turbine Shaft 2-1/4 turns.
6	UWPM-01-D	Cooling Water Supply Pump Motor	1	24 weeks	1. Spray motor bearings with protective oil layer, Tectyl 511 or equivalent through oil hole.
7	UWPM-01-E	Cooling Water Supply Pump Motor	1		
8	UWPST-01-A	Steam Turbine	1	4 weeks	1. Introduce small quantity of VG46 oil through bearing breather or filler plug and turn rotor 2-3 times to distribute oil around the journal and bearing surface.
9	UWPST-01-B	Steam Turbine	1		
10	UWPST-01-C	Steam Turbine	1		

Figure 5.A1: Managing equipment storage.

Appendix 5.B: Pickling procedure for reciprocating compressor suction piping

General Recommendations

1. The job should be executed by experienced people.
2. Operators must wear adequate safety equipment (gloves and glasses).
3. Accomplish entire pickling operation in as short a time as possible.

Preliminary Work

1. Install an acid-resistant pump connected to a circulating tank.
2. Provide 1–1/2 in or 38 mm – or greater– acid resistant hoses for the connections. Prepare a suitable assembly sketch.
3. For ensuring the filling of the system, flow must go upward and vents must be installed.
4. Provide method for heating the solutions, for example, a steam coil.

Pretreatment

Pretreatment is required only when traces of grease are present.
1. Fill the system with water at 90 °C or 194 °F.
2. Add 2% sodium hydroxide and 0.5 percent sodium metasilicate (or sodium or- thosilicate if cheaper). If these compounds are not available and only a small amount of grease is present 2% of NaOH and 3% of Na_2CO_3 may be used.
3. Circulate for 20–30 min at 90 °C or 194 °F.
4. Dump the solution and wash with water until pH = 7.

Acid Treatment

1. Fill the system with water at 50 °C or 122 °F.
2. Add 4% of Polinon 6A® or equal and circulate to ensure its complete distribution.
3. Add hydrochloric acid to reach the concentration of 7%.
4. Circulate intermittently for about 45 min or more until the pickling has been accomplished.

Notes

1. In order to avoid corrosion:
   (a) Keep the flowrate lower than 1 m/s.
   (b) Take samples of the solution and check for the Fe^{3+} content: if $[Fe^{3+}] > 0.4$ percent, dump solution.
2. In order to determine when the system has been adequately pickled, put a piece of oxidized steel in the circulation tank and inspect it frequently.

Neutralization

1. Add sodium hydroxide for neutralizing the acid, and water to avoid a temperature rise.
2. Circulate for 15–30 min.
3. Dump the solution and wash with water until pH = 7.

Note

The concentration must be calculated on the overall volume of the solution.

Passivation

1. Fill the system with water at 40 °C or 104 °F.
2. Add 0.5% of citric acid and circulate to ensure complete mixing.
3. Check the pH of the solution: if pH = 3.5, slowly add ammonia to raise pH to 3.5.
4. Circulate for 15–20 min.
5. Slowly add ammonia to raise pH to 6 in 10 min.
6. Add 0.5% sodium nitrite (or ammonium persulfate).
7. Circulate for 10 min.
8. Add ammonia to raise pH to 9.
9. Circulate for 45 min.
10. Stop the pump and hold the solution in the system for at least 3 h.
11. Dump the solution.

Cleaning of Large Compressor Piping: Method II

Cleaning of the piping may be done by commercial companies with mobile cleaning equipment or by the following recommended cleaning procedure. After hydrostatic tests have been made and the pipe sections have been cleaned as thoroughly as possible on the inside, the piping should be pickled by the following (or equivalent) procedure:

1. Remove all grease, dirt, oil or paint by immersing in a hot, caustic bath. The bath may be a solution of 8 ounces of sodium hydroxide to 1 gallon of water with the solution temperature 180–200°F. The time of immersion is at least 30 min, depending on the condition of the material.
2. Remove pipe from caustic and *immediately rinse with cold water.*
3. Place pipe in an acid pickling bath. Use a 5–12% solution of hydrochloric (muriatic) acid depending upon the condition of the pipe. Rodine inhibitor should be added to the solution to prevent the piping from rusting quickly after removal from the acid bath. The temperature of the bath should be 140–165°F. The time required in the acid bath to remove scale and rust will vary, depending on the solution strength and condition of piping; however, 6 h should be a minimum. The normal time required is about 12–14 h.
4. Remove pipe from acid bath and *immediately wash with cold water* to remove all traces of acid.
5. Without allowing piping to dry, immerse in a hot neutral solution. Add 1–2 ounces or 28–56 g of soda ash per gallon of water solution which may be used to maintain a pH of 9 or above. The temperature of the solution should be 160–170°F. Litmus paper may be used to check the wet piping surface to determine that an acidic condition does not exist. If acidic, then repeat neutral solution treatment.
6. Rinse pipe with cold water, *drain thoroughly and blow out with hot air until dry.*
7. *Immediate steps must be taken to prevent rusting,* even if piping will be placed in service shortly. Generally, a dip or spray coating of light water displacement mineral oil will suffice; however, if piping is to be placed in outdoor storage for more than several weeks, a hard-coating water displacement-type rust preventative should be applied.
8. Unless piping is going to be placed in service immediately, suitable gasketed closures must be placed on the ends of the piping and all openings to prevent entrance of moisture or dirt. Use of steel plate disks and thick gaskets is recommended for all flanges. Before applying closures, the flange surfaces should be coated with grease.
9. Before installation, *check that no dirt or foreign matter has entered piping and that rusting has not occurred.* If in good condition, then pull through a swab saturated with carbon tetrachloride.
10. For nonlubricated units, where oil coating inside piping is not permissible (due to process contamination), even for the starting period, consideration should be given to one of the following alternatives:
    (a) Use of nonferrous piping materials, such as aluminum.

(b) Application of a plastic composition or other suitable coating after pickling to prevent rusting.

(c) After rinsing with water in step six, immerse piping in a hot phosphoric bath. The suggested concentration is 3–6 ounces of iron phosphate per gallon of water, heated to 160–170 °F, with pH range of 4.2–4.8. The immersion time is 3–5 minutes or longer, depending on density of coating required. *Remove and dry thoroughly, blowing out with hot air.*

Caution: Hydrochloric acid in contact with the skin can cause burns. If contacted, *acid should be washed off immediately with water.* Also, if indoors, adequate ventilation, including a vent hood, should be used. When mixing the solution, always *add the acid to the water, never the water to the acid.*

Appendix 5.C: Machinery piping considerations

Designation: _____
Location: _____
Service: _____
Note: Cross out items that are not applicable

1. **Pipe hanger considerations during commissioning:** Consider the installation sequence which pipes away from the equipment nozzles.

   _____ (Yes or no)

2. **Pipe hanger considerations during commissioning:** Assure that the machinery is always aligned without the piping being attached.

   _____ (Yes or no)

3. **Pipe hanger considerations during commissioning:** Adjust the pipe for proper fit up, then connect.

   _____ (Yes or no)

4. **Pipe hanger considerations during commissioning:** Witness the alignment changes while the pipe is being installed.

   _____ (Yes or no)

5. **Pipe hanger considerations during commissioning:** Follow detailed checklist 16.1. For piping larger than 8 in NPS, sand bags or similar loads may be needed on the hanger adjacent to the machinery.

   _____ (Yes or no)

6. **Pipe hanger considerations during commissioning:** Pull stops on all system hangers.

   _____ (Yes or no)

7. **Pipe hanger considerations during commissioning:** Assure that indicators do not move out of "1/3 total travel" cold setting zone. If the indicator moves beyond the travel limit, a redesign of the piping may be necessary.

   _____ (Yes or no)

(continued)

8. **Pipe hanger considerations during commissioning:** Adjust hanger to return travel marker to "C."

_____
(Yes or no)

9. **Pipe hanger considerations during commissioning:** Record alignment of machinery, then re-install piping system hanger stops.

_____
(Yes or no)

10. **Final checks, prior to system operation:** Disconnect piping as necessary.

_____
(Yes or no)

11. **Final checks:** Flush and/or steam blow system.

_____
(Yes or no)

12. **Final checks:** Re-pipe and realign.

_____
(Yes or no)

13. **Final checks:** Weigh the hanger adjacent to the machinery.

_____
(Yes or no)

14. **Final checks:** Pull system pins, check "C" setting and fine tune as needed.

_____
(Yes or no)

15. **Final checks:** Verify that the travel stays within 1/3 of total "C"- Zone.

_____
(Yes or no)

16. **Flange joint practices:** Generally, follow a checklist format.

_____
(Yes or no)

17. **Flange joint practices:** Generally, follow a checklist format.

_____
(Yes or no)

18. **Flange joint practices:** Aim at avoiding flange leakage by making sure all bolt stresses are even and the flanges are within tolerance.

_____
(Yes or no)

19. **Flange joint practices:** Critical flanges should be defined as joints placed in services in excess of 500 °F or 260 °C and in sizes above 6 in or 15 cm in diameter.

_____
(Yes or no)

20. **Flange joint practices:** Critical flanges should be identified and records maintained as shown in Figure 5.A1.

_____
(Yes or no)

Appendix 5.D: Pipe flanges and bolt-up/torque application

Designation: _____
Location: _____
Service: _____
Note: Cross out items that are not applicable
1. Prior to gasket insertion check the following:
   Check conditions of flange faces for scratches, dirt, scale, protrusions; wire brush as needed.

_____
(Yes or no)

2. Verify gasket material in use, the bolt dimensions and the bolt grade.

_____
(Yes or no)

(continued)

---

3. Assure that only new gaskets are used. Check for damage on gaskets (spiral wound gaskets should not be loose and should have at least three evenly spaced spot welds on the I.D. or approximately one spot weld every 6 in or 15 mm).

          _____
          (Yes or no)

4. Use a straight edge to verify flatness of the flanges.
Mating flanges should be parallel within 1/32 in or 8 mm at the extremity of the raised face.

          _____
          (Yes or no)

5. **Torque application procedure:** Snug bolts up in crisscross pattern, applying the load evenly, at least one thread past the nut.

          _____
          (Yes or no)

6. **Torque application procedure:** Repeat crisscross operation until no. 1 bolt is not less than 80% of the initial setting after the last crisscross tightening sequence. See standard table values for torque application sequence.

          _____
          (Yes or no)

7. **Torque application procedure:** Set torque wrench to the next increment, repeating crisscross pattern until the final torque has been reached in all bolts.

          _____
          (Yes or no)

8. **Torque application procedure:** Check that after the last bolt is tight, no.1 bolt torque value is not less than the final torque.

          _____
          (Yes or no)

9. **Hot bolting:** Using the same crisscross pattern used initially, re-tighten all bolts to the final torque, restoring the bolt's original stress level.

          _____
          (Yes or no)

10. **Hot bolting:** If leakage is present, start at the point of leakage and proceed in a crisscross pattern.

          _____
          (Yes or no)

11. **Hot bolting:** If leakage does not stop, examine flange face for damage, distortion and so on.

          _____
          (Yes or no)

12. **Hot bolting:** Check flange alignment, if needed cut pipe and re-weld.

          _____
          (Yes or no)

13. **Hot bolting:** Reassemble joint.

          _____
          (Yes or no)

14. **Hot bolting:** Check support system for adequacy.

          _____
          (Yes or no)

Appendix 5.D1

Date: _____

Recorder: _____

Crew: _____

Line No.: _____

Joint No.	Flange one		Flange two		Joint alignment (3)					Gasket Condition			Bolts	
In flow Direction Sequence.	Face (1) Condition	Warpage Check (2)	Face (1) Condition	Warpage Check (2)	A B	A - B	C D			New Gasket ?	Proper Type?	Any Defects	Lubed ?	Length Checked
1														
2														
3														
4														

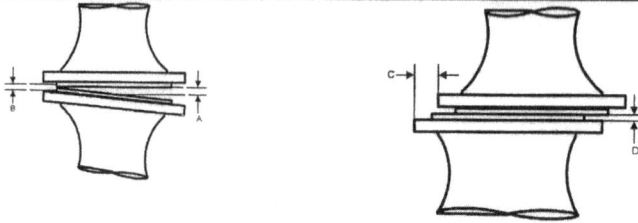

Figure 5.D1: Piping alignment record.

Appendix 5.E: Procedure for blowing steam turbine inlet lines

I Procedure for Blowing 110 psi and 600 psi or 7.6 bar and 41 bar Steam Turbines Inlet Lines

Notes

1. This blowing procedure has been found satisfactory in past startups. It conserves steam without unduly affecting overall results. Prepare a sketch of system before using similar procedure.
2. Steam systems have been successfully cleaned with suitable pigging procedures. Such procedures should be considered if proposed by an experienced contractor.

A. Status Before First Blow

1. This procedure covers only the main lines comprising the bulk of the pressure system. Individual small laterals — 1" to 3" or 2.5 to 8 cm diameter — are individually valved from the main and may be blown as required. However, steam jet ejector laterals and laterals to lube-oil-containing turbines will be blown as part of this procedure.
2. 24" or 61 cm _____ has been blown up to valves _____ and _____.

3. Close valves _____ group, and _____ group.
4. Open blocks and bypass around _____ . Also, open valves _____ group, and _____ .
5. Remove internals from double check valves in 24" or 61cm _____ at

   _____ .
6. Disconnect flange at _____ .
7. Remove _____ .
8. Replace _____ , with spool piece.
9. Disconnect lines at all equipment in portion of system to be blown (i.e., at turbine and ejector inlets).
10. Crack open all drains on portion of line to be blown.
11. Remove _____ .

B. Procedure
1. Long blow
   a) Allow blow to continue for 20 min.
   b) Allow line to cool for about 1 h.

2. Short blow
   a) Open controlling valve as quickly as possible.
   b) Blow for 5 min.
   c) Close valve and cool line for about 15 min.

C. Sequence
1. First blow
   a) Warm up 24" or 61 cm _____ and 8" or 20 cm _____ using valve

       _____ .
   b) "Long Blow" this portion of the line.
   c) Close valves _____ and _____ (open valve wide).
   d) Alternate "Short Blows" between valves _____ (for 8") _____,
       and _____ (for 24"_____)
   e) After fourth blow, close valves _____ and _____ .
   f) Replace _____ .
   g) Install target plate at _____ and blow for fifth time using valve _____ .
       Repeat if necessary.
   h) Open all valves in _____ group and initiate "Short Blows" using
       valve _____ . Check all openings in group _____ to assure good
       steam flow.
   i) After fourth blow, install target plates at _____ , and _____
       Repeat if necessary.

2. Second Blow

    a) Warm up 14" _____ using valve _____ and 6" _____ using valve _____.

    b) "Long Blow" these lines.

    c) Alternate "Short Blows" between valve _____ and _____ . Check all openings to assure good steam flow.

    d) After fourth blow install target plate at _____. . Repeat if necessary.

II Alternate Procedure for Blowing 110 psi and 600 psi or 7.6 bar and 41 bar Steam Turbines Inlet Lines

1. General. To avoid startup delays and unnecessary dirt and erosion problems with steam turbines, it is necessary to begin startup with a clean steam supply system. Special-purpose steam turbines are designed to run on super-heated steam free from inclusions such as sand, weld slag, rust and so on. Although the turbines will be provided with steam supply strainers, foreign particles of sufficient size may still pass through, causing severe erosion of nozzles and blades as well as sticking of TTVs due to dirt. Further, as the suction strainer collects dirt, the power of the turbine will decrease requiring the strainer to be cleaned.

    This procedure represents a method of steam system cleaning which has been used to produce acceptable cleanliness. Use of steam velocity, temperature cycles and targets to check cleaning results is required.

2. Steam Velocity for System Blowing. A minimum acceptable criterion for determining the steam velocity for blowing is one and one-half times the maximum service steam velocity through the lines. This should be considered a minimum since "the more steam the better." To minimize the consumption of steam for blowing, throttled steam should be admitted to the section of line to be blown. That is, expand the steam to only a few PSI or bar above atmosphere before admitting it to the line to be cleaned.

3. Steam Temperature for System Blowing. Steam blowing must be done at or near design temperature. In this manner, maximum use is made of the differential coefficients of expansion and the different rates of thermal growth between the piping and the scale, slag and other lodged foreign matters which are to be removed. Throttling of design steam upstream of the section to he cleaned will still provide nearly design temperature.

4. Preparations for Steam System Blowing. The system to be blown must be identified and broken down into:

- Steam mains or main supply headers from the generation source.
- Steam laterals or feeder lines which carry the steam from the main header to the equipment.

5. Responsibility for each system component must be clear. Next, the locations where blowing steam is to be exhausted must be identified. These locations should be selected at the end of single piping runs. Crossover lines in parallel systems may have valves open and closed or blinds used to create a single pipe run.

Once the exhausting locations are identified, provisions for directing the exhaust and mounting steam targets should be made. Directing the exhaust steam in a safe direction may require temporary elbows and chimneys which must be fabricated.

5. Fixtures. Brackets for holding steam targets should be prepared for each exhaust location. Brackets may be designed and welded to temporary elbows and chimneys or bolted to existing flanges. See Figure 5.5.

Steam targets should be square with an area to cover the exhaust steam path. The target bracket should hold the target about one pipe diameter away from the paint of the exhaust steam. Targets made of aluminum plate 1/8" or 3 mm thick have been used satisfactorily.

On steam mains where pipe areas are large, strip targets may be used to sample pipe cleanliness. Final blowing of laterals will determine actual cleanliness.

6. Blowing at Steam Turbines. Steam turbines are provided with steam strainers and elbows upstream of the TTV. These provide convenient locations for exhausting steam during blowing procedures. Often, the strainer can be removed and the elbow installed in its place to direct the exhaust steam away from equipment. All that remains is to provide a target and its bracket.

7. Procedures for Steam Blowing. Steam system cleaning discussed in this write-up applies the theory of thermal loosening of foreign material attached inside of steam lines and sweeping these particles away with the velocity of the steam. This will necessarily require several cycles of heating, blowing and cooling. A schedule which has been successful on insulated steam lines is as follows:

7.1 Warm steam line (10 min or as required).
7.2 Blow steam at 1–1/2 times maximum service velocity (1 h).
7.3 Shut down and allow line to cool (2–3 h).
7.4 Repeat Steps 7.1 through 7.3.
7.5 Mount target and repeat Steps 7.1 through 7.3.
7.6 Continue blowing with Steps 7.1 through 7.3 until target results are satisfactory.
7.7 After satisfactory target is obtained, allow line to cool thoroughly (10 h) and blow again to assure system cleanliness. A suggested standard of cleanliness is:
- Maximum number of impressions 12 per in^2 or 2 per cm^2.

- Maximum depth of impression in Type 6000 aluminum not to exceed 0.005 in or 130 µm.
- Maximum diameter of impressions 0.015 in or 380 µm.

7.8   When two successive targets with a thorough cooling period between blows are obtained, the line is indicated as clean.

Appendix 5.F: Lube oil system commissioning procedure

Petronta Chemicals	
	Leondorf Olefins Plant 1
Cold Ends – Refrigeration Area	Machinery/Commissioning
SOP 1033	*LBP25A/B Lube Oil System Commissioning Procedure*
Purpose	The objective of this procedure is to prepare and commission the quench water pump turbine lube oil systems prior to overspeed testing
Precautions: Safety, Health, Environmental Quality	– Proper oil containment should be utilized during the commissioning to prevent accidental spill/release of lube oil at location such as filters, reservoir and so on.
References:	Manufacturer supplied equipment documents under tag number LBP-25A/B P&IDs: LB-198-1045, 1502, 1507, 1509, 1511, 1513
Prerequisites:	1. The following utilities must be available and lined-up to the relevant systems: – 415 V 50 Hz electrical power to both lube oil pump motors (Both motors in manual OFF) _____ – Nitrogen cylinder hooked up to oil accumulator _____ – Cooling water available to lube oil coolers and turbine _____ – Instrument air available to steam turbine sealing air supply _____ 2. Control valves on lube oil system checked and calibrated _____ 3. Lube oil system has been flushed and verified to be clean for service, and all flushing screens removed from the system _____ 4. Electric motor drivers for lube oil pumps, LBP25A/B have been run-in for 4 h. Uncoupled vibration readings taken after units were warm, approximately 1 h, and again every hour during the duration of 4-h run. _____ 5. Lube oil reservoir has been filled to the maximum operating level indicated on the oil tank. _____ 6. Steam turbine (LBPST25A/B) sealing air is commissioned _____

(continued)

Materials Required:	Tools necessary for opening flanges and draining oil to remove air from lines and equipment	_____

Major activities:
A. Preparation for Startup
B. Venting of Lube Oil System
C. Performing Valve functional Tests and Pump Swing
D. Functional Tests of Oil System Alarms and Trips

Equipment List:

Equipment Number	Description	P&ID
XP21AP01A/B	LBPST25A Lube Oil Pumps	LB-175-1507
XP21BP01A/B	LBPST25B Lube Oil Pumps	LB-175-1511
XP21ATK01	LBPST25A Lube Oil Reservoir	LB-175-1507
XP21BTK01	LBPST25B Lube Oil Reservoir	LB-175-1511

Procedure Steps:

A. preparation for startup

1. Open suction and discharge valves on both lube oil pumps. _____
2. Open the 1" block valves and the bypass globe valve around the lube oil kickback or dump control valve (_____), to allow oil to flow back to the reservoir and avoid slugging of the dry oil filter cartridges with high-pressure oil _____
3. Swing the three-way transfer valve of lube oil coolers to the mid position and open the equalizing valve. _____
4. Open all three 1" block valves around the lube oil temperature control valve, (_____). _____
5. Swing the three-way transfer valve of lube oil filters to the mid position and open the equalizing valve. _____
6. Ensure that all globe valves on the impulse lines to the transmitters are opened. _____
7. Open the 1" block valve lining up the accumulator to the oil supply. _____
8. Commission nitrogen from N2 bottle and pre-charge the accumulator to a pressure of 130 kPag; close the ¾" N2 supply globe valve. _____
9. Open the 1" block valves on the lube oil supply pressure control valve, (_____); the control valve should be in full open position before sensing oil pressure and the 1" bypass globe valve should be closed. (Note: Always try to visually determine status of valves.) _____
10. Confirm the motors for the pumps have been energized before proceeding. _____

B. venting of lube oil system

1. Once lube oil skid is ready to operate, start the "A" lube oil pump. _____

(continued)

	NOTE: *This is the initial commissioning of the lube oil system so ensure that all the pressure gauges and transmitters are properly bled to remove all trapped air before reading the instruments.*
	Caution:
	Increase the oil pressure gradually while filling the filters. Read oil pressure from (_____) on main oil pump discharge line. Design pressure is 2000 kPag; sudden surge in oil pressure through the filters can collapse the filter cartridges leading to machinery bearing and shaft damage.
	2. Once the oil pump is running steady, slowly close the 1" bypass globe valve around the oil system kickback valve (dump valve), (_____), and watch the pump discharge pressure increase gradually to about 517 kPag on the pump discharge pressure gauge. _____
Instrument	3. Confirm that the oil pressure (PI _____) downstream of oil coolers, is at 517 kPag; if pressure is not at 517 kPag after the bypass valve is fully closed, adjust the back pressure control valve, (V _____), to the correct pressure (Ensure that the control valve reference line is commissioned.). _____
Instrument	4. Confirm that the lube oil supply pressure on (G _____) is at 103 kPag; if not, adjust pressure control valve, (V _____), as required. _____
Console	5. Confirm that the low and low–low-lube oil alarms are cleared and the lube oil permissive to start is satisfied. _____
	6. As the oil system fills, verify that the oil reservoir level remains in the normal operating level; fill if needed. _____
	7. Open the 3/4" high-point vent valves on both of the lube oil coolers and vent until oil is observed draining from the vent lines. _____
	8. Swing the three-way transfer valve to the "B" cooler and confirm that oil is overflowing through the vent line. _____
	9. Swing the three-way transfer valve to the "A" cooler and confirm that oil is still overflowing through the vent line. _____
	10. Close the cooler high-point vent valves. _____
	11. Open the cooling water supply and return valves to the lube oil coolers and vent the coolers using the 3/4" vent valves. _____
	12. Shut the cooler vent valves after all vapor is vented from the coolers. _____

(continued)

---

13. Open the ¾" vent valves on both of the lube oil filters and observe the oil overflow through the sight glass, (FI ____) on the vent line.   _____
14. Swing the three-way transfer valve to the "B" filter and confirm that oil is overflowing through the vent line.   _____
15. Swing the three-way transfer valve to the "A" filter and confirm that oil is still overflowing through the vent line.   _____
16. Close the filter housing vent valves.   _____
17. Verify flow in sight glasses at all bearing oil drains. _____
18. Perform a final check on the critical parameters listed as follow:
18.1 LO header pressure, (G _____) is 517 kPag   _____
18.2 LO supply pressure (G_____) is 103 kPag   _____
19. Bleed the impulse leg of all transmitters in the oil system and verify the readings with their respective pressure gauges.   _____
20. Verify that all alarms on lube oil system in DCS/local panels are cleared.   _____
21. Check main lube oil tank level, and top-up oil to normal operating level if required.   _____

C. Performing valve functional tests and pump swing

Objective

Perform functional checks on safety relief valves, pressure control valves and standby and emergency pumps
NOTE: *The following steps are performed to adjust the settings of the main oil pump discharge pressure relief valve to 1,035 kPag.*
1. Confirm that the selector switch on the "B" pump is in HAND position, not running, and the "A" pump is running; if the "B" pump is running, press the STOP button.   _____
2. Open the 1" bypass valve around the "A" oil pump discharge pressure relief valve.
NOTE: *The oil system's pressure alarms/trip may be activated once the bypass valve is opened. Temporarily ignore the alarms and proceed with the next steps.*   _____
Caution:
Do not increase the pump discharge pressure above 2,000 kPag while performing the PRV setting adjustment.
3. Start closing the running oil pump discharge 1" block valve and watch the pump discharge pressure increase on the local pressure gauge.   _____
4. If the discharge pressure cannot be raised to 1,035 kPag after fully closing the discharge block valve, start slowly closing the 1" relief valve bypass valve to raise the pump discharge pressure to 1,035 kPag.   _____

(continued)

5. If the pump discharge pressure exceeds 1,035 kPag, stop closing the bypass when the pump discharge pressure reaches about 1,100 kPag and adjust the relief valve until pressure drops to 1,035 kPag, then continue closing the bypass, repeating this step as necessary, until the valve is fully closed and the pressure is maintained at 1,035 kPag. _____

6. If the bypass valve fully closes and the discharge pressure does not reach 1,035 kPag, adjust the relief valve spring tension higher to increase the discharge pressure to 1,035 kPag. _____

7. Normalize the oil system by opening the main oil pump discharge block valve. _____

8. Shut down the "A" pump by pressing the STOP button and start the "B" pump by pressing the START button and repeat steps 2–7 for the relief valve on the "B" pump discharge. _____

9. Clear all existing oil pressure alarms, if any, once the system pressure is stable.
   Perform functional test of lube oil kickback pressure control valve (V_____).

10. With the "B" oil pump running, begin to test the lube oil kickback pressure control valve (V_____) by slowly opening the 1" bypass globe valve. _____

11. Confirm that oil kickback valve starts closing to maintain a pressure of 517 kPag downstream of the filters (F _____); continue opening the bypass until kickback valve is fully shut. _____

12. Verify that pressure on (G _____) is maintained at 517 kPag. _____

13. Begin closing the 1" bypass globe valve and observe the kickback valve starts opening while maintaining oil header pressure; continue until bypass valve is fully shut. _____
    Perform functional test of lube oil supply pressure control valve (V_____).

14. Verify that the pressure indicated on (G _____), downstream of control valve, is reading about 103 kPag. _____

15. With the "B" oil pump running, begin to test the lube oil supply pressure control valve (V _____) by slowly opening the 1" bypass globe valve. _____

16. Confirm that control valve starts closing to maintain the pressure on (G _____); continue opening the bypass until control valve is fully shut. _____

17. Close the 1" bypass valve. _____

(continued)

18.	Verify that lube supply pressure is maintained at 103 kPag. _____
	Perform pump swing tests for both pumps.
19.	Turn the "A" pump selector switch to AUTO. _____
20.	Manually turn off the "B" pump by moving the selector switch to HAND and pressing STOP and confirm the following: _____
20.1	Lube oil supply pressure low alarm for (LPA _____) and (LPA _____). _____
20.2	"A" pump starts on low-lube oil pressure. _____
20.3	Lube oil supply pressure low–low alarm is not activated. _____
21.	Inspect "A" pump for leaks, vibrations or abnormal sounds. _____
	Observations: _____
22.	Turn the "B" pump selector switch to AUTO. _____
23.	Manually turn off the "A" pump by moving the selector switch to HAND and pressing STOP and confirm the following: _____
23.1	Lube oil supply pressure low alarm for (LPA_____) and (LPA _____ and is activated. _____
23.2	"B" pump starts on low-lube oil pressure. _____
23.3	Lube oil supply pressure low–low alarm is not activated. _____
24.	Inspect "B" pump for leaks, vibrations or abnormal sounds. _____
	Observations: _____
25.	Restart the "A" pump and run both pumps in parallel. _____
26.	Confirm kickback valve LBP1587V (LBP1589V) handles additional flow with two pumps running, and maintains header pressure of 517 kPag on LBP1584G (LBP1589G). _____
27.	Shut down the "B" pump and confirm that none of the following are activated: _____
27.1	Low-lube oil supply pressure alarm _____
27.2	Low–low-lube oil supply pressure alarm _____
28.	Restart the "B" pump on manual, and confirm kickback valve LBP1587V (LBP1589V) handles sudden increase in flow with two pumps running, and maintains header pressure of 517 kPag on LBP1584G (LBP1589G) _____
29.	Confirm that the pump discharge pressure relief valves do not open. _____
30.	Shut down the A oil pump and confirm that none of the following are activated: _____
30.1	Low-lube oil supply pressure alarm _____
30.2	Low–low-lube oil supply pressure alarm _____

(continued)

---

31. Shut down the "B" pump and confirm the following: _____
31.1 Low-lube oil supply pressure alarm activates _____
31.2 Low–low-lube oil supply pressure alarm activates _____

D. Functional Tests of Oil System Alarms and Trips

List of Oil System Alarms and Trips

TAG	Description
LBP582T	LBPST21A Lube Oil Filter Differential Pressure High
LBP592T	LBPST21B Lube Oil Filter Differential Pressure High
LBP583I	LBPST21A Lube Oil Supply Pressure Low (standby pump autostart)
LBP584I	
LBP588I	LBPST21B Lube Oil Supply Pressure Low (standby pump autostart)
LBP589I	
LBP583I	LBPST21A Lube Oil Supply Pressure Low–Low (TRIP)
LBP584I	
LBP588I	LBPST21B Lube Oil Supply Pressure Low–Low (TRIP)
LBP589I	

Objective

Perform functional tests on all alarms and interlocks related to the oil system

1. Start the "A" lube oil pump as per procedure, and prepare to test the above oil system alarms and trips. _____
2. Switch the "B" pump to AUTO. _____

Caution:
The main turbine (LBPST25A/B) trip systems shall be tested during the interlock test of the oil system. As such, the TTV shall be opened slightly to observe valve trip action. During the test, the main inlet steam double isolation valves must be fully closed to prevent any steam entry into the turbine

Test Lube Oil Filter High Differential Pressure Alarm

3. Close the low-pressure side (downstream) block valve on the filter differential pressure transmitter and slowly bleed the low-pressure side. _____
4. As the differential pressure reaches 172 kPag, confirm that the filter high differential pressure alarm is energized. _____
5. Close the bleed valve and reopen the downstream block valve. _____
6. Confirm that the filter high differential pressure alarm can be cleared. _____

Lube Oil Supply Pressure Low and Standby Pump Auto Start

7. Close the block valve on the impulse line for lube oil supply pressure transmitters FKP583I (FKP588I). _____
8. Slowly open the bleed valve on LBP583I (LBP588I); when the pressure drops below 80 kPag confirm the following: _____

(continued)

8.1	The low-lube oil supply pressure alarm is activated.	_____
8.2	The "B" pump starts.	_____
9.	Ensure that steam turbine trip is not activated	_____
10.	Close the bleed valve and reopen the block valve on the impulse line to the lube oil supply pressure transmitter.	_____
11.	Shut down the "A" pump and put it to AUTO.	_____
12.	Close the block valve on the impulse line for lube oil supply pressure transmitters LBP584I (LBP589I).	_____
13.	Slowly open the bleed valve on LBP584I (LBP589I); when the pressure drops below 80 kPag confirm the following:	_____
13.1	The low-lube oil supply pressure alarm is activated.	_____
13.2	The "A" pump starts.	_____
14.	Ensure that the steam turbine low-lube oil pressure trip is not activated.	_____
15.	Close the bleed valve and reopen the block valve on the impulse line for lube oil supply pressure transmitter.	_____
16.	Shut down the "B" pump and leave hand switch in HAND position.	_____

Lube Oil Supply Pressure Low–Low *(TRIP)*

17.	Confirm that the double block valves on the steam turbine inlet is closed and there is no pressure on the inlet steam line pressure gauge, LBP1580G (LBP1593G).	_____
18.	Press RESET on each of the 3 Protech 203 overspeed modules.	_____
19.	Press the ALARM ACKNOWLEDGE on the Triconex governor to reset the solenoids.	_____
20.	Reset, latch and open the TTV.	_____
21.	Close the isolation valve on the impulse line for both lube oil supply pressure transmitters LBP583I and LBP584I (LBP588I and LBP589I).	_____
22.	Open the bleed valve on LBP583I (LBP588I) and drive its signal to 0 kPag.	_____
23.	Open the bleed valve on LBP584I (LBP589I) slowly and confirm the following when the pressure reading of the transmitter drops below 45 kPag:	
23.1	The low–low supply pressure alarm is activated.	_____
23.2	Solenoid dump valve, LBY581X (LBY583X) on main turbine trip system is de-energized and opened.	_____
23.3	TTV is shut under trip condition.	_____
24.	Close both the bleed valves to the oil pressure transmitters.	
25.	Open the isolation valves on the impulse line for each of the pressure transmitters	_____

** End of Procedure **

Revision History

Rev	Date	CHANGES	Initiated by:
Draft A	30/04/18	Original Draft	Nathan (NT) Little

Validation

Employee name: _____

Date: _____

Check the applicable box below.

☐ This procedure was performed/reviewed (circle one) as written and is accepted as valid and technically correct.

☐ This procedure, as currently written, requires the following revisions to be technically correct:

Page #	Comment
_____	_____
_____	_____
_____	_____
_____	_____
_____	_____
_____	_____
_____	_____
_____	_____
_____	_____
_____	_____
_____	_____

Employee signature: _____

Supervisor
signature: _____

Attn. mailroom:

Please return this sheet to the SOP – manufacturing coordinator.

References

1 American Petroleum Institute API RP 686, 2nd Edition, 2009, *Recommended Practice for Machinery Installation and Installation Design.*
2 Jerome E. Spear, *Plant Services*, Aug 07, 2007.
3 https://www.thirdcoastcertifications.com
4 USA – With the **Occupational Safety and Health Act of 1970**, Congress created the **Occupational Safety and Health Administration (OSHA)** to ensure safe and healthful working conditions for working men and women by setting and enforcing standards.
5 Canada – *Occupational Health and Safety Act* since 1979, with today's version based on significant amendments made in 1990. Enforced by the Ministry of Labor, the OHSA's overarching role is to ensure workers are protected from workplace hazards.
6 American Society of Mechanical Engineers/ American National Standards Institute.
7 Bloch, Heinz P., "Optimized Equipment Lubrication—Conventional Lube, Oil Mist Technology and Full Standby Protection," 2nd Edition (2022), De Gruyter, Berlin/Germany, ISBN 978-3-11-074934-2.
8 Bloch, Heinz P.; "OIL-MIST LUBRICATION HANDBOOK – Systems and Application," 1987, Gulf Publishing Company, Houston, Texas, ISBN 0-87201-640-4.
9 TMI Staff Contributors, *Protecting new equipment before installing offshore*, Turbomachinery, INTERNATIONAL, February 29, 2020.
10 See Reference 1.
11 Rules and recommended standards for machinery foundations: GMRC, ISO, DIN, EOTA.
12 Charlie Rowan, AVOIDING COSTLY MISTAKES (when grouting large ski-mounted equipment), COMPRESSORTech^{TWO}, August/September 2011, Pages 30 to 33.
13 Adapted by permission from Theo de Kok and Tom Hoekstra, Industrial Services at EMHA, theo.dekok@emhabv.nl, *If The Foundations Be Shaken, Addressing Inspection and Repair of Recip Foundations*, CTSS, 2018 Edition, Page 209 to 216.
14 Geitner, F.K. & Bloch, H.P. - Series *Practical Machinery Management for Process Plants, Volume III, Machinery Component Maintenance and Repair*, Fourth Edition, ELSEVIER, 2019, and preceding editions, 650 Pages.
15 Ibid, Pages 111 to 127.
16 See Reference 1.
17 Klaus Brun and Rainer Kurz, *Compressors are not pipe anchors*, Turbomachinery International • September/October 2010, Page 44.
18 See Reference 14, Pages 190 to 201.
19 Sarah Simons, Benjamin White and Francisco Fierro, Southwest Research Institute, *Applying the Energy Institute and GMRC/PRCI* Guidelines for the avoidance or reduction of vibration problems in small diameter piping branch connections," 2016 Turbomachinery & Pump Symposia.
20 Fabreeka.com (25/02/2001)
21 See Reference 8.
22 ISO Standard 4406:99.
23 See Reference 7.
24 Source: Heinz P. Bloch, Don Ehlert and Fred K. Geitner, *Optimized Equipment Lubrication, Oil Mist Technology and Storage Preservation*, Reliabilityweb, Inc., 2020, ISBN 978-1-941872-98-7.

Part II: **Execution**

After having laid out the prerequisites, we want to proceed with the details of commissioning and startup of specific major equipment. The following text will cover shaft-sealing elements and their support systems as a typical example for what is often referred to as auxiliary equipment that is nevertheless essential to the operation of plant process machinery.

Chapter 6
Shaft Sealing Systems

6.1 Introduction

The long-term reliable operation of rotating and reciprocating plant process machinery – turbocompressors for example – must be ensured by well-proven shaft sealing elements. Shaft seals eliminate or reduce fugitive emissions. Figure 6.1 shows a classification chart of such seals.

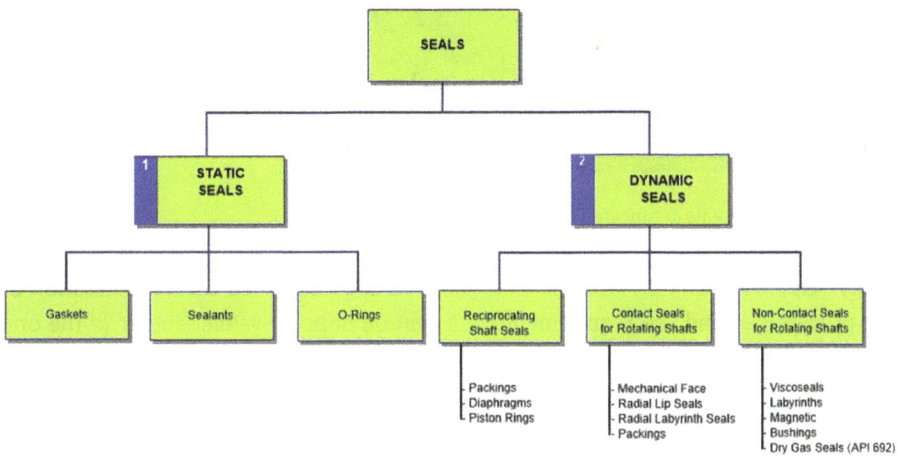

Figure 6.1: Seal classification chart.

Here, we would like to briefly look at contact and non-contact seals as they relate to commissioning and starting up process machinery – refer to Figure 6.2 to see where they are fitted.

One day, in the fall of 1988, our manager said to us: "You must have felt last night like one of the people launching the first space rocket – not knowing whether or not it would work." He was referring to our self-directed retrofit of gas-lubricated shaft seals[1] and a successful startup on two critical motor-driven 3,000 hp/2,240 kW ethylene refrigeration compressors. Our achievement was a first for our large multinational petrochemical company, even though, admittedly, gas transmission companies in North America were applying gas seals in the early 1980s, well ahead of the HC process industries.

The ice was broken in those days and many compressor owners and operators followed suit by replacing their cumbersome oil lubricated shaft seals with gas seals. Oil lubricated seals had been the standard shaft sealing device – see Figure 6.3 – for

https://doi.org/10.1515/9783110701074-006

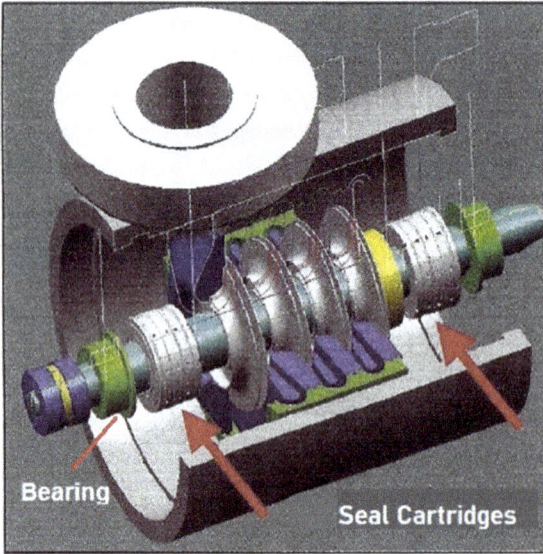

Figure 6.2: Where seals are located – centrifugal compressor.

many years. Albeit often reliable, their disadvantage has been oil contamination of the process gas and the requirement for a seal oil supply system similar to the one shown earlier in Figure 5.9.

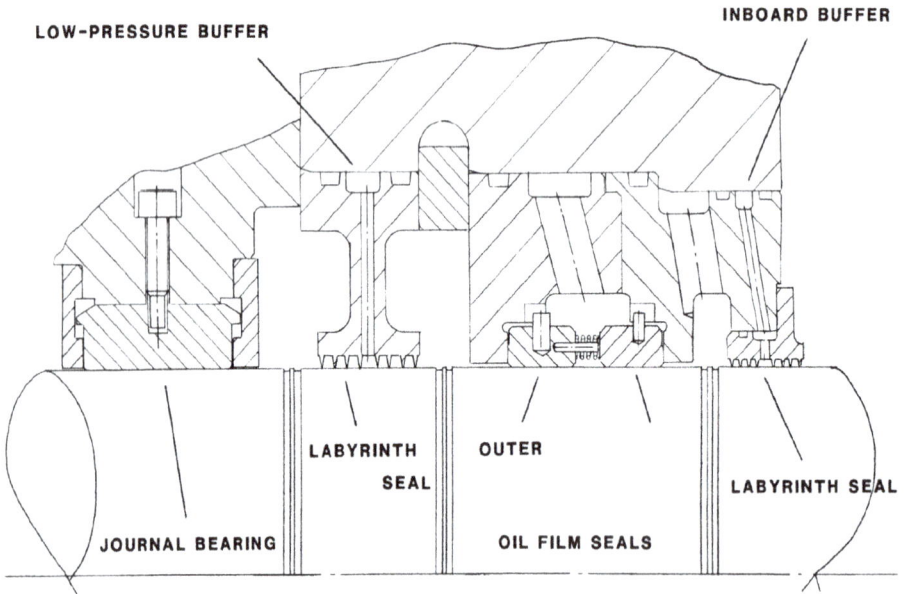

Figure 6.3: Cross section of a classic oil shaft seal assembly (source: Cooper Bessemer).

Today gas seals have become to be known as a reliable and cost-effective alternative to oil type seals. Their key characteristics are low leakage rates, wear-free operation and an extremely low level of power consumption.

Gas seals are non-contacting, dry running mechanical seals. Figure 6.4 shows the principal design features of such a seal. It is usually furnished as a cartridge containing a spring-loaded stationary seal face or sliding ring (1) sealed by an O-ring and a rotating seat or mating ring (2). The sealing faces slide over each other without contact. This results in almost no wear and a long seal life.

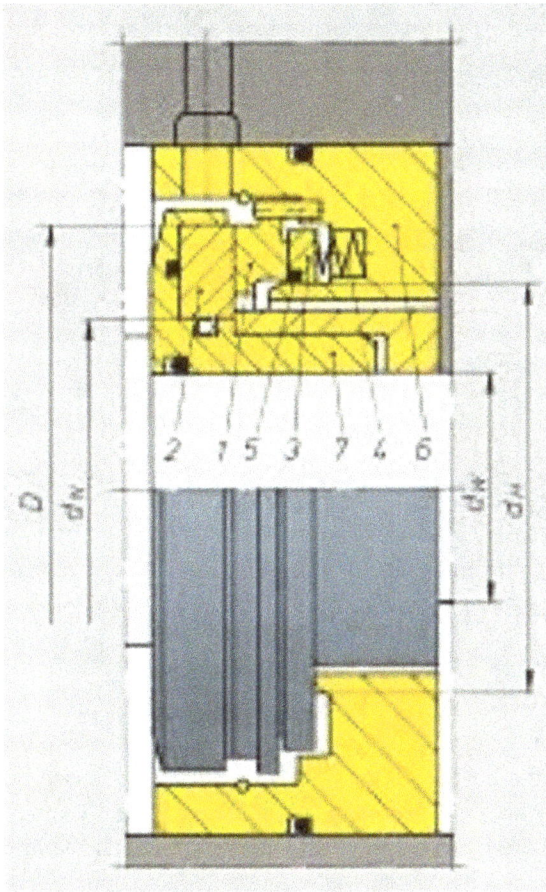

Figure 6.4: Cross section of a dry gas seal: 1, stationary seal face (sliding ring); 2, rotating seat (mating ring); 3, thrust ring; 4, compression spring; 5, O-ring; 6, housing; 7, shaft sleeve with cupped retainer; D, Outer diameter of the sliding face; d_N, Nominal diameter; d_W, Shaft diameter; d_H, Hydraulic (pneumatic) diameter. (source: *EagleBurgmann DGS Catalogue*).

The seal face of the rotating mating ring is divided into a grooved area at the high-pressure side and a dam area at the low-pressure side (see Figure 6.5). The stationary sliding ring is pressed axially against the mating ring by spring forces and sealing pressure.

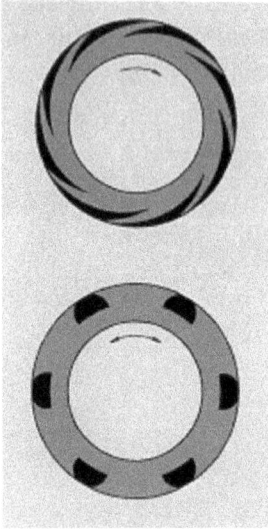

Figure 6.5: Mating ring V-grooves and U-grooves (note: arrows indicate sense of rotation).

The sealing gap is located between the mating ring and the sliding ring. For proper non-contacting operation, these two rings have to be separated by a gas film acting against the closing forces in the sealing gap. The gas film is achieved by the pumping action of the grooves and the throttling effect of the sealing dam. Groove geometry is critical for trouble-free operation of the seal.

6.2 Risk Profile

Failure rates of gas seals are in the neighborhood of 0.175 failures/year based on current experience,[2] meaning that we could expect a seal problem every 6 years or so. However, judging from the spare parts list offered in Appendix 3.C and failure reports,[3] it becomes evident that seals are considered vulnerable in a commissioning and startup environment. Their failure mode, failure mechanisms and failure causes are listed in Table 6.1.

Table 6.1: Failure mechanisms and causes for dynamic seals.

Failure mode	Failure mechanism	Failure causes
Leakage	Wear fracture	– Contaminants – Application (engineering, design), e.g., slow roll issues with DGS. – Misalignment – Surface finish – Shaft out-of-roundness – Inadequate lubrication – Reverse pressurization
Leakage	Dynamic instability	– Misalignment – Rotordynamics
Leakage	Embrittlement	– Contaminants – Fluid/seal compatibility – Thermal degradation – Idle periods

Table 6.2. represents a general risk assessment of a shaft sealing system before and during its initial startup.

Table 6.2: Risk assessment work sheet for a shaft seal system.

Phase	Risk	Most probable cause[1]	Suggested actions/mitigating measures
Rigging and lifting	Low	QP	– Not applicable. It is best to use OEM seal specialist for seal installation.
Handling, staging and storage protection	High – DGS packages vulnerable to inadequate storage protection resulting in rework and delay	OR, HK	– Provide proper storage space – dry cool environment.
Foundation and grouting	Low		– However, internal machine alignment will suffer, if unit foundation is not adequate.

Table 6.2 (continued)

Phase	Risk	Most probable cause[1]	Suggested actions/mitigating measures
Piping and tubing	**High** – No or improper cleaning – Ingestion of foreign objects after cleaning	OR, HK	– Cleanliness and assembly to the machine should follow practices established as part of the project execution agreement between OEM and the owner organization.
Shaft alignment	**High – indirect**	QP, DO	– Shaft alignment starts with leveling the unit. A minimum permissible deviation from the horizontal line would be customarily 0.002 in/ft. or 0.2 mm per linear meter. Adhering to this will extend seal life.
Lubrication system	**High – liquid oil film seals** – Contamination	QP, DO, CO	– Cleanliness of the lubrications system is of utmost importance. – Use one of the international cleanliness standards such as ISO 4406:99 or the OEM's recommendation for seals requiring seal oil.
Commissioning, startup and initial operation	**High – DGS** – Is clean, dry buffer gas not available? – The seal not protected from bearing oil. – How is the compressor pressurized or de-pressurized? – Settling out pressure not known. – How is the machine brought up to operating speed? – Are all involved personnel familiar with the machine train operating procedures?	DO	– Solves all outstanding issues or questions – especially settling-out pressure. – Thoroughly clean system by bypassing machine inlets and air blowing. There should be no foreign particles showing in test filter cloth. – Inspect and, if necessary, replace dual filter elements – Insist on "dry run" using temporary instrumentation such as manometers and rotameters. – Rotameters to stay in place until initial startup and run-in is completed.

Table 6.2 (continued)

Phase	Risk	Most probable cause[1]	Suggested actions/mitigating measures
	– The full control system not included and adequately described in the operating manuals. – Auxiliary seal gas supply not fully commissioned and tested. – DGS supply system insufficiently insulated. – Slow roll issues not resolved.		– Trace and insulate supply and reference gas line as required by site conditions. – Address slow roll questions early.

[1]Refer to Figure 2.12.

6.3 Commissioning and Startup

It is evident from the foregoing that shaft seals are highly susceptible to contamination by dirt and small foreign objects. Another important point pertaining to dry gas seals (DGSs) is the fact that they require a reliable clean and dry buffer gas supply, a condition that is not always guaranteed – particularly during the initial startup phase. It is therefore, important to assure that this supply cannot be interrupted. Figure 6.6 represents a simplified flow diagram illustrating the buffer gas supply to the machine seal cavities.

The startup team must be intimately familiar with such a system which, today, is much more complex, similar to the one shown in Figure 6.7. Here, the owner decided to employ a seal gas booster as an answer to the quest for a reliable conditioned gas flow into the seal.

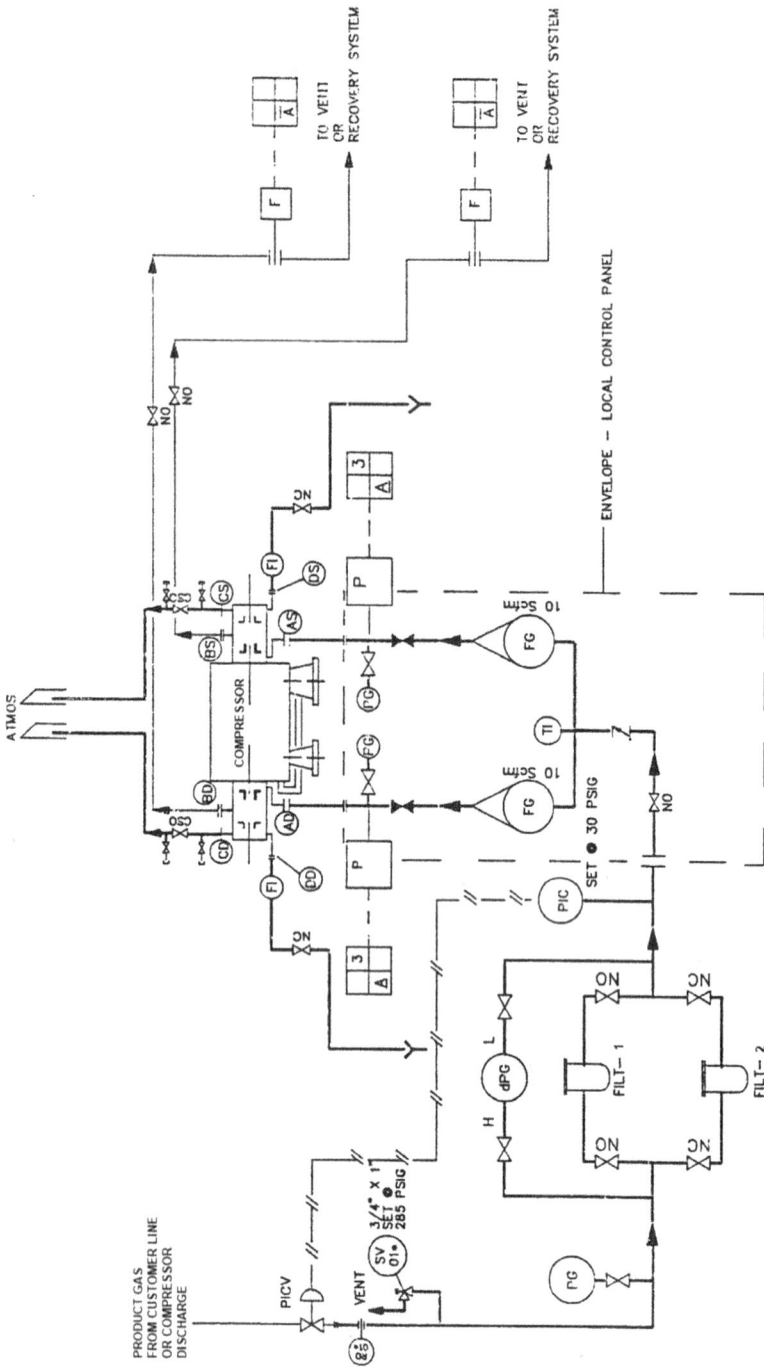

Figure 6.6: Simplified flow diagram for a dry gas seal (DGS) support system.

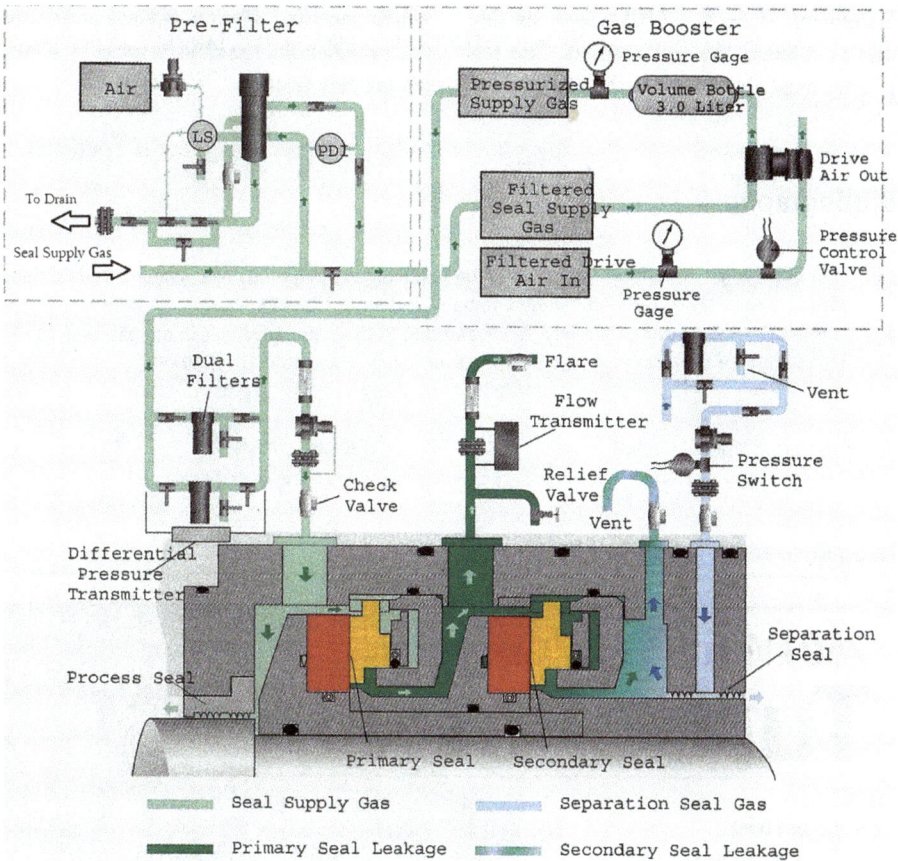

Figure 6.7: Tandem dry gas seal with gas supply booster (courtesy of Flowserve Co.).

The procedure for commissioning this system would be guided by the desire to end up with an almost surgically clean system. One would disconnect the supply piping at the equipment connection points and first flush the system with warm water and then dry it out with hot air. Additional hints about cleaning, commissioning and initial start can be obtained from the recently issued standard on DGSs.[4]

References

1 API 692 "Dry Gas Sealing Systems for Axial, Centrifugal, and Rotary Screw Compressors and Expanders, First Edition, June 2018.

2 $\lambda \sim 20 \times 10^{-6}$ [20 Failures/10^6 h].

3 S. Zardynezhad, Toyo Engineering Canada Lt., Calgary, Alberta, *Achieve successful compressor startup by addressing dry-gas seal failures*. HYDROCARBON PROCESSING, JANUARY 2015, Pages 65 to 69.
4 Refer to Reference 1.

Bibliography

[1]   H.P. Bloch and F.K. Geitner, An Introduction to Machinery Reliability Assessment, 2nd Edition, ISBN 0-88415-172-7, Gulf Publishing Company, Houston, TX, 1994, 242.
[2]   H.P. Bloch, H.P. Staff, C. Carmody, A. Plc and U.K. Rotherham, Advances in dry gas seal technology for compressors. Hydrocarbon Processing, December 2015, 69–73.

Chapter 7
Commissioning and Startup (C&S) of Electric Motors

7.1 Overview

7.1.1 Introduction

Electric motors (EMs) are the most common machines in process plants. They are an efficient prime mover and add flexibility to the design of process plants such as petroleum refineries, petrochemical plants and gas-processing plants. EMs can be built with characteristics to match almost any type of load. They can be designed to operate reliably in outdoor locations where they are exposed to weather and atmospheric contaminants.

Motors have to be correctly applied in order to assure reliable performance. Critical items to be considered are:
- Load characteristics for both starting and running conditions
- Load control requirements
- Power system voltage and capacity
- Plant site conditions that would determine the type of motor enclosure,
- Hazardous area classification such as hydrocarbon, hydrogen and dust, to name a few

Motor configurations are used in two major categories: Horizontal and vertical. The output shafts of horizontal motors are arranged horizontal to the ground or mounting base, while the output shaft of a vertical motor will be perpendicular to the conventional mounting base. Vertical motors are used as pump drivers in the petroleum, petrochemical and chemical industries, because the entire assembly requires a smaller footprint. In this application, they are labeled "Vertical Pump Motors" to indicate that they have been made suitable for this service by furnishing them with thrust bearings to accept the vertical down-thrust of the pump.

Both horizontal and vertical motors are available in a variety of designs and enclosures. They are classified as standard duty, large heavy-duty, special industry designs and special application designs.

If we look at an EM as a system, we come to the conclusion that there are more involved than just the motor. The systems to be considered are:
- The facility's power distribution system, which includes wiring, conduits and transformers.
- The motor control, which may include starters, variable frequency drives and other starter systems.
- The motor that uses a variety of voltages and phases – the most common being single phase and three phase.
- The load in form of driven equipment such as pumps, compressors and fans.

https://doi.org/10.1515/9783110701074-007

7.1.2 A-C Motor Type and Selection

One of the first considerations in motor selection is to make a choice between a squirrel cage induction and a synchronous motor. The induction motor has the advantage of simplicity. It is a rugged machine and has a stellar record for dependability. In general, it can accelerate higher load inertias than synchronous motors and usually will do so in less time. Induction motor control is simple and no excitation equipment is required. Its disadvantages are that it operates at lagging power factor and has higher inrush or starting currents. Induction motors of up to about 22,000 horsepower or 16.4 MW are the preferred choice for very large applications.

The constant-speed synchronous motor has inherent advantages so that it often becomes the prime mover of choice for industrial drive applications at lower power ratings. Load speed can be exact. Torque characteristics of the motor can be varied by design measures to match the requirements of the driven load and the available power supply.

7.1.3 Induction Motors

Squirrel-cage induction motors are the most widely used kind in the size range of up to 150 kW or 200 horsepower. A typical induction motor is shown in Figure 7.1. An induction motor is an alternating current device in which the primary winding, the stator, is connected to the power source and the secondary winding or "squirrel cage" is the other member. The rotor carries the induced current. There is no physical electrical connection to the secondary; its current is induced. Induction motors are simple, rugged and reliable because they have no rotating windings, slip rings or commutators.

Figure 7.1: Electric motor.

An induction motor operates below the synchronous speed that is determined by the power cycles and the number of poles in the stator. For example, a two-pole, 60-Hertz[1] motor has a synchronous speed of 3,600 RPM, and a four-pole motor operates at 1,800 RPM. Normal operating speeds for induction motors are 3,550 RPM and 1,750 RPM. This deviation from synchronous speed is called slip, and it varies with load. Full load – maximum – slip varies from 1% in large motors to as high as 5% in small units. Most motors have an average slip of 3%.

7.1.4 Synchronous Motors

A synchronous motor is usually easier to start than an induction machine. However, synchronous motors have less thermal capacity in their windings and may be more severely taxed when accelerating high inertia loads.

Synchronous motors are rarely applied at 3,000–3,600 rpm common in the processing industries. They would be expensive to build even though large two-pole synchronous motors exist where a squirrel cage induction motor would have been the more economical choice.

Once synchronized and running, synchronous motors present special system problems. They may tend to pull out of synchronism on voltage dips that induction motors can ride through. A gradually increasing load from 0% to 125% of rated load will be easily accommodated. A suddenly applied load of 125% can easily cause the motor to pull out of synchronism with the electrical system.

When matching large synchronous machines to a system, it is important to perform a transient load study as part of the asset acquisition process. This will help ascertain whether or not the electrical system is capable of supporting the motor demands under transient conditions.

A short explanation of component parts and their applications in the Appendix may help. Motor-related terms may be found in the Glossary.

Modern "Above NEMA[2]" oil and gas duty motors include many features to achieve optimal performance and reliability. Motor stators are manufactured with CS core plate electrical steel, with indexed lamination stacking, fully sealed insulation system with Class F vacuum pressure impregnation technology, and heavy-duty bracing of stator coils and end turns.

The rotor system must endure and transmit heavy loads, aid in cooling of the motor and provide electrical performance.

7.1.5 Motor Enclosures

The selection of motor enclosures is determined by the environmental conditions and electrical hazard area classification under which the machine is to operate. The type of enclosure also influences motor maintenance intensity as well as commissioning and startup efforts. The following enclosures are used by the processing industries:

- **Open drip proof** (ODP): Due to the fact that the chamber has open vents and air can flow directly over the windings, this ODP motor tends to be slightly more efficient because the windings are cooler. Although the ODP motor does not require external cooling, the open concept can allow airborne contaminants and dust to enter the chamber. To protect against anything in the environment, some manufacturers' ODP motors include laminated windings and sealed bearings and they are safe to be used in indoor industrial settings.
- **Drip proof:** Largely applied indoors or enclosed spaces where no environmental challenge exists.
- **Weather-protected type I:** This is essentially a drip-proof protected motor with heaters and bearing seals; it is sensitive to weather and atmospheric conditions.
- **Weather-protected type II (WPII):** This represents the more commonly used outdoor enclosure. More expensive than the weather-protected type I enclosure, it minimizes the ingestion of water and dirt. This enclosure admits chemical contaminants in gas or vapor from attacking motor parts that are vulnerable to them. Figure 7.2 shows a large WPII motor.

Figure 7.2: Large WPII motor – 16 MW (courtesy: Siemens AG).

- **Totally enclosed pipe ventilated** (TEPV): Sometimes referred to as TEFV – Totally enclosed forced ventilated motor. It is similar in construction to an ODP motor, but it has external cooling air brought to the motor through ducts or pipes. Typically, air is forced into the motor by an externally powered fan.
- **Totally enclosed forced ventilation (TEFV):** Totally enclosed motors used on such process equipment as cooling towers are classified as non-ventilated, fan-cooled (TEFC). Totally enclosed motors are recommended and used for locations where fumes, dust, sand, snow and high humidity conditions are prevalent, and they can provide a high-quality installation either in or out of the air stream provided the typical problems of mounting, sealing and servicing are properly addressed. TEFC motors are also the preferred choice in hydrocarbon processing facilities. TEFV enclosures can be used indoors or outdoors in dirty or hazardous environments. Since the motor cooling air is piped in from a separate source the influx of dirt and gaseous contaminants is minimized.
- **Totally enclosed water-to-air cooled** (TEWAC): The TEWAC machine uses an air-to-water heat exchanger to reject motor heat losses. This type of motor is not a new development. Such machines have long been common in Europe. Their usage in the United States has been less frequent, but growing emphasis on noise reduction is creating new interest in the inherently quiet TEWAC machine. Certain risks are being perceived by owners and operators: Review of current specifications and orders shows that many motor users, inexperienced in TEWAC applications, are uncertain how to deal with possible cooler leakage, water flow control and monitoring, effects of high-water temperature, motor rating versus ambient temperature, water chemistry and cooler construction and mounting. TEWAC are designed to breathe during shutdown and frequently breather filters are used to remove particulate contaminants. It is more efficient than a TEFC motor because it does not have the external fan to drive. Its first cost is greater than the cost of a WPII motor but less than what a TEFC motor would cost if one excludes the additional capital cost for a cooling water system. ESD[3] controls are required for the water supply system.
- **Totally enclosed fan cooled** (TEFC): See also Figure 7.3. This is the highest degree of enclosure for an air-cooled machine. This motor has a closed-off chamber which requires the use of an external fan to blow air over it. A benefit to having the motor totally enclosed is that it is protected from the outside environment, including dust, airborne contaminants and many weather disturbances such as wind-driven rain. A TEFC motor is often the preferred choice in the hydrocarbon-processing facilities. In large sizes, the TEFC motor has an air-to-air heat exchanger. Internal motor air is recirculated around the outside of the tubes while outside air is driven through the tubes by a shaft-driven fan. These motors are quite expensive especially in large sizes because of the high volume of cooling air required relative to motor size. These motors are indicated for use in very dirty or hazardous locations

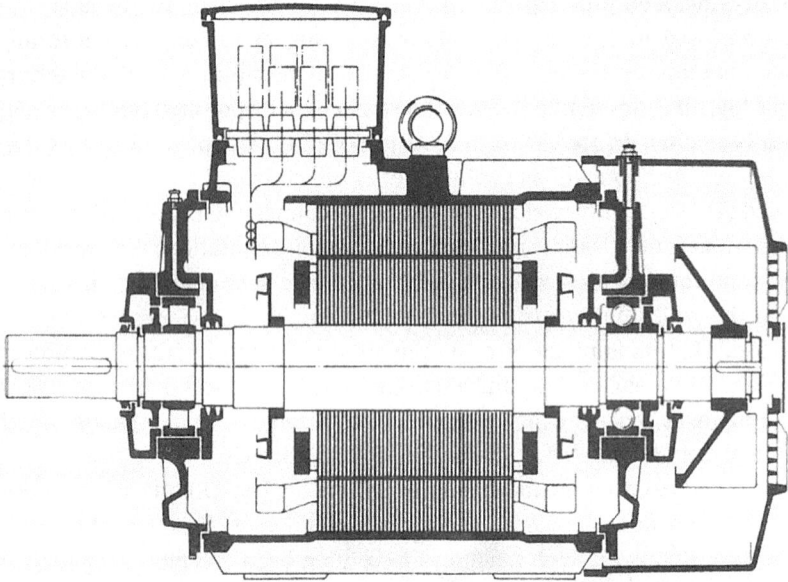

Figure 7.3: Cross section of a totally enclosed fan-cooled (TEFC) motor (source: Siemens AG).

– **Explosion proof:** An explosion-proof machine is a totally enclosed machine whose enclosure is designed and constructed to withstand an internal explosion. It is also designed to prevent the ignition of combustibles – gas, vapor and dust – surrounding the machine by sparks, flashes, or explosions which may occur within the machine casing.

Motors are designed for compliance to specific requirements of the hazardous area classification; the choice of enclosure must match the requirements.

7.1.6 Motor Service Factor

Motor service factor is an indication of the machine's maximum allowable continuous power output as compared to its nameplate rating. A 1.0 service factor motor should not be operated beyond its rated horsepower at design ambient conditions, whereas a 1.15 service factor motor will accept a load 15% in excess of its nameplate rating. Usually, motor manufacturers will apply the same electrical design to both motors but will use class B insulation on 1.0 service factor motors and class F insulation on 1.15 service factor motors. Class B insulation is rated at a total temperature of 130 °C and class F is rated at 155 °C.

More importantly, a 1.15 service factor motor operates at a temperature from 15 to 25 °C lower – compared with the temperature rating of its insulation – than does a 1.0 service factor motor operating at the same load. This, of course, results in longer insulation life and, therefore, longer service life for the motor. For this reason, many equipment manufacturers will recommend the use of 1.15 service factor motors for loads at or near nominal horsepower ratings.

Since increased air density raises the load on air movers, an added attraction for using 1.15 service factor motors is that there is less chance of properly sized overloads tripping out this equipment category during periods of reduced heat load and low ambient temperatures.

7.1.7 Insulation and Heating

Insulation used in quality EMs is considered to be non-hygroscopic. However, the insulation tends to slowly absorb water and, to the degree that it does, its insulation value is reduced. Also, condensed moisture on insulation surfaces can result in current leakage between pinholes in the insulation varnish. It is, therefore, advisable to keep the inside of the motor dry on installations exposed to high humidity by using heaters.

They serve to keep the temperature inside the motor 5–10 °C higher than the temperature outside the motor. Motors in continuous service will be heated by motor losses, but idle motors in storage require the addition of heat to maintain this desired temperature difference. It is recommended to always provide electric space heaters, sized and installed by the motor manufacturer.

7.1.8 Motor Torques and Speeds

Most process machines do not require high starting torque motors. Normal torque motors perform satisfactorily for pumps, fans, blowers and other driven equipment, thus imposing far less stress on the driven components. Normal torque motors should be specified for the bulk of single-speed applications, and variable torque in the case of two-speed applications.

There are five points along a motor speed–torque curve that are important to the operation of many machines:
(1) locked-rotor torque,
(2) pull-up torque (minimum torque during acceleration),
(3) breakdown torque (maximum torque during acceleration),
(4) full-load torque and
(5) maximum plugging torque (torque applied in reversing an operating motor).
    Based on full-load torque, the average percentage values of the other torques

are as follows: locked-rotor torque = 200%; pull-up torque = 100%; breakdown torque = 300%; and plugging torque = 250%.

We stated earlier that EMs are constant speed prime movers. Their individual common speeds, for the 60 Hz system, are 3,600, 1,800, 900 and 450 rpm. However, variable speed motors have become more and more important as they allow for an economical throughput by flow and pressure variations of modern process machinery. Moreover, large EMs have been built that operate at high speeds to compete with other prime movers such as gas and steam turbines.

7.1.9 Motor Controls

Purchaser and owner are usually responsible for control devices and wiring which can also be subject to demanding service situations. Controls serve to start and stop the motor and to protect it from overload or power supply failure. It therefore becomes part of the motor service reliability chain. Control device and wiring are not routinely supplied as a part of a motor procurement contract, however, because of their importance to the system, the need for a thoroughness in the selection and wiring of these components cannot be overstressed. In all cases, motors and control boxes must be grounded. Grounding practices must follow OEM and plant site specific standards and also meet area electrical hazard classification.

The various protective devices, controls and enclosures required by most electrical codes are described in Appendix 7.A.

7.1.10 Wiring System

The design of the wiring system for the numerous process machines, fans, compressors, pumps and controls is the responsibility of the owner's engineer in compliance with the applicable site and national codes. Although the average installation presents no particular problem, there are some systems that require special consideration if satisfactory operation is to result.

Conductors to motors must be sized both for 125% of the motor full-load current and for voltage drop. If the voltage drop is excessive at full load, the resultant increased current can cause overload protection to trip. (Although motors should be operated at nameplate voltage, they can be operated at plus or minus 10% of nameplate voltage.)

In a normal system with standard components, even the larger machines will often attain operating speed in less than 15 s. During this starting cycle, although the motor current is approximately 600% of full-load current, the time delays in the overload protective devices prevent them from breaking the circuit.

It can never be stressed enough that the commissioning team, particularly the electrical SME, must double check conduit sealing, one of the finishing touches to electrical conduit that is often overlooked. Conduit sealing is to provide a barrier between hazardous areas; the location of seal fittings should be clearly identified on the drawings.

7.2 Commissioning and Startup

7.2.1 Risk Profile

EMs, like other process machinery, are vulnerable to ingestion of foreign objects or simply dirt. Dirt can harm bearings and winding resulting in failure. Windings frequently fail from overload but also from too frequent starts and finally, from old age. The mean-time-between failure of a pipe ventilated essentially ODP 10 MW motor has been experienced as 25 years of basically 24/7 operation in a best-of-class chemical processing plant. The general experience of a small-to-medium motor population is reflected in the risk profiles presented in Figures 7.4 and 7.5.

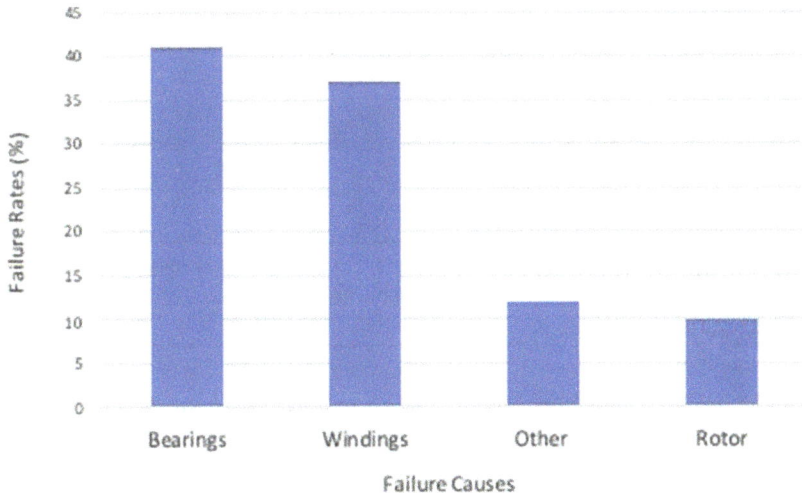

Figure 7.4: Risk profile of medium to large motors – 1983 EPRI failure rates.[4]

Large motors receive a shop test, and as mentioned in other chapters, we would like to point out the importance of having a good understanding of what went on during the factory acceptance test (FAT) at the OEM's shop. The startup team must be familiar with following items:

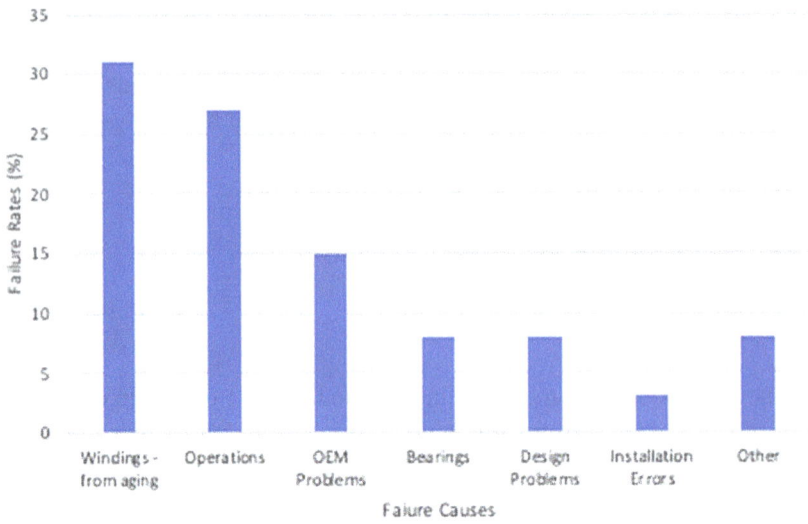

Figure 7.5: Risk profile of large motors – general field experience (2015).

- Inspector's report
- Shop test report
- vendor drawing and data requirement (VDDR)

Potential risks are being evaluated as shown in Table 7.1.

Table 7.1: Electric motor potential risk analysis record.

Phase	Risk	Most probable cause[1]	Suggested actions/mitigating measures
Rigging and lifting	**Medium** – Large motor inadequately rigged	QP	– Assure lifting plan has been prepared
Handling, staging and storage protection	**High** – Moisture contamination of windings	OR, HK	– Provide proper storage space – dry cool environment. – Provide storage oil mist for bearings – Energize space heaters – Periodic rotation of the rotor

Table 7.1 (continued)

Phase	Risk	Most probable cause[1]	Suggested actions/mitigating measures
Foundation and grouting	**Medium** – Cracks, voids – Edge lifting	QP, EN	– Have grout manufacturer rep at site during pour – Use grout with the proper exotherm formulation suited for the climate conditions at site.
Piping and tubing	**High** – No or improper cleaning – Ingestion of foreign objects after cleaning	OR, HK	– Cleanliness and assembly to the machine should follow practices established as part of the project execution agreement between OEM and the owner organization.
Shaft alignment	**High** – Magnetic center not recognized – Limited end float not understood	QP	– Shaft alignment starts with leveling the unit. A minimum permissible deviation from the horizontal line would be customarily 0.002 in/ft. or 0.2 mm per linear meter. – Initiate "toolbox talks"
Lubrication system	**Medium** – Wrong oil spec. – Mix-up – Contamination – Improper slope of drain line	QP, DO, CO	– Because external piping must be fabricated and installed as part of the lubrication system, the cleanliness obtained during FAT helps but should be discounted, therefore, the system should be flushed again to achieve cleanliness standard per ISO 4406:99.
Pre-commissioning commissioning	**Medium** – Conduits not sealed – Improper slope of drain line	DO, CO, QP	– Paint seals must be a commissioning item. – Involve technical group in final checkout.
Startup and initial operation	**High** – Improper phase winding – Re-acceleration issue not understood	QP	– Electrical SME must be involved – Check for reverse rotation during uncoupled run – Check hunting of rotor for magnetic center during uncoupled run

[1] Refer to Figure 2.12

7.2.2 Commissioning and Startup

Effort and duration of the C&S phase are singularly a function of motor size. Smaller motors, from fractional power to approximately 500 kW or 670 hp, require little attention even though they should be conscientiously cared for during the project phases of handling, receiving inspection, storage protection and installation including alignment.

Large motors, approximately from 500 kW or 670 hp up to 100 MW or 134,000 hp, must receive special attention. Meticulous tracking of all project phases is required. Important project phases are:

- **Rigging and lifting.** Rigging and lifting of heavy load such as large motors is subject to state or local government inspections and permits. We covered the subject in Chapter 5, and a rigging and lifting plan has to be submitted prior to commencing the motor lift.
- **Receiving, inspection and storage protection.** In general, it is advisable to follow In Reference [5], however, we remind the reader that our comments and experiencebased recommendations pertain to the machines found in modern industry. Storage protection of small motors has to take into account that they usually have grease lubrication where the bearing style can be told from code letters on the motor's nameplate. Not all motor manufacturers use the same letter code. The internet is a valuable source of information, as are the marketing departments of motor manufacturers.

When using oil mist storage preservation on small motors, it is best to follow these eleven guidelines:

1. If the code letters indicate sealed bearings, they have been provided with lifetime lubrication and cannot be regreased.
2. Unless it is possible to connect to an oil mist line, disregard the oil mist labels at all four locations in Figure 7.6.
3. At the two bearing housing's outlet ports, connect pipes or tubes exactly as shown in Figure 7.6. These ports will always remain open, both while in storage and in actual operation.
4. Thread grease fittings – that is, Zerk® fittings – into each of the two bearing housing inlet ports.
5. Use the exact type of EM grease that was previously supplied with the motor and apply it with a grease gun.
6. Slowly apply grease to the two inlet Zerk fittings until the grease leaves from the horizontal pipe or tubing portion at each of the two outlet ports.
7. With heavy-duty pliers, squeeze the two open outlet ports into oval shapes with minor diameter openings about 0.100 in or 2.5 mm; the grease in the horizontal pipe or tubing becomes a grease plug that will not allow dirt or water vapors to reach the bearing.
8. Do not rotate the motor. Do not regrease the motor, regardless of storage duration.

9. Seal the shaft openings with silicone rubber caulking. Choose a dark color, like black to discourage pilfering.
10. Coat all exposed machine surfaces with Product C, Appendix 5.A.

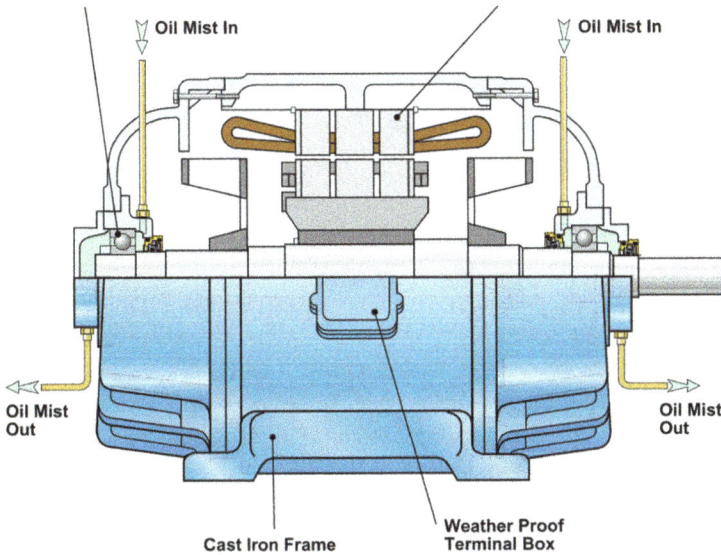

Figure 7.6: Small motors are almost always provided with rolling element bearings (source: AESSEAL, Inc., Rotherham, UK and Rockford, TN).

EMs should not be given a periodic regreasing while in storage, regardless of storage duration. The reasons for this experience-based statement become clearer when looking at typical ball bearings in their respective housings.[6] Motor-bearing arrangements differ from each other in various ways, and incorrect regreasing can do damage. Therefore, giving the bearings uniform storage protection by sealing them off (and *not* first filling them with grease) is a sensible preservation strategy. However, such sealing off, done by simply leaving the motor and its bearings alone, except for placing a bead of caulk around the shaft opening at the motor end plate, is not as trustworthy as oil mist blanketing. With blanketing, oil mist reaches and protects every interior surface of the bearing housing. Oil mist intruding into the motor windings is of little consequence.

An examination of proper motor lubrication illustrates that, unless the same style of bearing is used on all the plant's EMs, there is no substitute for following a proper work execution procedure. This procedure will be different for different styles of bearings. Good supervision and experience-backed lubrication management prevent failures and generate higher profits. But, again, for storage preservation of EM bearings in situations where oil mist is not palatable for reasons we cannot really fathom, try to follow these recommendations in conjunction with Figure 7.6.

7.2.3 Large Motors

Assuming that a large motor has sleeve or journal bearings, the following procedures should be followed for oil mist preservation:

1. Blank the oil return line – that is, insert blind.
2. Seal the shaft openings with silicone rubber caulking and tape.
3. Fill the bearing housing with Product C, Appendix 5.A.
4. Install a valve standpipe so that the inlet is higher than the bearing's housing.
5. Coat all exposed machine parts with Product C.
6. Do not rotate the motor shaft.

– **Foundation, Base Plate and Grouting.** A very important step is placing a large motor on its foundation. Soleplates can often be mounted to the motor feet after the machine has been lifted and lowered onto safe temporary supports. It can subsequently be placed onto the foundation. The soleplates will then be supported by foundation jack bolts while the assembly of motor feet and baseplate are being bolted down to the foundation bolts. Once the unit is level, the sole plate can be grouted in. After the grouting has cured the jack bolts must be withdrawn and the voids filled with grout. Refer to Figure 7.7.

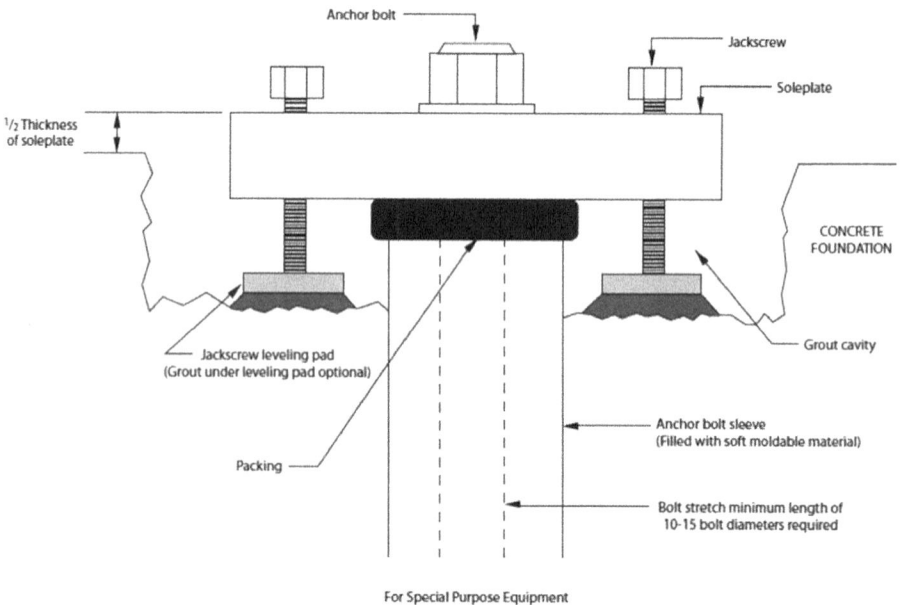

Figure 7.7: Typical soleplate installation (source: reference[5]).

- **Piping.** If piping for bearing lube oil and cooling water (e.g., TEWAC) is present, there has to be assurance that procedures for piping cleaning and installation established by the project team were followed.
- **Shaft Alignment.** At this stage, the motor has been set according to the outline drawings and the driven equipment is being aligned to the motor. The team must assure correct relative growth data pertaining to the transmission equipment and has been provided by the OEM and accepted by the commissioning team. Further, the owner's engineer, together with the OEM field service representative, should have agreed on the method of turning or barring over the motor shaft without causing damage to the bearings or bearing seals. Similarly, the alignment procedure accepted by the project team will be followed and records will be established accordingly.
- **Lubrication.** As stated before, larger motors are preferred to have oil film type or journal bearings. Bearing housings are usually of split type design to allow inspection and replacement without complete disassembly. Bearing housings contain:
  1. The oil reservoir and settling chamber
  2. Bearing housing oil reservoir drain valves
  3. Oil rings
  4. Constant level oiling device,
  5. Externally mounted sight gauges marked with the correct oil level.

As part of the commissioning procedure, any existing oil has to be drained and replaced by fresh oil according to the plant lubrication specification list.

7.2.4 Couplings

Large horizontal motor couplings have to take into account that sleeve bearings are not generally equipped with thrust bearings. The motor rotor is permitted to float, and as it will seek its magnetic center, an axial force of rather small magnitude can cause it to move off this center. Sometimes it will move enough to cause the shaft collar to contact and possibly damage the bearing cap. To avoid this, a limited end float coupling is used between the driven equipment and large motors to restrict axial movement of the motor rotor. Limited end float couplings usually allow for an end float of ¼ in. or 6 mm. When installing limited end float couplings, it is important to understand Figure 7.8. It tells us that the motor should be located so its shaft is in the center of rotor float meaning that dimensions "Y" should equal dimensions "Z." It is important to provide sufficient clearance between shaft ends on the hubs of the flexible coupling to allow for adequate thermal expansion of the shafts, as specified by the coupling manufacturer.

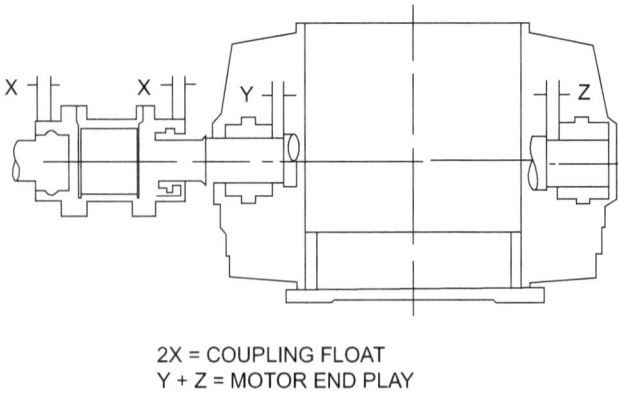

2X = COUPLING FLOAT
Y + Z = MOTOR END PLAY

Figure 7.8: Determination of endplay on a limited end float coupling.

The motor magnetic center is set in the OEM shop and comes with a small tolerance range that assures no undue axial forces are generated. The axial position of the rotor is determined by the fixed position indicator shown in Figure 7.9.

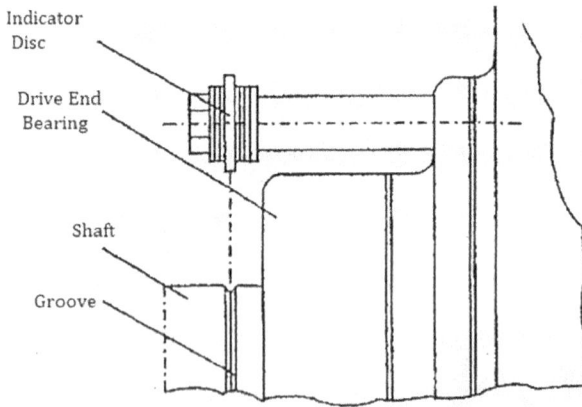

Figure 7.9: Motor rotor position indicator.

7.2.5 Startup Preparation

The visual inspection steps that should be performed are listed as:
– Checking nameplate data against purchase specifications – such as power rating – hp or kW – speed, voltage and service factor.
– Checking rotation arrow against driven equipment sense of rotation.
  Performing electrical inspection and tests.

1. Open each terminal box. Check for indication of the presence of moisture, free water and correctly made connections. Inspect not only workmanship of joints but make sure that proper leads are connected on dual voltage motors. It is well known that damp terminal leads are a common cause of low megger readings.
2. Spot check a few grease-lubricated motors by removing shaft and bearing caps. Look closely at grease and make sure it has not started to harden. Check odor. If the grease looks or smells bad, one should think about checking and possibly re-greasing all motors.
3. Make sure motor turns easily.
4. Keep terminations unfinished so that equipment is not subject to test voltage. Insulation should be penciled and otherwise prepared for completion of the terminations to the greatest extent possible for protection of the cables, including making up and grounding of stress cones for shielded cables of higher voltage motors.
5. Measure insulation resistance of motor windings before connecting power cables to the motor. Repeat measurements after power cable terminations are completed. A good guidance is provided in reference.[7]
6. Measure the 60:30 dielectric absorption ratio of motor windings rated above 600 V. Measure the insulation resistance by taking readings at the end of 30 s and again at the end of 60 s. Minimum acceptance value is 1.4.
7. Measure the 10:1 min polarization index of windings on motors rated above 1,500 hp or 1,100 kW. Measure the insulation resistance taking readings at the end of 1 min and again at the end of 10 min. Minimum acceptable value is 3.0.
8. Inspect cable termination, armor termination and terminal connection. Also, check terminal box covers insuring explosion proof flanges are new and comply with the electrical area classification.
9. Make a final check of conduit seals.
10. Minimum resistance shall be obtained as shown in Table 7.2.

Table 7.2: Minimum resistance values.

1,000 hp or 745 kW and below – Class A and B Insulation		
400 V	2,300 V	5,000 V
1.5 meg	3.0 meg	5.0 meg

Should the megger readings indicate that some of the motors require drying out before being placed in service, then one of the following methods should be followed:

a. Apply external heat – strip heaters or light bulbs – under a canvas cover, well vented. Heating should not go beyond 203 °F or 95 °C, as water must not be reaching the boiling point in the insulation.
b. Circulate current in the motor by applying a low voltage to the windings. Current should only be a fraction of full-load current and the temperature limitation of 203 °F or 95 °C should be observed.
– Follow and work through the checklist contained in Table 7.3 leading to the solo run.

Table 7.3: Motor commissioning checklist.

Designation: _____
Location: _____
Service: _____

Note: Cross out items that are not applicable

Pre-operational Checks

#		
1.	Mechanical assistance present	_____ (Yes or no)
2.	Electrical assistance present	_____ (Yes or no)
3.	Instrumentation assistance present	_____ (Yes or no)
4.	Process operations present	_____ (Yes or no)
5.	Arranged for power company inspection	_____ (Yes or no)
6.	Are applicable operation and maintenance manual, parts book, special tools and spares available?	_____ (Yes or no)
7.	Have the vendor drawing and data requirements (VDDR) been fulfilled?	_____ (Yes or no)
8.	Are the items/documents listed in the project VDDR form available?	_____ (Yes or no)
9.	Are there any outstanding items from shop inspector's reports?	_____ (Yes or no)
10.	Receiving, storage protection, foundation, grouting, piping and alignment checklists completed and attached.	_____ (Yes or no)
11.	Control loops functionally tested and correct, all set points set and verified.	_____ (Yes or no)
12.	New or calibrated gauges supplied.	_____ (Yes or no)
13.	All vendor/OEM and site requirements read and understood, checklists completed.	_____ (Yes or no)
14.	Make sure bearings isolation is not bridged – if applicable. Bearing isolation is indicated by special information plate on motor.	_____ (Yes or no)

Table 7.3 (continued)

15. Ascertain motor control center (MCC) starter unit settings are all correctly sized for induction motors.	_____ (Yes or no)
16. Check that dust filters – if any – are installed correctly.	_____ (Yes or no)
17. Check that space heater – if any – are installed correctly.	_____ (Yes or no)
18. Motor is bolted to soleplate and foundation to specifications.	_____ (Yes or no)
19. Solid single shims installed – part of alignment but double check.	_____ (Yes or no)
20. On pipe-ventilated motors, make sure external fan has been commissioned and is operating.	_____ (Yes or no)
21. On TEFC motors, check fan is in place – if accessible for inspection.	_____ (Yes or no)
22. Bearing temperature RTDs are wired up and active.	_____ (Yes or no)
23. Bearing temperatures are displayed in control room panel	_____ (Yes or no)
24. Bearing temperatures are displayed local.	_____ (Yes or no)
25. For water-cooled motors (TEWAC): Motor cooler is charged from cooling water side, vented and water pressure is > 57 psi or 4 bar – or as specified.	_____ (Yes or no)
26. Check with OEM if motor can be solo run without cooling water.	_____ (Yes or no)
27. Breaker was exercised – "cold trial" – from MCC. Information: Each unit is a combination motor starter that comprised a fused disconnect or molded case breaker with a contactor or electronic controlling device. Contactor is operated by push buttons on motor control unit or from control room or and also from local push button.	_____ (Yes or no)
28. Large motors only: low-voltage power circuit breaker raked in and in TEST/NORMAL mode.	_____ (Yes or no)
29. Local push (START/STOP) button is operative and released.	_____ (Yes or no)
30. Validate STOP button lockout.	_____ (Yes or no)
31. All motor protection devices (ESDs) were tested and are active	_____ (Yes or no)
32. Verify all vibration transducer calibration is correct. Include readout and alarm indication.	_____ (Yes or no)
33. Verify all vibration transducer calibration is in accordance with API 670. Include readout and alarm indication.	_____ (Yes or no)
34. All required vibration analysis equipment has been calibrated and ready for use.	_____ (Yes or no)

Table 7.3 (continued)

Bearing Commissioning – Oil-Lubricated Bearings

1. Bearing housings filled with specified lubricant and ready for operation.
   _____ (Yes or no)

2. Sight glasses at specified level: _____ in. or mm from top
   _____ (Yes or no)

3. All oil levels set to specified level: _____ in. or mm from bottom
   _____ (Yes or no)

4. Rotor magnetic center indication identified and clearly marked if required.
   _____ (Yes or no)

Bearings – Grease Lubricated
   _____ (Yes or no)

Note: Grease is normally used in bearings of motors ranging from fractional power to 500 kW or 650 hp.

1. Grease specification sheet available.
   _____ (Yes or no)

2. Remove grease inlet plug and check for grease.
   _____ (Yes or no)

3. Grease presence verified and applied following Plant guide lines: _____
   (see Chapter 4).
   _____ (Yes or no)

Oil Mist Lubrication

1. Oil specification sheet available.
   _____ (Yes or no)

2. Oil piping clean and blown – if applicable.
   _____ (Yes or no)

3. Oil mist re-classifiers properly installed. Orifice sizes identified and correct. All connection points and drains installed in correct locations.
   _____ (Yes or no)

4. Mist lubrication system in operation a minimum of 16 h prior to equipment operation.
   _____ (Yes or no)

Cooling Water

1. Cooling water piping flushed and connected – where applicable.
   _____ (Yes or no)

Vents and Drains

1. All vent and drain plugs installed with similar material of construction.
   _____ (Yes or no)

2. Double check location – high point vents, low point drains.
   _____ (Yes or no)

Coupling Safety Area

1. Coupling guard available and fitted.
   _____ (Yes or no)

2. Ensure that the coupling area is properly barricaded for solo run.
   _____ (Yes or no)

3. Ensure no visiting spectators are present within barricaded area.
   _____ (Yes or no)

Table 7.3 (continued)

Motor Solo Run – Typical for Large Motors	
1. Lockout-tagout (LOTO): Verify that all LOTO procedures have been followed prior to energizing the system.	_____ (Yes or no)
2. Ensure that motor is disconnected from driven equipment. Coupling spacer removed.	_____ (Yes or no)
3. Verify that the motor's coupling half is capable of a solo run; certain types require adapters.	_____ (Yes or no)
4. Check axial clearance – "bump-to-bump": a. Antifriction bearings: +0.005 in. or 127 µm b. Sleeve bearings according to magnetic center mark, not more than +0.250 in. or 6.0 mm	_____ (Yes or no)
5. Assure motor coupling half is securely clamped.	_____ (Yes or no)
6. Torque all adapter fixture bolts to a specified value – if applicable.	_____ (Yes or no)
7. Arrange for electrician to "bump" the motor by depressing start, then stop button to check for correct rotation. Correct if necessary.	_____ (Yes or no)
8. Commence 2- to 4-h solo run.	_____ (Yes or no)
9. Verify magnetic center indication.	_____ (Yes or no)
10. Monitor motor bearing temperature during solo run. If bearing temperatures are not instrumented, check them frequently with laser instruments.	_____ (Yes or no)
11. Inspect oil thrower ring operation in each bearing – there are usually inspection openings on oil lubricated sleeve-bearing motors.	_____ (Yes or no)
12. Monitor motor vibration during solo run.	_____ (Yes or no)
13. Monitor motor amp (all three phases) during motor solo.	_____ (Yes or no)
14. Monitor motor winding temperatures (winding RTDs) during solo run.	_____ (Yes or no)
15. Upon completion of solo run, usually of two to four hours duration, enter required data in progress tracking sheet – Figure 7.10.	_____ (Yes or no)

7.2.6 Initial Startup

Initial startup of the motor is for its solo run according to the checklist presented in Table 7.3. Another pre-start list is presented in Table 7.4. Motor startups are swift compared to other prime movers. Once the starter button is activated, it often takes no more than 30 s for the machine to reach its operating speed. Figure 7.10 shows a log sheet used by the startup team to track the progress of the project's motor initial startup activity.

Table 7.4: Checkout and run-in instructions for motors > 75 kW or 100 hp.

Note: Cross out items that are not applicable.

1. Are specific operation and maintenance manuals, parts books, special tools and spares available? _____ (Yes or no)
2. Have the vendor drawing and data requirements (VDDR) been fulfilled? _____ (Yes or no)
3. Are the items/documents listed in the project VDDR form readily available? _____ (Yes or no)

A. **Installation Inspection**

1. Receiving, storage protection, foundation and grouting checklists completed and attached. _____ (Yes or no)
2. Control loops functionally tested and correct all set points set and verified. _____ (Yes or no)
3. Alignment with driven equipment has been accepted including adequate shims. Hold-down bolts are tight, run-out and axial end play ok. _____ (Yes or no)
4. Inspect bearings, for cleanliness and lubrication.
5. Oil mist system (if necessary) has been commissioned. Bearings were greased. Grease spec: _____ _____ (Yes or no)

B. **Pre-startup Checks**

1. Megger motor and push button leads (100 megs acceptable) check for proper grounding of motor. _____ (Yes or no)
2. Coupling run-in plates are installed if required. _____ (Yes or no)
3. Roll motor rotor by hand to ensure it is free. _____ (Yes or no)
4. Check axial end clearance. _____
   a. Anti-friction bearings: +0.005 in. or 127 µm (Yes or no)
   b. Sleeve bearings: ~±0.250 in. or 6.3 mm

C. **Startup and Run-in**

1. Bump motor for rotation check. _____ (Yes or no)
2. Start motor for 4-h run, standing by to observe vibration levels and bearing temperatures. _____ (Yes or no)
3. Record data requested on motor run-in sheets at beginning and at end of motor run. _____ (Yes or no)
4. For sleeve bearing motors, mark magnetic center on shaft (should be within 3/32 in. or 2.4 mm of geometric center). _____ (Yes or no)
5. At end of run, check coupling fit and installation. (Record on driven equipment log sheet.) _____ (Yes or no)

UNIT:_____

	PRE-STARTUP CHECKS										
Motor Number Designation	Megger Motor/ Pushbutton	Coupling Run-In Plates	Bearings Flush & Lubricated	Motor Rolls Free	Rotor End Clearance	Bump Rotation OK	4 Hour Run Start Time	Bearing Vibration Recorded	OB Bearning Temp.	IB Bearning Temp.	Witness Signature/ Date

Figure 7.10: Motor startup progress tracking.

7.2.7 Motor Operation

Once started up and proven in an initial operating run, EMs demand very little attention. Winding temperature rise is frequently an area of concern, and one should note and remember the industry rule of thumb that states, each 10 °C of reduction in motor temperature will result in a doubling of motor winding life.

Appendix 7.A: Motor protective devices, controls and enclosures

1. **Fusible Safety Switch or Circuit Breaker (Small Single-Phase Motors):** This device provides the means to disconnect the controller and motor from the power circuit. It also serves to protect the motor-branch-circuit conductors, the

motor control apparatus and the motors against overcurrent due to short circuits or grounds. It must open all ungrounded conductors and be visible – not more than 50 ft or 15 m away – from the controller or be designed to lock in the open position. The design must indicate whether the switch is open or closed, and there must be one fuse or circuit breaker in each ungrounded conductor. A disconnect switch must be power rated or must be able to carry 115% of full-load current and be capable of interrupting stalled-rotor current. A circuit breaker must also be capable of carrying 115% of full-load current and be able to interrupt stalled-rotor current.

2. **Non-fused Disconnect Switch:** This switch is generally only required if the fusible safety switch or circuit breaker either cannot be locked in the open position or cannot be located in sight of the motor.

3. **Manual and Magnetic Starters:** These controls start and stop the motor. They also protect the motor, motor control apparatus and the branch-circuit conductors against excessive heating caused by low or unbalanced voltage, overload, stalled rotor and frequent cycling. Starter requirements are determined by the basic horsepower and voltage of the motor. Overloads in a starter are sized to trip at not more than 125% of full-load current for motors having a 1.15 or higher service factor, or 115% of full-load current in the case of 1.0 service factor motors. Single-phase starters must have an overload in one ungrounded line. A three-phase starter must have overloads in all lines. If a magnetic controller is used, it may be actuated by devices sensing certain process fluid parameters. Temperature sensors sensing cooling water temperature would be a typical example.

4. **Control Enclosures:** NEMA has established standard types of enclosures for control equipment. The types most commonly used in conjunction with process plant machinery are as follows:

   a. NEMA type 1 – general purpose: Intended primarily to prevent accidental contact with control apparatus. It is suitable for general purpose applications indoors, under normal atmospheric conditions. Although it serves as a protection against dust, it is not dust-proof.

   b. NEMA type 3 – dust-tight, rain-tight and sleet-resistant: Intended for outdoor use and for protection against wind-blown dust and water. This sheet metal enclosure is usually adequate for use outdoors on a cooling tower. It has a watertight conduit entrance, mounting means external to the box, and provision for locking. Although it is sleet resistant, it is not sleet proof.

   c. NEMA type 3R: This is similar to type 3, except it also meets UL requirements for being rainproof. When properly installed, rain cannot enter at a level higher than the lowest live part.

   d. NEMA type 4 – watertight and dust-tight: Enclosure is designed to exclude water. It must pass a hose test for water and a 24-h salt spray test for corrosion. This enclosure may be used outdoors on a cooling tower. It is usually a gasketed enclosure of cast iron or stainless steel.

e. NEMA type 4X: similar to type 4, except it must pass a 200-h salt spray test for corrosion. It is usually a gasketed enclosure of fiber-reinforced polyester.

f. NEMA type 6 – submersible, watertight, dust-tight and sleet resistant: Intended for use where occasional submersion may be encountered. Must protect equipment against a static head of water of 6 feet or 1.83 m for 30 min.

g. NEMA type 12 – dust-tight and drip-tight: Enclosure intended for indoor use. It provides protection against fibers, flying, lint, dust, dirt and light splashing.

h. NEMA type 7 – hazardous locations – Class I air-break: This enclosure is intended for use indoors in locations defined by the National Electrical Code for class I, division 1, groups A, B, C or D hazardous locations.

i. NEMA type 9 – hazardous locations – Class II air-break: Intended for use indoors in areas defined as class II, division 1, groups E, F or G hazardous locations.

References

1 Hz = cycles per second.

2 The **National electrical manufacturers association** (NEMA) defines standards used in North America for various grades of electrical enclosures typically used in industrial applications. Each is rated to protect against personal access to hazardous parts, and additional type-dependent designated environmental conditions.

3 ESD – emergency shutdown device.

4 EPRI – Electrical Power Research Institute, USA.

5 American Petroleum Institute API RP 686, 2nd Edition, 2009, *Recommended Practice for Machinery Installation and Installation Design.*

6 Bloch, Heinz P., "Optimized Equipment Lubrication—Conventional Lube, Oil Mist Technology and Full Standby Protection," 2nd Edition (2022), De Gruyter, Berlin/Germany, ISBN 978 3 11 074934 2.

7 IEEE Standard 43-2013, "The Recommended Practice for Testing Insulation Resistance of Rotating Machinery," or just IEEE 43.

Bibliography

[1] H.P. Bloch and F.K. Geitner, Series Practical Machinery Management for Process Plants, Volume IV, Major Process Equipment Maintenance and Repair, Second Edition, an imprint of Gulf Professional Publishing., at www.elsevier.com 1997, and preceding edition, Hardcover ISBN: 9780884156635, eBook ISBN: 9780080479002, 700.

[2] https://embed.widencdn.net/pdf/plus/megger/0vqvivnnu5/Motor-testing-with-MTO106_AN_en.pdf?u=ac6ctk

Chapter 8
C&S of Mechanical Drive Steam Turbines

8.1 Introduction

A steam turbine generates mechanical power from steam at high temperature and pressure. It can deliver constant or variable speed and is capable of close speed control. Drive applications include pumps, compressors, electric generators and many more. Steam supplies power to about 20–30% of all pumps and compressors in the process industries and related facilities.

Steam turbines have been classified by mechanical arrangement as single casing, cross-compound – more than one shaft side-by-side or multicasing tandem compound, two or more casings in a single train. They have also been characterized by steam flow, namely axial direction for most and radial direction for a few. Additionally, steam turbines can be categorized by steam cycle, condensing, non-condensing, automatic extraction or reheat.

In a steam turbine, steam flows through directing devices and impinges on curved blades mounted along the periphery of the rotor. By exerting a force on the blades, the steam flow causes the turbine rotor to rotate. Unlike a reciprocating steam engine, a steam turbine makes use of kinetic rather than the potential energy.

Steam turbines have developed toward multistage axial designs, where the expansion of steam is attained in a row of sequentially arranged stages. Staging allows increase in power output, while preserving high-speed capability required for direct coupling of driven equipment. Mechanical drive turbines power output ranges from some 25 hp (18 kW) or less to about 150,000 hp (110 MW) in, for example, air separation plants. They are frequently used in refinery and petrochemical plants because their processes produce excess heat which in turn is converted to drive steam.

8.2 The Steam Cycle

The cycle by which a steam turbine generates power is described by Rankine.[1] It is the process that consists of a heat source – boiler or heat recovery unit – that converts water to high-pressure steam. Water is pumped into the boiler at elevated pressures using boiler-feed water (BFW) pumps. Boilers can be medium to high pressure depending on the size of the unit and the temperature to which the steam is heated. Steam is then further heated to the boiling point corresponding to the pressure. Most frequently, it is superheated above the temperature of boiling.

In a multistage steam turbine, the pressurized steam is expanded to a lower pressure, and exhausted either to a condenser at vacuum and condensing conditions, or

https://doi.org/10.1515/9783110701074-008

into an intermediate-temperature steam distribution system – non-condensing – that provides steam to other applications. The condensate is returned to the BFW pumps for continuation of the cycle. It is not uncommon to have smaller units exhaust to atmosphere.

A steam turbine usually consists of a stationary set of blades, called nozzles, and a moving set of adjacent blades, called buckets or rotor blades, installed within a casing. The two sets of blades work together, such that the steam turns the shaft of the turbine and the connected load. The stationary nozzles accelerate the steam to high velocity by expanding it to lower pressure. A rotating bladed disc changes the direction of the steam flow, thereby creating a force on the blades that manifests as torque on the shaft on which the bladed wheel is mounted. The combination of torque and speed is the output power of a steam turbine.

8.3 Special-Purpose Steam Turbines

The hydrocarbon processing industry, for example, differentiates between two distinct turbine categories: General-purpose[2] and special-purpose turbines.[3] Figure 8.1 shows a typical general-purpose turbine, Figure 8.2 a medium size turbine and Figure 8.3 a large special-purpose turbine.

Figure 8.1: General-purpose steam turbine (courtesy of Siemens AG).

Figure 8.2: Medium-sized multistage special-purpose steam turbine (courtesy of Siemens AG).

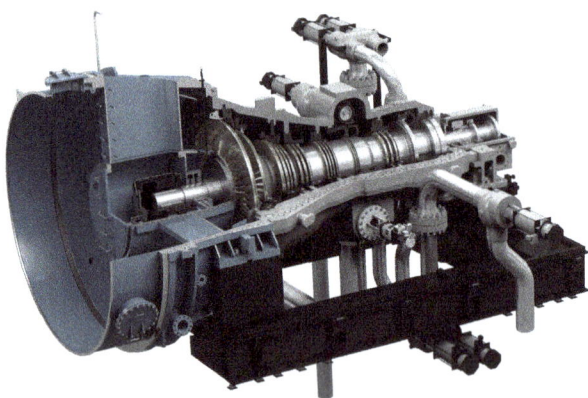

Figure 8.3: Large special-purpose steam turbine (courtesy of Siemens AG).

Table 8.1 presents turbine types and their application in process plants.

Table 8.1: Mechanical drive steam turbine types and their application in process plants.

Turbine type	Typical steam conditions	Applications
Non-condensing – back pressure	ISP: 1,200–140 psig or 82–9.6 bar IT: 900–450 °F or 482–232 °C EX: 500–50 psig or 34.5–3.4 bar	Process steam demand greater than steam power demand Various steam pressure levels required by process
Atmospheric exhaust	ISP: 135 psig or 9 bar IT: 400 °F or 200 °C	Off-site pump drives Fans and blower drives

Table 8.1 (continued)

Turbine type	Typical steam conditions	Applications
Condensing	ISP: 600–15 psig or 41–1 bar IT: 750–250 °F or 400–121 °C EX: 4 in. or 102 mm Hg abs.	Steam power demand greater than LP process steam demand Minimum live steam raising No high-pressure steam available Turbocompressor, blower and generator drives
Mixed pressure or induction	Combination of above	When excess steam exists at an intermediate pressure
Extraction a. controlled b. uncontrolled	Combination of above	Intermediate pressures(s) demand, specifically when variation in quantity is required Large process demand with fluctuations Small process demand with little change in demand

ISP, inlet pressure; IT, inlet temperature; EX, exhaust condition.

The above referenced standard[4] defines general-purpose steam turbines as vertical or horizontal turbines used to drive equipment that is usually spared, is relatively small in size (power) or is in non-critical service. Steam conditions will not exceed a pressure of 48 bar or 700 psi or a temperature of 400 °C or 750 °F or where speed will not exceed 6,000 revolutions per minute (rpm). The requirements for special-purpose steam turbines may be found in API 612. In both cases, the focus is on mechanical drive turbines in contrast to the much larger turbines utilized in electrical power generation.

8.4 General-Purpose Steam Turbines

General-purpose steam turbines are simple, compact units with small footprints. They do not require any utilities other than a live steam supply. More often than not, pressure lubrication is not needed because their bearings are lubricated by shaft-mounted oil rings dipping into an oil bath that has to be maintained at a specific level.

8.5 Risk Profiles

Pre-commissioning and commissioning of steam turbines must take into account that they, like other fluid machinery, are sensitive to ingestion of foreign particles.

General-purpose steam turbines are very likely to be contaminated by dirt as they are small and often get lost in storage facilities prior to installation. If they are not connected to a dedicated storage oil mist system, they tend to lose their protective flange covers and plugs. It would stand to reason to charge a conscientious millwright attached to the startup team to pay special attention to the project's machinery by tracking and accompanying individual turbines through receiving, lifting, base plate mounting, pipe connecting and alignment. As a rule, if any foreign matter is found inside turbine nozzles or the general appearance of the machine suggests it is contaminated by dirt, it must be disassembled, cleaned and reassembled.

Vulnerabilities of steam turbines are indicated in Figure 8.4a and b.

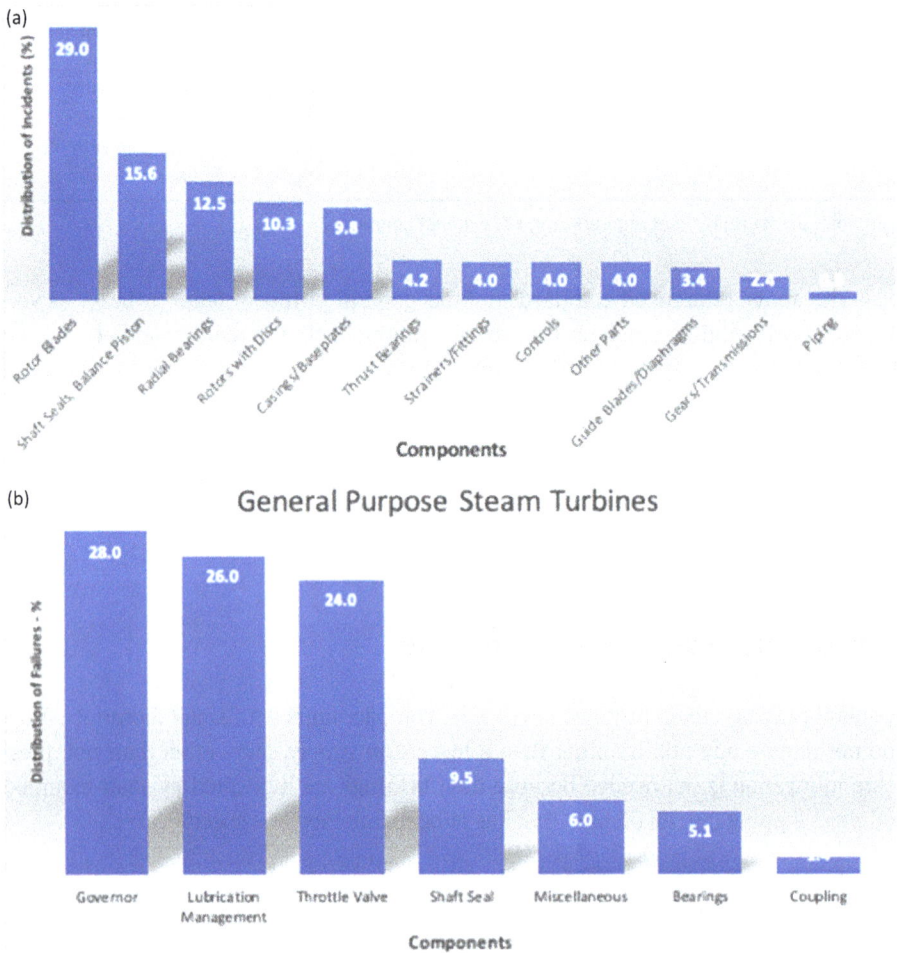

Figure 8.4: (a) Risk profile of general-purpose steam turbines. (b) Risk profile of special-purpose steam turbines.

Based on our experience, control system and associated linkage failures should be included in these distributions. However, we are unable to produce any data reflecting frequency of occurrence.

The result of an intrinsic-risk assessment is shown in Figure 8.5. It is based on the natural gas compressor train sketched in Figure 8.6. Feed gas compression trains are critically important due to high forced outage costs.

1	2
	StmT driven
	Turbocomp.
Rel. Risk #	294
Rel.Risk Level:	MEDIUM
Complexity*	21
1.7	7
2.3	2
4.3	6
5.1	1
6.5	5
7.0	0
Severity	14
Table 2.4.1	5
Table 2.4.2	5
Table 2.4.3	2
Table 2.4.4	2

Figure 8.5: Relative intrinsic-risk assessment of a steam turbine driven compressor train. *Refer to Tables 2.3 and 2.4.

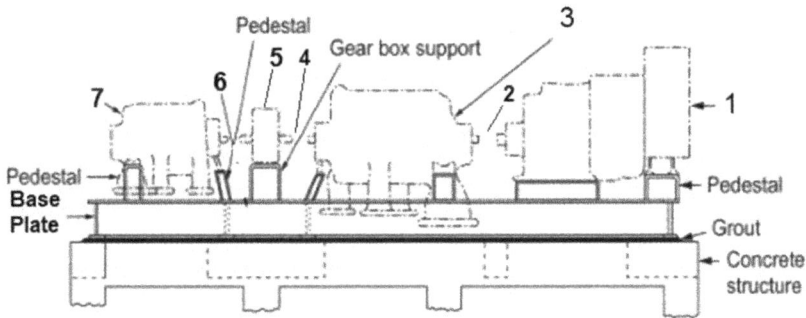

Figure 8.6: Sketch of a feed gas compressor train: 1, 8 MW steam turbine; 2, coupling; 3, LP centrifugal compressor at 10,500 rpm; 4, coupling; 5, speed increaser (gear box); 6, coupling; 7, HP centrifugal compressor at 12,500 rpm.

8.6 Risk Mitigation

A large portion of potential startup failure risks of steam turbines can be mitigated by tracking them and caring for them during the six project phases alluded to in preceding chapters of this book. As a reminder, we are listing them again in Table 8.2 to record the deliberations of the startup team.

Table 8.2: Project related risks and their mitigation – steam turbines.

Phase	Risk	Most probable cause[1]	Suggested actions/mitigating measures
Rigging and lifting	**High** – Large SP sets	QP, HS	– Review lifting plan – Have OEM fabricate spreader bar for lifting that meets local regulations, and have it approved by a local government inspector (if required)
Handling, staging and storage protection	**High**	OR, HK	– Provide proper storage space – dry cool environment – Prevent moisture intrusion: use vapor phase inhibitors (VPI), desiccants, dry instrument air as a purge where possible – Fill bearings and hydraulic circuits with the proper oil
Foundation and grouting	**Medium** – Cracks, voids – Edge lifting	QP, EN	– Have grout manufacturer rep at site during pour – Use grout with the proper exotherm formulation suited for the climate conditions at site
Piping and tubing	**High** – No or improper cleaning – Ingestion of foreign objects after cleaning	OR, HK	– Cleanliness and assembly to the machine should follow practices established as part of the project execution agreement between OEM and the owner organization – Supervised and certified piping cleaning and conscientious flange joint practices – steam piping[i]

Table 8.2 (continued)

Phase	Risk	Most probable cause[1]	Suggested actions/mitigating measures
Shaft alignment	**High** – Incorrect thermal growth predicted	QP	– Shaft alignment starts with leveling the unit. A minimum permissible deviation from the horizontal line would be customarily 0.002 in. per foot or 0.2 mm per linear meter – Conscientious machinery internal and external alignment routines – see alignment procedures and reference[ii] – Involve experienced personnel
Lubrication system	**High** – Wrong oil specification – Mix-up with other lubricants – Contamination	QP, DO, CO	– Have lubrication expert involved – Because external piping must be fabricated and installed as part of the lubrication system, the cleanliness obtained during FAT helps but should be discounted, therefore the system should be flushed again to achieve cleanliness standard per ISO 4406:99
Startup and initial operation	**High** – Improper startup sequence, for example, disregarded critical speed range	QP, DO	– An experienced startup team, and proper operator training

[1] Refer to Figure 2.12

[i]Bloch, H.P. & Geitner, F.K., Series *Practical Machinery Management for Process Plants*, Volume III, *Machinery Component Maintenance and Repair*, Fourth edition, ELSEVIER, 2019, and preceding editions, 650 pages, Chapter 4.
[ii]Ibid, Pages 46–48.

One activity deserves further elaboration, namely receiving and storage protection. An arguably effective storage and staging protection method is described in Appendix 8.A. Another long-term preservation process for steam turbines is outlined in Appendix 8.B.

Steam turbine piping cleaning and flange-up deserve special attention. While governor problems on special-purpose steam turbines have faded into the background due to advanced digital controls, lubrication problems are still to be reckoned with. As things go, lubrication systems have to undergo meticulous scrutiny for cleanliness during commissioning. It is well known that once these systems are clean and a reliable supply of lube oil to the bearings is assured, we should expect a long operating life from the steam turbine unit.

Prior to the commissioning phase the startup team will have reviewed the contractor's or EPC's drawings. As stated earlier, this is the last chance of bringing about any reliability enhancing changes. Table 8.3 shows a reasonable approach to that review. While a good many of the items listed will have been subject to prior discussions, meetings and decisions, there would be no harm in reviewing those items once more because, again, this effort would be the last window of opportunity for reliability enhancing modifications before the project solidifies as a completed plant.

Table 8.3: Contractor's drawing review – steam turbines.

Note: The following questions are in two parts. Those raised under "General" apply to all turbines. Those under "Condensing Turbines" are additional for this type of machine.

_____

_____

_____

_____

_____

General

1. Is there a warm-up vent (at least 11 in or 280 mm) on the inlet?	(Yes or no)
2. Does the inlet block have a 1 in or 25 mm bypass for line warm up?	(Yes or no)
3. Does the exhaust valve have a 1 in or 25 mm bypass for warm up?	
(Back pressure turbines only).	(Yes or no)
4. Is there a trap and bypass upstream of the trip and throttle valve?	(Yes or no)
5. Is there a trap and bypass on the steam chest of single valve turbines?	(Yes or no)
6. Is there a trap and bypass on the low point of the exhaust casing?	(Yes or no)
7. Is there a low pressure seal vent line on both seals?	(Yes or no)
8. If the vendor (OEM) has specified a pressure for this vent, has satisfactory	
control been provided?	(Yes or no)
9. What devices cause a trip of the turbine other than the built-in ones? _____	
Is this as specified?	(Yes or no)

Table 8.3 (continued)

10. What does the overspeed trip pressure switch actuate?	
Is this as specified?	(Yes or no)
11. If there is no built-in strainer in the trip and throttle valve, is there a Y-type	
strainer in the inlet line? We prefer to have both!	(Yes or no)
12. Has an exhaust line safety valve been provided between the turbine and the	
exhaust block valve if the exhaust pressure is above 75 psig or 5.2 bar?	(Yes or no)
Is the safety valve setting at or below the maximum design exhaust system	
pressure (including exhaust casing)?	(Yes or no)
13. Back pressure turbines with labyrinth seals must have an eductor and	
condenser. Are these shown?	(Yes or no)
14. Has the following instrumentation been provided?	
a. Inlet and exhaust TIs	
b. Inlet and exhaust PIs	
c. Steam chest PI (only on single valve units)	
d. First-stage PI	
e. Steam flow FIR	(Yes or no)
15. Have turbine washing facilities been provided?	(Yes or no)

Condensing turbines

1. Has a suitable shaft seal system been provided?	(Yes or no)
Usually the sealing steam will be taken from a pressure tap on the casing, but for startup, a pressure-controlled live steam supply must be provided to keep water out of the seals.	(Yes or no)
2. Is there an exhaust pressure safety device to relieve pressure in the event of cooling water failure?	(Yes or no)
Is this safety device water sealed?	(Yes or no)
3. Is there a minimum flow recycle arrangement on the condensate pumps?	(Yes or no)
4. Is the condenser level control and bypass (if valved) arranged for maximum pump out in case of air failure?	(Yes or no)
5. Are the pump glands sealed by discharge pressure?	(Yes or no)
6. Is the pump suction chamber vented back to the condenser steam space?	(Yes or no)
7. Has the following additional instrumentation been provided?	
a. Vacuum gauge on inlet and interstage of ejector	(Yes or no)
b. Seal steam pressure gauge	(Yes or no)

Piping layouts

1. Is the inlet steam taken off the top of the main header and is there a trapped dead leg on the header downstream of the turbine take-off?	(Yes or no)
2. Does the inlet slope continuously between the washing desuperh eater connection and the machine flange with no pockets of any kind to trap water?	(Yes or no)
3. Can the inlet pipe be readily diverted outside for initial blow-out?	(Yes or no)
The piping not to be blown must be capable of being thoroughly inspected internally.	
4. Can the trip and throttle valve be manipulated easily from the main platform?	(Yes or no)
This valve is used to start up the turbine and control its speed when out of governor range.	
5. The turbine exhaust safety valve must be removable for testing with the machine in service. Is it located so that if it is dropped it will not cause damage to other equipment?	(Yes or no)

Table 8.3 (continued)

6. On air blower drivers, are all steam vents on both inlet and exhaust lines away from and above the air intake hood?	(Yes or no)
7. Has the inlet and exhaust pipe been provided with sufficient direction anchors so that all piping growth will be away from the turbine?	(Yes or no)
8. Are the inlet and exhaust pipe supports, restraints, guides and anchors, as described in the piping stress calculations?	(Yes or no)
Has sufficient allowance for friction been made in the stress calculation?	(Yes or no)
9. Before startup, the operator must blow all steam line and casing drains. Are all these valves accessible for him or her to do this?	(Yes or no)
These drains are normally taken to a funnel. Is the location of the funnel such that the steam venting from it will not interfere with other equipment nearby or cause a hazard?	(Yes or no)
10. Piping must not run unnecessarily over parts which must be removed for maintenance, that is, bearing covers, top half casing, governor and trip and throttle valves. Has this problem been avoided and have crane capacity and lifting height been checked?	(Yes or no)
11. Is there a place where the casing top half can be set down during maintenance without interfering with maintenance?	(Yes or no)
Can it be moved to this location without passing over other equipment which might be running?	(Yes or no)
Can the rotor be removed to the maintenance shop without passing over running equipment?	(Yes or no)
12. Is the exhaust steam trap adequate to dispose of all the water required to make the inlet steam 1% wet?	(Yes or no)
Is it located on the low point of the exhaust?	(Yes or no)
If the casing will readily drain into the line, the trap should be on the line. Otherwise, it should be on the casing.	

Operability

1. Are all instruments clearly visible?	(Yes or no)
2. Does the operator have safe and easy access to all the bearings?	(Yes or no)
3. Has he or she clear access to the manual trip?	(Yes or no)
4. Can he see clearly the tachometer from the governor overspeed device?	(Yes or no)
5. Can he see clearly the tachometer from the trip and throttle valve?	(Yes or no)
6. Run through a starting sequence. Can all the operations required be done by one man?	(Yes or no)
7. Can operating personnel safely open the inlet and exhaust block valves?	(Yes or no)
8. Will the operators be able to manipulate the turbine washing system safely and in a controlled manner?	(Yes or no)

Similarly, the startup team should concern themselves with what happened to the machine during factory acceptance test (FAT) often referred to shop test and the preceding inspection report as per API 612.[5] In the case of a special-purpose machine, it would be important to know whether or not the turbine had received an ASME Power Test Code[6] performance test or just a mechanical run. The startup team should be aware of any problems, deficiencies or unsettled issues that might have been raised during these tests. Again, such deficiencies, possibly hidden until now, always have the potential of delaying commissioning and startup.

The FAT for steam turbines consists of a mechanical run test. This test is basically a means of checking rotor balance, controls, safety trips and searching for leaks. The basic procedure should be to run up to speed with steam conditions as close to design as possible. When conditions stabilize, including bearing and lube oil temperatures, the turbine should be operated for a period of 1 h, with no further rise in bearing and lube oil temperatures. On condensing turbines, steam inlet temperature or test duration may be reduced to prevent excessive temperature in the casing. Table 8.4 lists the questions a conscientious owner's engineer should be asking to become familiar with the machine at hand.

Table 8.4: Steam turbine shop test protocol.

	Conditions	
	Design	Test
1. Steam inlet pressure		
2. Steam inlet temperature		
3. Steam exhaust pressure		
4. Steam exhaust temperature		
5. Lube oil pressure		
6. Lube oil inlet temperature		
7. Speed		
(a) Maximum continuous		
(b) Normal operating		
(c) Trip		
(d) Calculated critical		
8. Maximum vibration		
(e)		
(f)		
(g)		
(h)		

Note: Cross out items that are not applicable

Table 8.4 (continued)

1. Did conditions 1, 2, 3, 5, 6, 7 and 8 on test match the design conditions to your satisfaction?	(Yes or no)
2. Is the turbine half coupling (with adaptor, if necessary) fitted for the test?	(Yes or no)
3. Is there a steam strainer on the inlet?	(Yes or no)
4. Is the actual critical speed within 5% of the calculated value?	(Yes or no)
5. Bearing temperature: Is the temperature rise across each bearing less than 60 ° F or 30 °C?	(Yes or no)

6. Vibration

Probe location	Magnitude (mils or m)	Frequency (cpm)

(a) Is there an absence of vibration at critical frequency, also at frequencies between 35% and 50% of running speed?

(b) Shaft and bearing housing vibration attenuation:

	A. Shaft	B. Housing	Attenuation
I.B. Bearing			
O.B. Bearing			

Is the attenuation less than 4? (If not, (Yes or No)
state actual figures in the report.)

(c) Vibration readings (mils or m):

	Just > Trip N	Max. cont. N	Difference
I.B. bearing			
O.B. bearing			

Is the difference in vibration levels at these speeds less than 20%? _____ (Yes or no)

7. Overspeed trip
The overspeed trip must be actuated at least three times. Is the difference between the highest and the lowest trip speeds less _____ than 0.5% of the highest? (Yes or no)
Note any problems with adjustment of trip, and operation of the trip values.
Note the actual final setting of the trip. _____rpm
Is this approximately 110% of maximum continuous speed? (Yes or no)

8. Are all auxiliary trips being tested – low-lube oil pressure and so on? _____ (Yes or no)

9. Is a thorough check being made for leaks, steam, oil, and air to governor (if applicable)? Finally satisfactory? _____ (Yes or no)

10. Is the machine stable at all speeds? (Hunting within 0.5%.) (Yes or no)

11. Is the speed control operating satisfactorily? (Yes or no)

Table 8.4 (continued)

12. (a) Are the data being taken, including the linearity of speed versus control signal?	(Yes or no)

Record

Signal

Speed

(b) On loss of control air, is the result as specified?	(Yes or no)
13. Bearing inspection: Do bearing surfaces show normal wear patterns?	(Yes or no)
14. If there is a spare rotor to be run, request an internal inspection (normally specified).	
(a) Is there an absence of rubbing?	(Yes or no)
If rubbed, demand a clearance check.	
15. Check the internal alignment and clearance data from final assembly drawing (if available). Okay?	(Yes or no)
(a) Is alignment good; clearances within tolerances?	(Yes or no)
(b) Likewise for a spare rotor if it has been fitted.	(Yes or no)
16. Have copies of vendor's log sheets, and final internal clearance diagrams been obtained?	(Yes or no)
17. Test witnesses (owner or OEM representatives) should have tried to find out if any difficulties occurred in preparing for the test. Such problems could be repeaters.	

8.7 Commissioning General-Purpose Steam Turbines

Once we are assured of having followed rigid pre-commissioning and commissioning routines, we can prepare for the initial startup. A pre-start checklist in Table 8.5 represents a review of what must be accomplished in order to begin starting up the turbine for its uncoupled or "solo" run.

Table 8.5: Pre-start checklist for general-purpose steam turbines.

	Comments
1. Are there any outstanding items from inspector's checklist (API 611)?	
2. Are there any outstanding items from the completeness summary – are they punch listed?	
3. Are applicable operation and maintenance manual, parts book, special tools and spares available?	
4. Have the vendor drawing and data requirements (VDDR) been fulfilled?	
5. Are the documents listed in the project VDDR form readily accessible?	
6. Are cross-sectional drawings and critical internal clearance data handy?	
7. Receiving, storage, foundation, grouting, piping and alignment checklists are completed and have been reviewed.	
8. Vibration analysis equipment has been calibrated and ready for use – if applicable. As a minimum, handheld vibration measurement instruments should be handy.	

Table 8.5 (continued)

	Comments

10. Have solid shims been placed under the turbine feet – instead of multiple shims?
11. Slacken all bolts connecting steam piping to turbine. Note any excessive springing of pipe.
12. Disconnect coupling between turbine and driven equipment.
13. Check rotating elements for freedom.
14. Record alignment readings between turbine and driven machine.
15. Check that gaskets are installed at turbine flanges.
16. Check that Y-strainers with a valved blowdown are installed in steam turbine inlet piping.
17. Reconnect piping to turbine and
    a.   Check freedom of ration using strap wrenches
    b.   Record final alignment data
17. Review steam piping for adequate support and effect on equipment. Ascertain exhaust stack and traps are constructed in keeping with personnel protection considerations.
18. All steam inlet lines must be adequately trapped to prevent condensate accumulation.
19. Turbine exhaust lines must have steam trap if turbine is remote or automatic start or if condensate if going to a condensate system. Casing may also need a steam trap.
20. Assure steam traps have been checked and allow blowdown to sewer.
21. Check that safety guards are provided and ascertain they are strong enough to resist deflection if contacted by personnel.
22. Check soundness of grouting – review grouting checklist.
23. Ensure cooling water piping had been inspected for cleanliness and flushed on site – if applicable.
24. Check that cooling water to the bearings and oil coolers is supplied and connected following OEM instructions. Each bearing housing should have its own separate parallel cooling water piping – if applicable.
25. Check that inlet and exhaust gate valves are accessible and operable.
26. 1/2 in. or 12 mm is minimum acceptable auxiliary size. Correct any deviations.
27. Check that valved blowdowns are installed on casing drains. For turbines with automatic starting, these blowdowns should be trapped since startup would be unattended.
28. Carbon and labyrinth packing vents are usually unrestricted as per vendors' recommendation. Make sure these vents are installed as per OEM instructions – for seals operating under atmospheric or positive pressure.
29. The inlet of each parallel cooling water path should have a valve for flow regulation.
30. Each parallel path of a closed cooling water system should have a sight flow indicator – if applicable.
31. Check if sentinel warning valves are installed and set to specifications. Note: Most turbines are equipped with a sentinel valve and not a safety valve. A sentinel valve is only a warning device that indicates that pressure is too high in the turbine case. They are set at approximately 75% above exhaust pressure. To relieve pressure on the turbine case, it is necessary to manually block the steam supply valve and open the case drains.

Table 8.5 (continued)

	Comments

32. At above 150 psi or 10.3 bar inlet steam pressure or if the exhaust casing has lower design pressure, a safety valve is required between the turbine casing and exhaust block valve.
33. Check for unplugged openings.
34. Any separate lube system should have
    a. Oil filter
    b. Oil reservoir breather
    c. Oil cooler
    d. Temperature indicator (TI) in and out of cooler
    e. Pressure gauge (PI) for each pressure level
    f. Pressure gauges (PI) before and after filter

Note that lube oil systems must be very carefully reviewed for potential intrusion of gland-leakage steam into the lube oil. When in doubt, provide reliable water removal equipment or reliably purged bearing housings.
35. Check procedure for purging steam lines. Were targets used?
36. Was steam line purging – blow-out, thermal cycling – supervised by owner's engineer?
37. Inspect lube oil pumps and drivers – is there an alignment record for these pump sets – if applicable?
38. Flush out lube oil systems and blow out oil mist lines.
39. Check out instrumentation.
40. Check that a speed indicator is handy – when panel mounted, visible to the operator – or hand held.
41. Replace lube oil filters – If applicable.
42. Drain and refill reservoir.
43. Commission lube oil system and oil mist system. Verify operation.
44. Make sure the turbine-coupling half is secured. Some designs require an adapter plate or strap.
45. Exercise the T&T valve prior to steam admission.
46. Certain turbines require carbon shaft seal break-in. If so, follow break-in procedure described in Appendix 8.D.
47. Follow GP turbine startup and shutdown steps listed in Table 8.6.
48. Run turbine uncoupled. Trip on overspeed three times. Check governor control.
49. Check bearings for excessive temperature.
50. Stop unit and inspect – clean if required – Y-strainers.
51. Connect turbine to driven equipment.
52. Check hot alignment.
53. Install lagging on turbines.
54. Check automatic start feature where provided and demonstrate operation.

Having completed these with the premises, we can begin with the startup. We recommend to proceed according to the sequence listed in Table 8.6, leading to the initial run of the turbine.

Table 8.6: Startup and shutdown procedures for general-purpose steam turbines.

A. Startup
1. Check the area to see if it is clean and free of foreign materials.
2. Open case drains and check for water. Check steam traps for proper operation. Be sure all condensate is completely drained from inlet and exhaust lines and from turbine case before proceeding further.
3. Check lubrication system to make sure oil level is okay and oil is clean and free from water.
4. Be sure driven equipment such as pumps, blowers, compressors or other driven equipment – is lined up[1] correctly. See appropriate operating procedures for centrifugal pumps or other driven equipment.
NOTE: Does not apply to solo runs.
5. Make sure exhaust valve is open into the low-pressure system. Exhaust valve should be painted yellow and car-sealed open (CSO).
CAUTION: If exhaust valve has been closed, be sure that condensate is drained
from the exhaust system before opening the exhaust valve.
6. Manually reset and trip the manual hand trip at least twice to assure operability. Reset the manual hand trip,
CAUTION: Never operate a turbine if the manual hand trip is not operable.
7. Slowly open the bypass line around the inlet block valve and let steam blow freely through the drain valves at a high velocity to clear all water from lines.
NOTE: If turbine exhausts to atmosphere, watch for condensate buildup in the exhaust line when starting turbine. If no bypass line is available, crack the supply line valve slowly a small amount.
8. When the turbine is free of water and warmed up, close all drain valves and bypasses around traps.
9. Slowly open steam supply line. Assure the governing system has assumed control by using a tachometer if possible. If no tach is available, tell by change in sound of turbine and by observing the movement of the governor linkage.
10. **CAUTION:** If there is an indication that the governor has not assumed control of the turbine at operating speed, immediately shut down the turbine and have the governor checked.
11. After the governor has assumed control, open the steam supply valve fully.

B. Shutdown
CAUTION: Never close the exhaust valve on a steam turbine without venting the turbine case to the atmosphere.
1. Shut turbine down by tripping the manual hand or local trip. Turbine should come to a complete or near stop. If not, it should be checked by mechanical personnel.
2. Close the block valve on supply steam. Make sure the bypass valve around the block valve is closed.
3. Open case drains and supply line bleeders between the inlet valve and the turbine.
4. Close the exhaust steam valve to the low-pressure steam system and finish opening the remaining drain valves.
CAUTION: Never close the exhaust steam valve until you are sure the supply line is blocked. Never leave the turbine until all pressure is completely bled off.
NOTE: if turbine is to be shut down and left lined up and in standby for automatic startup, see specific turbine operating procedures.

---

[1]Jargon for activating a piping system

One important step to get the machine ready for its initial run, the solo run, is following the safety critical procedure outlined in Appendix 8.C. It forces us to check the overspeed trip mechanism by means of the hand-tripping device, and be sure it is working properly. Then we reset it.

8.7.1 Carbon Ring Run-in

General-purpose turbines usually feature carbon rings as shaft seals. Carbon rings are tight fitting components hugging the shaft in order to prevent steam from escaping into the atmosphere. Figure 8.7 is an illustration of a packing box assembly housing the carbon rings. Carbon rings require break-in. If this has not already been accomplished during a FAT, or if damaged seals were discovered before the initial run, the replacement carbon rings require break-in. The incentives are:
-   Longer life expectancy
-   Increased turbine efficiency
-   Protection of bearings and journals by keeping condensing steam out of the oil
-   Lower vibration levels

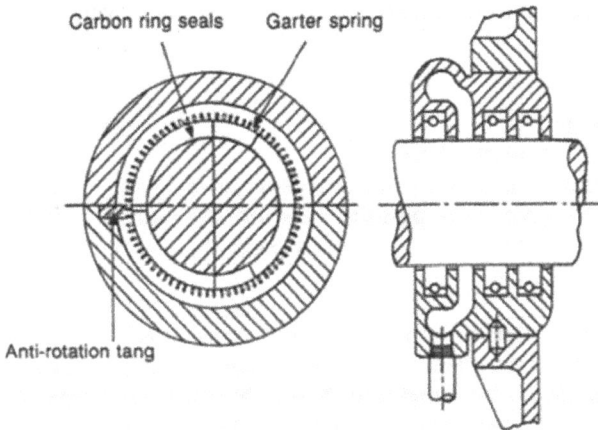

Figure 8.7: Packing box for general-purpose steam turbines.

A detailed carbon break-in procedure is shown in Appendix 8.D.

8.7.2 Casing Preparation

A common requirement for all types of steam turbines it to see that the turbine casings and connecting pipes are drained at all times. During operation, any accumulation of water cools the adjacent metal and causes distortion which, if severe, may cause blade rubs or vibration. During shutdown periods, accumulation of water causes excessive corrosion that impairs the efficiency of the turbine.

For multistage special-purpose turbines, the turbine casings are provided with built-in drains from each zone to the next lower pressure zone and finally to the exhaust. Orifices are provided for continuous drainage during normal operation, and hand-operated bypasses – where necessary – for use during starting and shutdown periods.

Similar drains must be provided from all connecting pipelines. These include the steam inlet line and the atmospheric relief line. On condensing machines, all drains – except from the high-pressure steam inlet – should connect to the condenser or a vacuum trap because, when starting or operating at light load, vacuum may exist in the entire back end.

Follow-up and overview of the commissioning and startup of steam turbines in connection with a major project is achieved by completing the tracking sheet shown in Figure 8.8.

8.7.3 Commissioning Example

A typical example of commissioning a general-purpose steam turbine driven pumping service is presented in the following paragraphs. It is to demonstrate how detailed a startup procedure must be in order to ensure success and it also shows that it is important to mark-up flow sheets either by numbering or red-tracing the lines and components that are involved in the procedure.

Example of a Commissioning and Startup Procedure

General-Purpose Steam Turbine in Automatic Service.

Plant description: Two vertical in-line 110 kW or 147 hp pump sets in propylene recycle feed service.

P-2827 steam turbine-driven set designated as main pump.

P-2828 motor-driven set designated as standby pump.

Premises: Prime movers have been commissioned and solo run. Couplings and coupling guards have been installed and checked for safety. The piping system has been cleaned and pressure tested. All connections have been tightened and leak tested. Instructions contained in Tables 8.5 and 8.6 are being followed.

Pumping system commissioning and startup – refer to Figure 8.9. Call-out numbers correspond to procedure step numbers. P-2828 is not shown.

1. With the automatic steam admission valve P275V and the 6 in. valve upstream of it closed, commission the MP steam system (135 psi or 9.3 bar) by blowing down condensate and starting up the steam traps. Open 6 in. valve (CSO) downstream of P275V and open ¾ in. bypass or warm-up valve.

Process Unit ____________

	PRE-STARTUP CHECKS														
Turbine Number Designation	Supply Line Blown	Gov./Trip Mechanism Is Free	Bearings Are Clean	Steam Strainer Installed	Axial End Clearance	Throttle Valve Stroke	Run-In Start/Stop Time	Minimum Governor RPM	Maximum Governor RPM	Overspeed Trip, RPM 1	2	3	Vibration Recorded	Hot Alignment Recorded	Witness Signature/Date

Figure 8.8: Typical general-purpose steam turbine pre-startup and run-in log sheet.

2. Open 6 in. valve to vent LP steam system, 15 psi or 1 bar, to vent to atmosphere, drain condensate. Allow turbine to warm up over a period of 8 h by passing a small amount of steam through the turbine. Set valves appropriately. Make sure O/S trip valve is open and armed.
3. Admit propylene to pump casing by opening valve Z270V. Make sure pump seal cavity leak-off line to vapor space of D-2827 is wide open. Check if seal is holding. Pump may be isolated during turbine warm-up.

Startup after the Turbine Is Warmed
4. Open blowdown connections in steam piping upstream of turbine.
5. Make sure pump is filled with liquid. Open 6 in bypass valve to get turbine to break away and settle out at a run-in speed of 500 rpm.
6. Close 6 in. valve to atmospheric vent of the LP steam system, if turbine slows down, increase speed by opening the 6 in bypass valve by several additional turns on the hand wheel. Allow turbine to run at 500 rpm for 1 h.
7. Prepare for automatic turbine startup: Switch H271S on "auto" and open 6 in valve upstream of automatic steam admission valve. This valve is to be CSO after completion of initial startup.
8. Assure ¾" blowdown valves in drip legs just upstream of turbine are still open. Switch off PM-2828, the field switch for the motor driven pump. The automatic steam admission valve should open slowly and bring P-2827 on line.
9. Close steam blowdowns.

8.8 Special-Purpose Turbines

The important differences between large special-purpose turbines and smaller general-purpose turbines should be kept in mind as being:
- Frequently unspared critically important machinery.
- Usually operated for longer periods of time – it is not uncommon to see them operated for 5 years on a "24/7" basis without a mechanical outage.
- Equipped with comparatively more complex protective devices, electronic speed governing and control systems.
- Provided with different type of trip and throttle, T&T, valves.
- Connected to the driven machine by high-speed special couplings – see Chapter 11.
- Equipped with labyrinth casing-to-shaft seals with steam leak-off and seal connections.
- Connected to a pressure oil system that not only supplies lube oil to the turbine and driven machine bearings but also to the hydraulic governing and steam flow control system – see also Chapter 5.

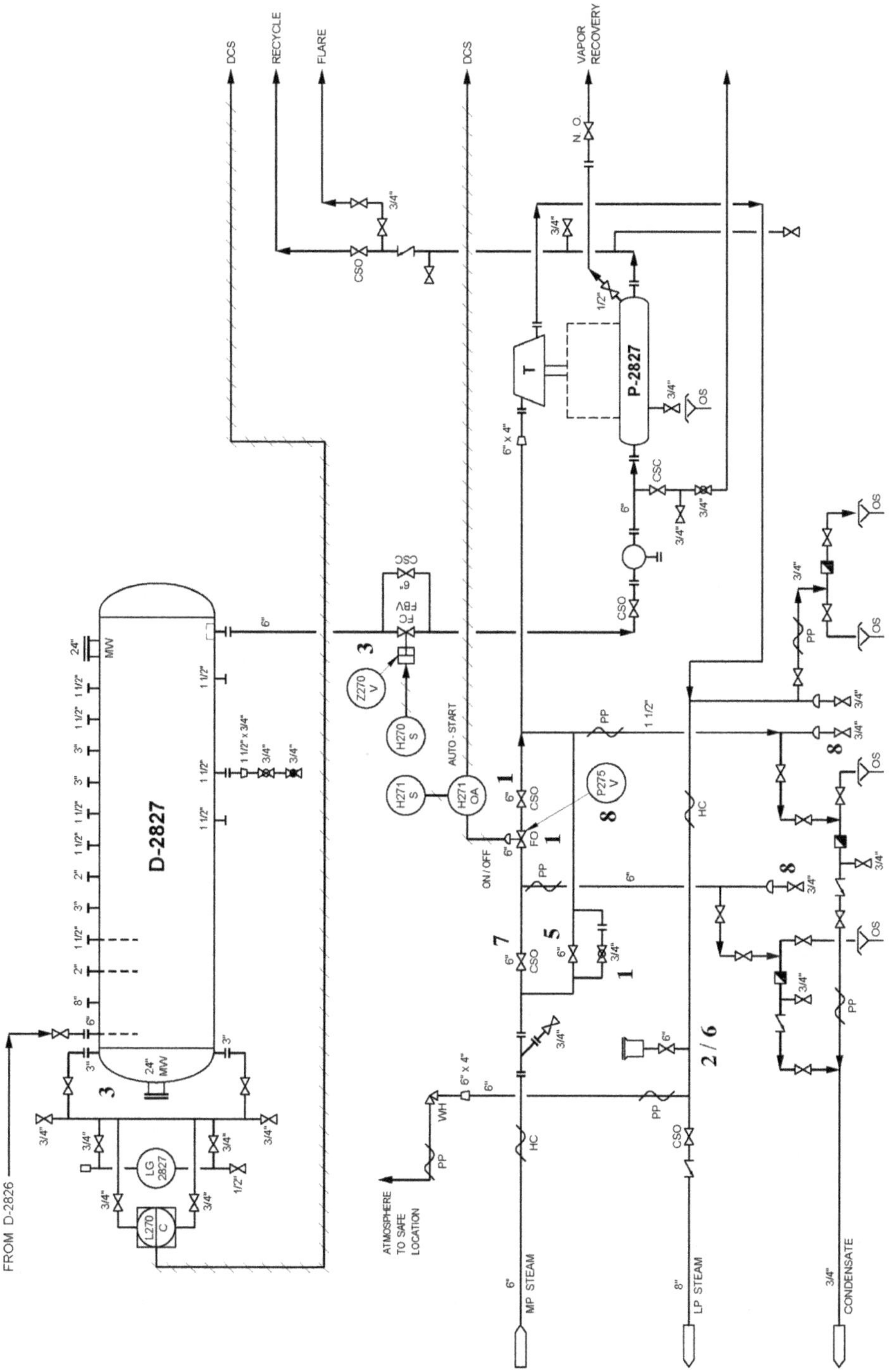

Figure 8.9: Simplified flow diagram of a propylene recycle pumping service.

– Provided with instrumentation and HMI displays to indicate and record temperatures, pressures, speed, shaft movement and vibration for interfacing with IoT or SCADA technologies.

8.8.1 Commissioning and Startup – Special-Purpose Steam Turbines

The project phases of receiving, storage, erection and piping will eventually lead to mechanical completion and commissioning of the large steam turbine. Commissioning consists mainly of making sure steam piping is clean, and we elaborated on this phase in Chapter 5. A checklist similar to Table 8.5 should be used to review and assert conscientious turbine pre-commissioning and commissioning efforts. Such a checklist should be composed by the OEM field service representative – FSR – and the owner's engineer working together for transparency and good communication.

Earlier we talked about the need for cooperation of owner and vendor representatives. Here, in the startup phase of a major critically important machine, personnel harmony is especially called for as painful delays can result if there is discord and miscommunication. At one turbine startup occasion we saw the complaint " . . . don't feel the love . . . " entered in an OEM FSR's log book.

Generally, there are seven steps to prepare a large mechanical drive condensing steam turbine for either overspeed check or a slow roll. They are shown in Figure 8.10, which could be a lead in for a commissioning and startup procedure.

8.8.1.1 Starting a Special-Purpose Condensing Steam Turbine

As we approach the initial startup of a special steam turbine, operating personnel should have been trained to be abreast of the contents of the machine's operating manual. It is available in either hard copy or on the plant's information system. As a minimum, it should contain the following operating procedures:
– General description of the turbine and drive
– Emergency equipment and procedures
– Initial startup
– Normal startup
– Emergency shutdown
– Major components, care and feeding

8.8.1.2 Safety Precautions

Under no circumstances should the trip valve be blocked or held open to render the trip system inoperative. Overriding the trip system, and allowing the turbine to exceed the rated (nameplate) trip speed may result in catastrophic asset losses. Here are some words of caution:

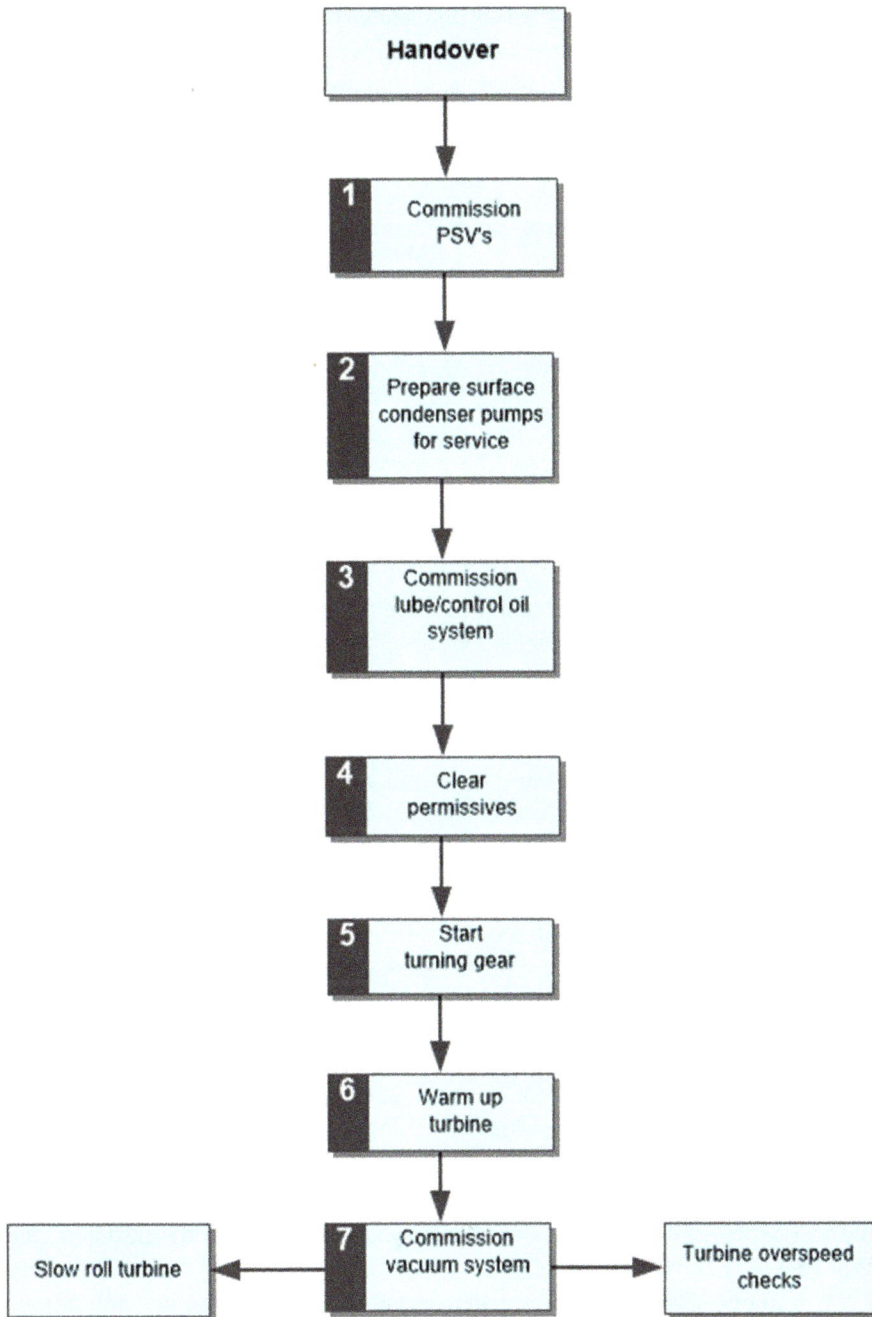

Figure 8.10: Cold pre-startup steps for a major steam turbine.

1. Do not operate the turbine if inspection shows that the rotor shaft journals are corroded.
2. Be sure the rotor is not rubbing any stationary parts and rotates freely by hand before starting.
3. Check that all piping and electrical connections are made before operating the turbine.
4. Ensure that all valves, controls, trip mechanisms and safety devices are in good operating condition.
5. Modern large steam turbines have advanced control systems that should be commissioned and be available for the initial startup. See Figure 8.11. They include the driven machine, for example, a compressor. Their features are:
   - Speed control
   - Extraction steam pressure/flow control
   - Inlet and/or backpressure control
   - Electronic overspeed protection and trip triggering
   - Startup sequencing (auto roll up)
   - Critical speed avoidance
   - Compressor surge control and protection
   - Compressor load and capacity control
   - Compressor inlet temperature control
   - Drum level control
   - Lube oil monitoring
   - Compressor seal monitoring
   - Alarm and trip functions
   - Compressor valve sequencing
   - Emergency shutdown logic (ESD)

8.8.1.3 Testing the Overspeed Trip

When the turbine is first started after installation, it is very important to test the overspeed trip by actually overspeeding the machine. This may be done by use of the overspeed test routine as part of a modern electronic governing system. Refer to the governor instructions for operation.

Modern mechanical drive steam turbines today feature well-engineered and highly reliable electronic overspeed prevention systems. These systems offer end-users the security and dependability of a two-out-of-three voting system with the flexibility and ease of do-it-yourself configuration.

A typical overspeed prevention system monitors turbine speed and will initiate a trip command to prevent overspeed events. Systems of this type incorporate three identical speed modules that individually measure a frequency-input signal from a passive or active magnetic pickup sensor. A supervisory module continually monitors

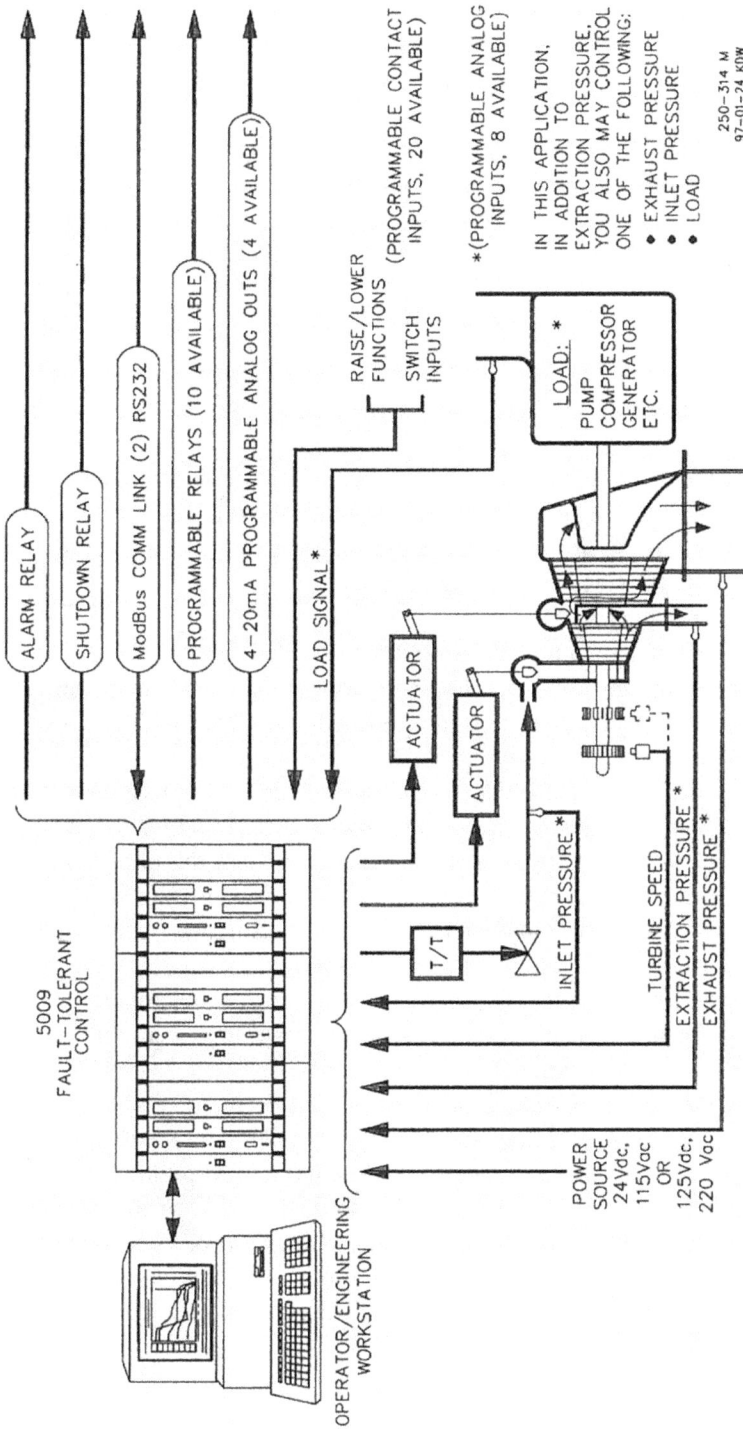

Figure 8.11: Control system for special-purpose steam turbines (source: Woodward Governor Co.).

the three speed modules for proper operation, which helps to eliminate unnecessary downtime and increases system availability.

As shown in Appendix 8.C, regardless of what type of governor has been furnished, great care should be taken to have only personnel essential for the activities present – no spectators should be allowed since there is always the risk of the overspeed protection system not working as it should and the turbine speeding away. Overspeed of a steam turbine can result in self-destruction with potentially devastating consequences for asset and life as the turbine disintegrates and debris scatters about.

8.8.1.4 Startup Sequence

1. Make sure the turbine coupling half is securely strapped and that there is no safety concern.
2. Be sure that the oil supply to the turbine is operating. See that ample oil pressure is established at the bearings and in the control system.
3. See that the turbine casing drains, the extraction line drains, the gland leak-offs are open, and that the steam line is free of water.
4. Open the exhaust valve.
5. Establish water circulation through the condenser.
6. Vent the incoming steam line to atmosphere until the incoming temperature is at the minimum recommended operating temperature. This could take several hours working with the utility section of the process plant.
7. Open the throttle valve a sufficient amount to start the rotor immediately, then close it and open it again just enough to keep the rotor rolling 200–400 rpm. Listen for rubs or other unusual sounds, especially when the rotor is rolling with the steam shut off, for at this time a foreign noise can be heard easily.
8. Start the condensate pump and operate intermittently, if necessary, to maintain level.
9. As soon as the rotor is in motion, turn on the water to gland condenser and steam to the ejector. Close all atmospheric casing drains to prevent drawing air through the drains when partial vacuum is established.
10. Start the second-stage air ejector or the priming ejector if one is used.
11. Keep the turbine rolling at low speed – approximately 200–400 rpm – to allow the parts to become partly heated. Maintain this slow rolling for at least about 20 min. The duration of the rolling is determined by specific operating instruction issued by the OEM. They should be adhered to. In case the unit had been started and run before and has shut down for any reason the rotor might be bowed and continued rolling at low speed will heat the rotor uniformly and straighten it.

12. At the end of the rolling period, bring the unit up to speed slowly, taking 10–15 min to reach full speed. After the unit comes up to speed, reduce the speed again and slow roll the turbine for a somewhat longer period of time.
13. Shut off the priming ejector/hogging ejector, if one is used. When the maximum vacuum is obtained with the second-stage ejector, start the first-stage ejector.
14. Make sure that the governor properly controls the speed of the turbine with full steam pressure and vacuum.
15. Close the drains from pressure zones when it is assured that all water has been removed and condensation stopped.
16. Make certain that the temperature of the oil supply to the turbine is maintained between 110 and 120 °F or 43 and 49 °C. The temperature of the oil leaving the bearings should not exceed 160 °F or 71 °C.
17. Open throttle valve fully.
18. Make sure that the governor properly controls the speed of the turbine with full steam pressure.

8.8.1.5 Shutting Down
This involves the following steps:
1. Decrease the load to about 20% of full load; except in an emergency shutdown, load should be removed gradually.
2. Then remove all loads and shut down the turbine by tripping the overspeed trip mechanism manually through an ESD loop such as low oil pressure.
3. Be sure that the oil supply is maintained until the machine becomes relatively cool. If this is not done, the heat conducted along the shaft from inside the turbine may injure the bearings.
4. When the turbine comes to rest, close the exhaust valve and open all drains between the throttle valve and the exhaust valve.
5. Shut off the air ejectors.
6. Open the vacuum breaker if one is provided.
7. Shut off the water to gland condenser and steam to the ejector.
8. Shut down the condensate pump.
9. Open all blowdown drains.

While HMI displays on modern special-purpose steam turbines facilitate surveillance and monitoring tasks around a steam turbine-driven train, it is of interest to look at a traditional log sheet as shown in Figure 8.12. It tells us the important operating parameters for steam turbines. Today, operators are tracking these parameters by incorporating them into modern comprehensive shift logs.[7]

SYSTEM	PARAMETER	Once per Shift		Twice per Shift		Weekly		Monthlly		Notes
		Check	Log	Check	Log	Check	Log	Check	Log	
LUBE OIL SYSTEM										
Oil Reservoir	Level		✓							
	Temp.		✓							
Main Pump	Disch. Pressure		✓							
	Bearing Oil Level		✓							
	Slow Roll (Turbine)	✓								
	Swing Pumps							✓		
Supply Header	Pressure		✓							
	Temperature		✓							
Cooler	Oil Outlet Temp.		✓							
	CW Inlet Temp.		✓							
	CW outlet Temp.		✓							
Filter	Differential Press.		✓							
Bearings (Each)	Temperature		✓							
	Pressure		✓							
Accumulator (if furnished)	Level		✓							
	Pressure					✓				
Oil Conditioner/Purifier	Operation		✓							
	Reservoir Level						✓			
	Filter Δ P						✓			
Control Oil Supply Header	Pressure		✓							
Lube Oil Analysis									✓	
STEAM SYSTEM										
Sealing steam	Pressure		✓							
Gland Seal	Pressure		✓							
Gland Eductor Supply	Pressure		✓							
Gland Codenser	CW Inlet Temp.		✓							
	CW Outlet Temp.		✓							

Figure 8.12: Traditional operating log sheet – steam turbines.

8.8.2 Water Washing

It is possible that steam quality has not come up to par during the plant's startup period. Turbines start to indicate internal fouling and exhibit loss of power. Water washing will often restore the machine to its original condition.

The potential problems of water washing mechanical drive steam turbines are:
- Misalignment due to piping stress as the temperature is reduced.
- Water slugging.
- Loss of clearance due to differential contraction between rotor and stator.
- Vibration due to nonuniform deposit removal.
- Thrust bearing failure due to almost complete plugging of one stage.
- Damage to blading if hit by water retained in the exhaust casing.

There can be no doubt that the above problems hinge on the technique used for water washing.

One method that has been successful in the past is the following:[8]

On most machines, the misalignment due to pipe stress will not be significant. After all, we are only disposing of the superheat, whereas during startup, the machine is exposed to a temperature change at least two times as large. However, if piping strains are a problem on startup, one must make sure all sliding supports are free before attempting to wash.

The risk of water slugging can be avoided by two measures. We use a venturi nozzle for desuperheating. Also, we insist that the piping fall continuously between the desuperheater and the machine inlet. Even if the water is not broken up into droplets in the desuperheater, it will pass harmlessly through the turbine as a constant stream. Refer to Figure 8.13.

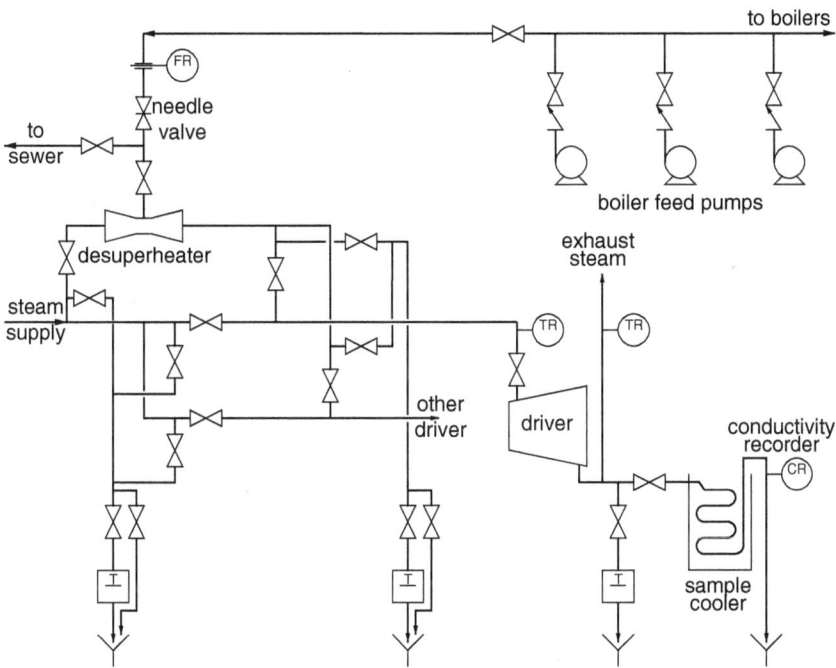

Figure 8.13: Simplified flow diagram for steam turbine water wash.

To prevent loss of clearance, we always limit the rate of temperature change to 180 °F or 90 °C per hour. The greatest hazard would be failure of the injection pumps when at maximum injection rate. Such a failure would produce a very high rate of change of temperature and would most likely result in an axial rub. To guard against this, we try to use boiler feedwater, since these pumps are the most reliable in the plant.

We attempt to reduce the chances of nonuniform deposit removal by halting the increase of injection rate whenever a deposit is actually being removed. This condition is detected by measuring the conductivity of the exhaust condensate.

It is alleged that thrust bearing failures have occurred because of stage plugging when an upstream wheel has shed its deposit before a downstream one. We have not found this to be the case.

One of the criteria we use to check a turbine design before purchase is "Can all the condensate be removed from the exhaust?" Some turbine designs are such that the blading is within about 1 in. or 25 mm off the bottom of the casing. Others do not have a casing drain at the lowest point. Others have a ½ or ¾ in., respectively, 12.5 or 19 mm casing drain. All of these designs are suspect unless the condensate can drain freely out of the exhaust.

We consider that a stage is washed adequately when the condensate conductivity falls to half its peak level. When this point is reached, the water injection rate is again increased. The wash is considered completed when the inlet steam is saturated and the exhaust conductivity is down to 200 micromhos.

After the wash is completed, the inlet temperature is raised to normal at a maximum rate of change of 180 °F or 90 °C per hour.

Initiation of the normal steam flow by bypassing the desuperheater, is the last hazard. We have found that water builds up in the line upstream of the valve, even when the bypass is left open. We now always leave the main valve cracked open to prevent the water buildup.

We conclude that on-load washing is safe provided reasonable care is exercised. We have, however, observed that deposit solubilities vary considerably between subsequent washes on the same machine. This same variability has been observed on two machines supplied by steam from the same source for the same time period. Some machines can be successfully cleaned without making the inlet saturated but most have required a wet inlet.

During our earliest washes, we believed that condensate was essential as the desuperheating medium. We reasoned that any other water would leave salts behind during the temperature-increasing phase. Five of the machines have now been washed using boiler feedwater, without any observable problems or deterioration of the cleaning.

Also, during our earliest wash endeavors we noted an apparent accelerated fouling rate during the first few days after a wash. We have no reasonable explanation of this phenomenon. It levels out quickly and does not appear to affect either the maximum mass flow or the efficiency. Currently, we merely warn the operators to disregard it if it is observed.

The steam supply should be free from moisture and preferably superheated. A receiver-type separator with ample drains should be provided ahead of the shut-off valve to prevent water from entering the turbine. When a separator is not provided, a continuous drain must be connected to the lowest point of the steam inlet piping.

Appendix 8.A: Field storage and staging procedure for steam turbines

1. A blind flange must be provided at the steam inlet flange of the trip and throttle valve (T&T valve).
2. Blind flange(s) must be provided after the extraction check valve(s) (Exhaust check valve in the case of a backpressure steam turbine.)
3. Blind flange(s) must be provided at the air vent pipe for the after-cooler of the air ejector condenser(s) and steam supply line for the steam ejector(s).
4. Blind flange(s) must be provided at the condensate water outlet flange(s).
5. Close all the drain valves for T&TV, turbine casing, steam, piping and so on.
6. Dry nitrogen or another inert gas must be supplied by using the pressure leading pipe for and extraction pressure regulating valve(s) pressure measurement pipe(s) – refer to Figure 8.A1.
7. Internal pressure must be observed by providing a U-tube manometer on the piping of the pressure gauge.
8. Seal tapes must be attached on the valve rod leak parts of the trip throttle valve (s), regulating valve(s) and extraction pressure regulating valve(s).
9. The blind flanges must be provided at the bearing vent flanges.

Cautionary Note: Various countries have different safety regulations guiding the use of nitrogen for industrial purposes. In the United States, Occupational Safety and Health Administration (OSHA) tasks industrial manufacturers to maintain a safe working environment for employees. For example, OSHA 29 CFR 1910.146 sets guidelines for confined spaces that could contain higher than normal concentrations of nitrogen gas.

Regardless of industrial location, it is vital that all production process managers conduct nitrogen gas risk assessment exercises to determine their level of exposure, and institute adequate preventive measures. Also, all personnel must be trained on the proper use of personal protective devices as well as proper actions to carry out in case of accidental hazardous exposure.

Appendix 8.B: Long-term preservation steps for steam turbines

1. Seal the shaft openings with silicone rubber caulking – select a dark color, like black to discourage pilfering. Then apply tape to the shaft's surface.
2. Dry out with instrument-grade (i.e., −40 °F or −40 °C dew point) air.
3. Some owners will fill the turbine casing, including the steam chest, with oil containing 5% rust prevention concentrate. Hold the governor valve open as necessary to ensure the chest is full. We recommend instead of oil, VPI (vapor phase inhibitors) bags at the turbine flanges. Before shipment, the internals should be sprayed with VPI for a 6-month storage.

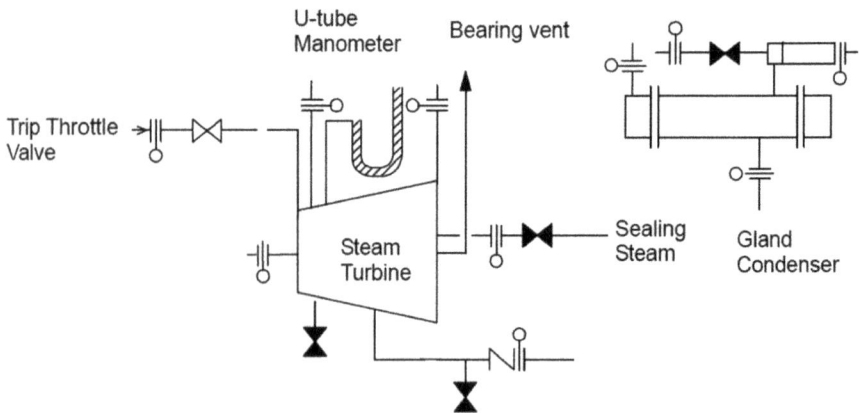

Figure 8.A1: Field storage arrangement for steam turbines (Mitsubishi).

4. Vent the casing as required to remove trapped air. Fill the trip and throttle valve with oil.
5. Coat all external machine surfaces, cams, shafts, levers and valve stems with product C^9.
6. Coat the space between the case and shaft protrusion with product A. Cover the space with black tape.
7. Fill the bearing housing with oil.
8. Coat the casing bolts with product C.

Appendix 8.C: Procedure to safely perform the solo run of a GP steam turbine

PETRONTA *Chemicals* **Process Unit**	
Process	Machine: (*Describe Function*)
Procedure no.	*Example:*
Purpose	The purpose of this procedure is to safely perform the solo run of a GP Steam Turbine. The primary objective of a solo run is to test the mechanical overspeed trip mechanism using a bypass line. The secondary objective is to confirm turbine mechanical integrity in preparation for unit startup. The Operator Technician must be thoroughly familiar with the entire procedure before starting the turbine for the first time.

(continued)

Safety, health environmental (SHE) consideration	– Caution must be exercised when working near rotating equipment. Extra precaution should be taken around all steam piping. This will be the initial commissioning of all steam piping; therefore, personnel should be alert to potential steam leaks.
	Caution must be exercised when draining hot condensate while warming steam lines and commissioning steam traps to prevent exposure to condensate and steam.
	Manual control of the steam turbine speed will be accomplished by throttling steam using the globe valve in the bypass line around the inlet steam block valve. The inlet steam block valve shall remain closed during the entirety of this procedure.
	During the overspeed test, the turbine shall be covered with a Kevlar blanket to prevent flying debris in the event of a turbine overspeed incident.
	During the overspeed test, no personnel shall be located in the line of fire of the turbine or coupling.
	Turbine coupling shall be removed and the coupling guard secured in place during the test.
References:	– Work permit system manual
	– OEM operation and maintenance manual
	– Task risk assessment document
	– Pipe and instrumentation diagram: XYZ
Responsibility and assistance	– Trained operator technician to perform turbine warm up and startup.
	– Trained machinery technician to perform turbine overspeed trip test.
	– Machinery FLS shall coordinate the work permit and necessary resources with Process Manufacturing Coordinator covering equipment preparation, steam commissioning, equipment handover, OEM operation & maintenance manual.
	– Machinery FLS/ET shall collate the test results and file them in the equipment file for base line reference data.
Prerequisites:	– Work permit should be approved prior to start of work
	– Standard PPE required for this task to be used
	– The area to be barricaded and a sign board to be displayed
	– Only ETs who have fulfilled the competency block test can perform the inspection and trip test
	– ETs should possess the TRA document and the procedure during the testing phase
	– Confirm that the coupling spool between the turbine and the pump has been removed and coupling guard properly in place during the test.
	– To ensure the remote digital tachometer is functional and calibrated
	– Cooling water is lined up to the turbine.
	– Steam for inlet at 660 psig and 750 °F or 4,550 kPag and 399 °C
Materials/tools to be used	– Remote digital tachometer
	– Kevlar blanket
	– Handheld tachometer for confirmation of speed at low rpm.

Activities listed above must be executed in the order documented in this procedure. Failure to execute this procedure in the correct order could lead to an uncontrolled turbine overspeed event and potential damage to equipment and injury.

Major steps

A: Trip mechanism Functionality checks
B: Warming up of steam Turbine
C: Confirm trip valve Positive shut-off
D: actual turbine Overspeed trip test

Procedure steps

	A: **Trip mechanism functionality checks**	
Purpose	Confirm the integrity of the tripping mechanism (trip valve, latch and linkages) for the reliable overspeed protection of the steam turbine.	
	CAUTION: Check driven machine and verify that coupling spool has been removed. Verify that the coupling guard has been reinstalled for safety purpose.	
	1. Visually check the <u>governor linkage</u> condition for wear and looseness. Grease the linkages. 2. Visually check the <u>trip valve linkage & spring</u> condition for wear and looseness. Grease the linkages.	_____ _____
	Warning: Do not apply lube oil or any anti-rust containing hydrocarbon liquids to the trip or governor valve stem. Fire hazard! [] Applied thin layer of MOLYKOTE™ or equal	_____
	3. Check the trip and governor valve stems are free from rust and able to move freely.	_____
	Warning: AFTER ENGAGING THE TRIP LATCH TO THE FULLY OPENED TRIP VALVE, THE VALVE HANDLE MUST BE TURNED FULLY TO THE ORIGINAL POSITION IN ORDER FOR THE TRIP VALVE TO FUNCTION DURING EMERGENCY. IF THIS IS NOT DONE, THE TRIP VALVE CANNOT CLOSE DURING A TURBINE OVERSPEED	_____
	4. Manually reset the trip valve by turning the handle clockwise until the trip latch can be engaged. 5. Turn the handle counterclockwise fully and ensure that the trip valve stays in the open position. 6. Pull the hand trip lever and confirm that the trip valve goes to full close position. 7. Repeat the above two steps to confirm repeatability of the trip valve's reaction	_____ _____ _____ _____

(continued)

	8.	Contact machinery engineer if any of the following malfunction criteria are observed: _____ [] Trip valve does not shut [] Trip valve fails to close completely (stem bent, internal seizure, etc.) [] Sticking or binding of linkages preventing the trip valve from closing [] Sticking or binding of linkages slowing the trip valve closure
	9.	Check oil level in bullseye and Trico Oiler; fill as necessary using oil _____ specified (e.g., VG46).
	10.	Apply reflective tape on turbine shaft for purpose of confirming _____ tachometer speed with an external handheld meter.
	11.	Install Kevlar blanket over the turbine casing. _____

B: Warming up of steam turbine

Purpose	Remove any steam condensate from the turbine casing and inlet line, and check all flanges/connections for steam leaks. **CAUTION:** **This Will Be the Initial Commissioning of all Steam Piping and Casings. Personnel should be Alert to** **Potential Steam Leaks.** **Warning:** **Never open the turbine main inlet steam block valve, or its bypass, unless the turbine exhaust valve is completely open** **Opening the main steam inlet block valve, or its bypass, while the exhaust valve is closed could lead to turbine casing overpressure and potential damage to equipment and injury.**

SHE Risk!	1. With the main inlet steam double isolation valves and exhaust valve _____
	    closed and steam turbine under trip condition, open the casing and
	    exhaust drain valves (both upstream and downstream of exhaust
	    isolation valve) to drain all condensate from the casing and exhaust
	    line.
	2. Verify the proper operation of steam traps and all condensate is _____
	    completely drained from the exhaust line and from turbine case before
	    proceeding further.

CAUTION:
Be Sure Condensate Is Drained from Exhaust System before Opening the Exhaust Valve
Failure to Do so May Result in Equipment Damage During Startup

	3.	Once all condensate is confirmed to have been drained, slowly open _____ the exhaust isolation valve fully to allow LP steam to enter the turbine casing for the purpose of warming up.
	4.	Continue to blow condensate from the system until dry steam is _____ observed and casing is heated near the saturation temperature of the exhaust LP steam (290 °F or 145 °C).

(continued)

	5.	Partially close all low point drains and leave them cracked open to allow condensate to escape	_____
	6.	On the small-bore bypass line across the main isolation valves, keep the ¾" gate valve fully closed. Open the ¾" globe valve and count the number of turns to fully open the globe valve. Record number of turns in the section B of attached commissioning/overspeed record sheet	_____
	7.	Then shut the ¾" bypass globe valve fully and open the ¾" gate valve fully. The gate valve on the small-bore bypass can be kept fully open until the end of the test.	_____

Warning:
BEFORE OPENING THE INLET BLOCK VALVE OR ITS BYPASS FOR THE PURPOSE OF WARMING UP, ENSURE THAT THE TRIP VALVE IS FULLY CLOSED AND THE INLET PIPING DRAIN VALVES ARE OPENED TO PREVENT HP STEAM FROM ENTERING THE TURBINE AND TURNING IT.

SHE Risk!	8.	Confirm that the trip valve is fully shut	_____
	9.	Open all condensate drain valves from inlet piping and trip valve. Prepare to drain condensate from the steam inlet line.	_____
	10.	With the ¾" bypass gate valve fully opened, gradually crack open the globe valve to allow a controlled amount of steam across the main block valves for the purpose of warming up the inlet piping – turbine should not turn! Remove condensate through the drain valves between the isolation valves and the turbine	_____
	11.	Use only the small-bore bypass for warm up until there is no condensate discharge from the drain valves.	_____
	12.	Close the bypass globe valve and drain off the remaining steam in the piping using the drain valves. Ensure that the steam pressure within the inlet piping before the trip valve is zero (XYZ G)	_____
	13.	Partially close all low point drains and leave them cracked open to allow condensate to escape	_____

Warning:
Ensure that the Steam Pressure Within the Inlet Steam Piping is Zero before Resetting the Trip Valve. Resetting the Valve with Full Inlet Steam Pressure Can Cause Uncontrolled Turbine Acceleration or Trip Valve Stem to be Bent
Trip Valve Handle and Hand Trip Lever Will Be Hot once Steam Enters the Turbine and Piping. Ensure PPE Is Used During Handling.

(continued)

	14. Arrange the governor inlet valve position to fully open by:
	   – Disconnecting the linkage between the valve and the Woodward
	     TG-13 governor
	   – Manually pulling the valve stem outwards to the fully open position
	(Test shall be conducted by throttling bypass globe valve only)
		_____

SHE Risk!	15. Reset the trip valve by turning the handle clockwise, engaging the trip latch and then turning the handle counter clockwise. Now steam turbine is ready to operate.	_____

C: Confirm Trip Valve Positive Shut-off

Purpose	– **Bring the steam turbine to slow roll (500 rpm) and manually "Trip" the trip valve to confirm that it closes sufficiently to bring the uncoupled turbine to a complete stop.** – **Confirm calibration of digital tachometer with handheld non-contact tachometer at low speed.**	

Warning:
STEAM MUST BE ADMITTED SLOWLY INTO THE TURBINE THROUGH THE BYPASS GLOBE VALVE
OPENING THE VALVE TOO QUICKLY COULD LEAD TO UNCONTROLLED ACCELERATION OF THE TURBINE AND AN OVERSPEED EVENT

**Note:** Prepare to count the number of turns the ¾" bypass globe valve must be turned in the next step, before the turbine shaft begins to move.	_____
1. Open the small-bore bypass globe valve gradually for turbine to slow-roll at 500 rpm.	_____
2. Record the number counted above in section C of attached commissioning/overspeed record sheet.	_____

Warning:
HANDHELD TACHOMETER CAN ONLY BE USED TO VERIFY TURBINE SPEED DURING SLOW ROLL (<1000 RPM)
If the speed readings between the digital and handheld tachometers are not the same, trip the turbine and contact the machinery engineer.
INCORRECT SPEED INDICATION CAN RESULT IN UNEXPECTED TURBINE OVERSPEED AND CAUSE INJURY

3. Use non-contact tachometer to confirm calibration of the electronic tachometer installed at site. Ensure that the digital tachometer and handheld is reading within 10 rpm.	_____
4. Check there is no abnormal noise and vibration. Allow for warming up of steam turbine and continue for draining of condensate.	_____
5. After running the turbine at 500 rpm for 15 min, manually trip turbine to check the integrity of the tripping mechanism and the shut off valve.	_____

(continued)

---

6. Contact machinery engineer if any of the following malfunction criteria _____
are observed:
[] Turbine does not come to a complete stop.
[] Trip valve fails to close completely (stem bent,
internal seizure, etc.).
[] Sticking or binding of linkages preventing the trip
valve from closing.
[] Sticking or binding of linkages slowing trip valve
closure.
[] Trip valve does not shut.
7. Close small bore piping globe valve. _____

D: Actual turbine overspeed trip test

Purpose | **Test the safety critical mechanical overspeed trip mechanism using the globe valve in the bypass line around the inlet steam block valve to manually increase the turbine speed to the actual designated trip speed.**

Equipment detail:
- **Tag number:** *ST0001A*
- **Area:** *Process Unit Area XYZ*
- **Governor type:** *Woodward TG-13*
- **Rated speed:** 1,450 rpm
- **Design trip speed:** 1,752 rpm
- **Allowable trip range:** 1,717–1,787 rpm

---

1. Prepare the turbine for restart following steps B8–B14 _____
*Note: Prepare to count the **total** number of turns the bypass globe valve requires (from full closed) to allow turbine to reach rated speed, in the next step.*
2. Open the bypass globe valve gradually to increase speed in steps of _____
200 rpm until rated speed.
3. Record the total number of turns the bypass globe valve requires (from _____
fully closed) to allow turbine to reach rated speed.
4. Observe no abnormal noise and vibration throughout the speed _____
increasing stage.
5. Run the turbine at rated speed for at least 2 h. Record bearing _____
horizontal, vertical and axial vibrations and bearing housing
temperatures every 15 min on the attached record sheet. Continue to
run the turbine until the bearing temperatures stabilize. Notify
machinery engineer if:
[] Vibrations exceed 3 mm/s
[] Bearing housing temperatures exceed 85 °C
Trip turbine if:
[] Vibrations exceed 13 mm/s
[] Bearing housing temperatures exceed 130 °C
*Note: Prepare to count the **total** number of turns the bypass globe valve requires (from full closure) to bring the turbine to trip speed in the next step.*

(continued)

Warning:
DO NOT EXCEED 1,827 RPM AT ANY TIME
EXCEEDING THIS SPEED COULD RESULT IN TURBINE FAILURE AND AN SHE
EVENT

SHE Risk! 6. Proceed with the overspeed trip test. Increase the turbine speed by   _____
slowly opening the bypass globe valve.
The speed should be increased slowly (to avoid sudden trip)
until 1,752 rpm is reached.

7. Record the trip speed   _____
[] _____ Trip speed

8. Record the total number of turns the bypass globe valve requires (from   _____
full closure) to bring the turbine to trip speed:
[] _____ Number of turns

9. Close the bypass globe valve and open all the drain valves (casing,   _____
governor valve, trip valve, piping, etc.) to depressurize both the
turbine and piping.

10. Contact machinery engineer if any of the following malfunction criteria   _____
are observed:
[] Turbine does not come to a complete stop
[] Trip valve fails to close completely (stem bent,
internal seizure, etc.)
[] Sticking or binding of linkages preventing the trip
valve from closing
[] Sticking or binding of linkages slowing trip valve
closure
[] Trip valve does not shut
[] Overspeed trip occurring outside the acceptable
range of trip speeds for that particular turbine

11. If the trip speed is not within the target speed range, isolate the steam   _____
turbine on both inlet and exhaust. Adjust the spring tension according
to the OEM manual.
Note: Tripping speed drop approximately x rpm per 0.1 mm reduction
in shim thickness

12. Repeat the test from step No. B1 until the desired speed is obtained.   _____
Confirm by tripping the turbine three times.

13. Record all three trip speeds in the attached commissioning/overspeed   _____
record sheet and inform the machinery engineer to review the test
data.

14. Reconnect the governor valve linkages for normal operation.   _____

15. Visually inspect coupling flexible elements for any signs of cracking   _____
prior to re-installation.

(continued)

16. Check alignment, install coupling spacer & guard. Prepare for equipment handover to process.	_____

Note: Governor speed fine tuning shall be done together with unit operator during commissioning.

17. Completely filled up and sign off both attachments for filing purposes. Trip test data shall be retained for at least 5 years.	_____

END OF PROCEDURE

Appendix 8.D: General-purpose steam turbine carbon seal break-in

GP Steam Turbine Carbon Break-In

The carbon break-in procedure consists of the following steps:

1. Heat the lubrication oil to a minimum of 100 °F or 38 °C before beginning slow roll. The operating oil temperature target is usually 110–120 °F or 43–49 °C. Mount dial thermometers on the gland seal housing, mid-turbine case and exhaust casing. These temperatures are used to determine the steady-state temperature point of the turbine prior to slow roll.

2. Open all casing drains, trip and throttle valve and steam line drains leading to the turbine and begin slowly admitting warm-up steam. Do not start slow roll until the turbine is hot. Larger condensing turbines, particularly partial admission turbines, may require a special manufacturer's recommended startup procedure to avoid localized rotor bowing.

3. Slow roll at 500 rpm at least 1 h. Open sealing steam line and establish 5–8 psi or 35–55 kPa pressure.

4. As the turbine gets hotter or the vacuum increases, it will speed up rapidly using the same steam flow due to the increased availability of energy.

5. Close case drains as appropriate.

6. Record vibration readings at both ends of the turbine.

7. Raise the speed to 1,000 rpm and immediately record vibration levels. Stay at 1,000 rpm for 1 h minimum. At about 1,200 rpm, the bearing oil film is carrying the rotor and the shaft has established a reasonably stable orbit in the bearing. Assuming that the rotor has relaxed its thermal bow, the "first" reading you will get at 1,000 rpm is primarily residual rotor unbalance. After about 15–30 min, you will observe an increase in vibration – about .25 to 1.0 mil or 6 to 25μ. Gradually, the vibration will drop nearly back to the first reading you took at 1,000 rpm. This

is what you have been looking for; a slight carbon seal ring rub followed by a return to steady state.

8. Raise the speed to 3,500 rpm in 500-rpm increments, repeating the sequence of immediately taking "new-speed" steady-state vibration levels and watching for the vibration increase and decrease cycle caused by the carbon rings breaking in.

9. At about 2,500–3,500 rpm, the new carbons are fairly well glazed and nearly run in. This is also the point where most people destroy their packings by assuming that the job is complete.

10. From 3,500 rpm, raise the speed by increments of 1,000 rpm up to running speed going rapidly through the critical speeds.

11. After running smoothly, if a sudden severe jump in vibration occurs, immediately drop the speed to 2,000 rpm or less for about half an hour. The carbon ring seals were grabbing the shaft. After a 1-h cool down, return the turbine to operating speed.

12. Run at normal maximum running speed for 1 h prior to checking the overspeed trip and coupling up.

13. The key to a successful and long-life carbon ring break-in is patience and the continuous presence of an operator or technician through the entire procedure to handle any contingency.

14. Upon successful break-in of carbon seal rings, it would be well to check the machine's steam strainer as stated in Table 8.3. It protects the turbine from large particles of scale, welding beads, and other foreign matter. This removable strainer does not guard against abrasive matter, boiler compound, acids or alkaline substances, all of which may be carried over in the steam. These substances may corrode, erode or form deposits on the internal turbine parts, thus reducing efficiency and power. It is necessary that feed water treatment and boiler operation be carefully controlled to ensure a supply of clean steam, if prolonged satisfactory operation is desired.[10]

References

1 https://en.wikipedia.org/wiki/Rankine_cycle (02/26/2021)

2 API 611 (R2014), General-Purpose Steam Turbines for Petroleum, Chemical and Gas Industry Services, Fifth Edition, March, 2008.

3 API 612, Petroleum, Petrochemical and Natural Gas Industries – Steam Turbines – Special Purpose Applications, Seventh Edition, August 2014.

4 See Reference 2.

5 See Reference 3.

6 ASME PTC 6-2004 (R2014)

7 Ibm.com/MAXIMO Shift Logs for Oil and Gas.

8 Conversations with Brian Turner, P. Eng. († 2009).

9 For Recommended Preservation Products see Appendix 5.A.

10 Bloch, H.P. & Geitner, F.K., Series *Practical Machinery Management for Process Plants*, Volume IV, *Major Process Equipment Maintenance and Repair*, Second Edition, an imprint of Gulf Professional Publishing, at www.elsevier.com 1997, and the preceding edition, Hardcover ISBN: 9780884156635, eBook ISBN: 9780080479002, 700 pages.

Chapter 9
C&S of Gas Turbines

9.1 Overview of Key Systems

9.1.1 General

Gas turbines are used as prime movers for numerous classes of process machinery. As mentioned before, they are referred to as mechanical drives as opposed to electrical generator drives with much higher power ranges. Mechanical drive turbines are generally in power ranges of 700–80,000 hp or 520 kW to 60 MW. Units within the output range of 20–30 MW seem to be the most common. They are used when process facilities do not have a steam system that can be used to provide power to steam turbines or the process facility is not using an electrical utility to provide electricity for the facility. In the midstream business of the petrochemical industry, such as gas pipeline transmission service, they have all but replaced internal combustion engines as compressor drivers.

They are usually packaged. Major package components and systems include the gas turbine, the starting system, the lube oil system, control system, on skid electrical wiring, a skid with drip pans, piping and manifolds, air inlet filter and silencer, exhaust ducting and silencer, optional package enclosures with ventilation system and combustible gas detection – see Figure 9.1.

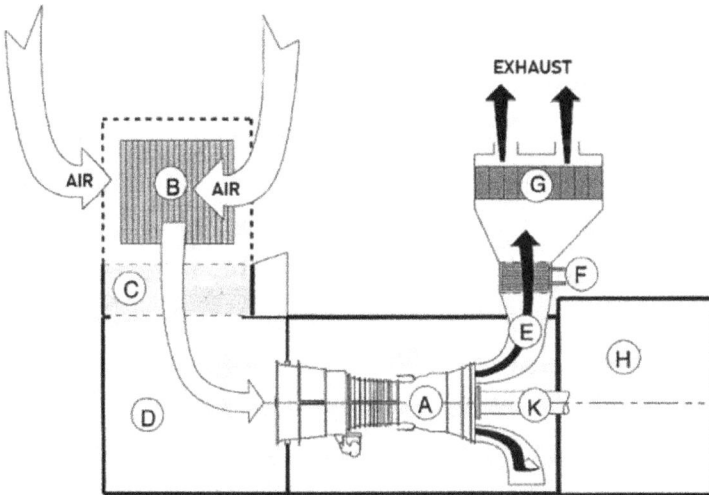

Figure 9.1: Typical gas turbine package: A, gas turbine; B, combustion air filter; C, combustion sound baffle; D, intake air plenum; E, exhaust gas collector; F, exhaust heat exchanger; G, exhaust sound baffle; H, driven machine (load); K, drive shaft (PTO) and coupling (adapted from reference[1]).

https://doi.org/10.1515/9783110701074-009

Two gas turbine concepts must be understood in our context. One, the single shaft machine, mostly used for electrical generator drives, as well as mechanical drives, and two, the industrial two-shaft mechanical drive gas turbine. The advantage of a two-shaft machine is that it satisfies the requirement of the process industry to provide variable speed capability and thus flow and pressure control when driving fluid machinery. Figure 9.2a and b illustrates the concept.

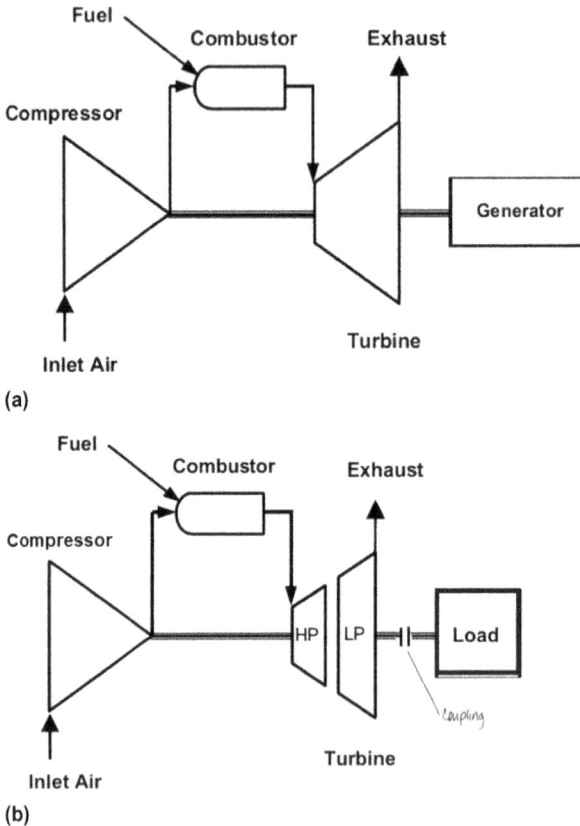

(a)

(b)

Figure 9.2: (a) Single-shaft gas turbine concept and (b) two-shaft gas turbine concept (source: GE).

Gas turbines are highly engineered machines with complex monitoring systems, instruments and controls applied to support the systems and the gas turbine itself.

A control concept is outlined in Figure 9.3.

Interconnected protective systems initiate alarm or shutdown signals based on unsafe conditions such as:
- Overspeed
- Excessive temperature (bearings; exhaust)

- High vibration
- Problems revealed by flame detection and combustion monitoring
- Low-lube oil pressure

It further provides fully automatic sequential operation of the turbine, its auxiliaries and its load. It provides automatic control by varying speed and load of the gas turbine based on a process setpoint, for example, pressure or flow when driving a pump or compressor.

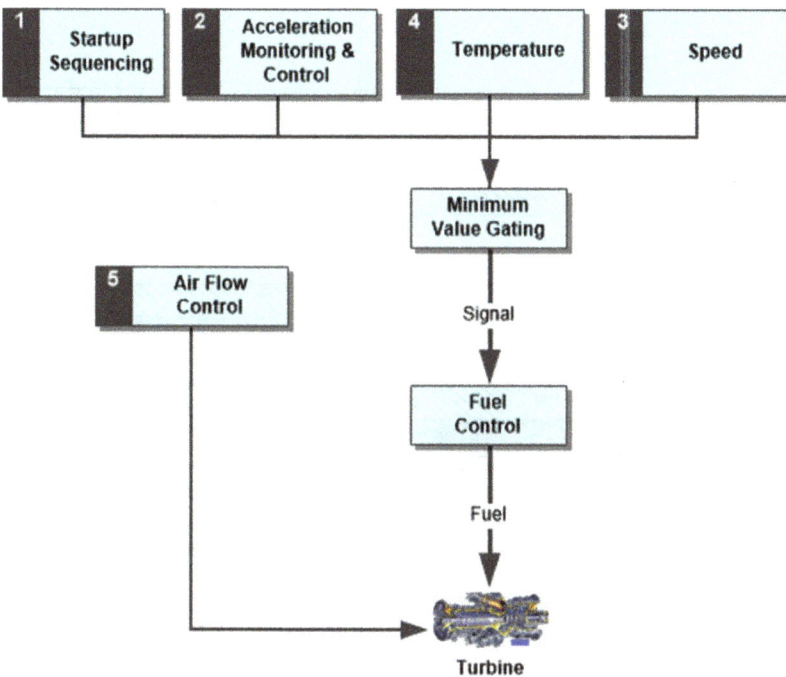

Figure 9.3: Basic gas turbine control concept.

A more complex overview of modern control systems includes:
1. Human–machine interface systems (HMI), dedicated computer systems provided by the gas turbine manufacturer and not to be confused with the operating owner's distributed control system (DCS). Generally, the HMI communicates with the DCS to provide valuable information for the Process Operators. Fuel Supply Control System controls the fuel gas valve and monitors fuel gas supply pressure.
2. The combustion control system for modern gas turbines is intricate. Its function is to control emissions from the combustion process by ensuring a lean mixture of fuel with the inlet air as well as controlling the time the combustion process

occurs so that the production of NO_x, CO and volatile organic compounds (VOC) is minimized, thereby ensuring the lowest emissions possible.

3. The startup control system is used to start the gas turbine from zero speed and ramp up the gas turbine speed to achieve the minimum speed required to initiate combustion as illustrated in Figure 9.4. Once the combustion system control parameters are satisfied, the startup system is disengaged and control is handed off to the acceleration, speed and temperature control systems.

Gas Turbine Starting Sequence

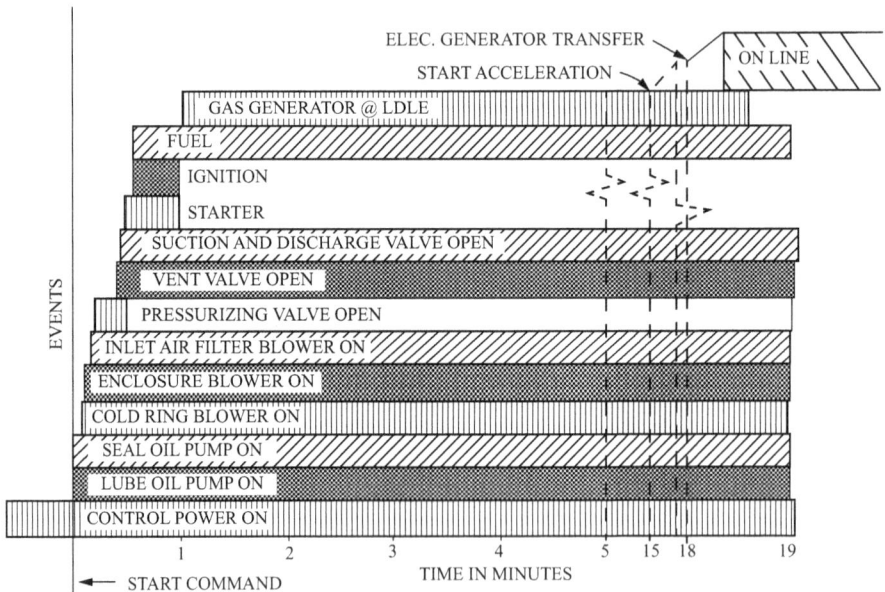

Figure 9.4: Example of a single-shaft gas turbine starting sequence.

4. Acceleration control system accelerates the gas turbine from minimum speed to initiate combustion to the minimum operating speed for the gas turbine.

9.1.2 Fuel Supply Control System

Fuel gas supply pressure and flowrate are continuously monitored and controlled to ensure they are suitable for the reliable operation of the gas turbine. These control systems are generally the first systems to be commissioned. All the associated instruments, filters, valves and flanges must be confirmed to be installed according the P&IDs and the gas turbine manufacturer's requirements. The system must be checked for leaks before fuel gas is introduced.

During the pre-commissioning of the fuel gas supply system the fuel control system will execute an integrity check of the fuel gas stop valves. In fact, this test is automatically performed before every start of the gas turbine. This is a critical safety check and should never be omitted. These valves stop the flow of fuel gas and are required to stop the gas turbine.

9.1.3 Combustion Control System

The combustion control system ensures adequate power is available to the rotating assembly, it controls the low emission performance of the combustion system and adjusts the combustion system in response to load changes. This is a complex control system that must be tuned in situ to ensure it provides stable and low emission combustion. The gas turbine manufacturer's commissioning team should be used to complete the tuning of the combustion control system during both the solo operation of the gas turbine and its coupled and loaded operation. This should be followed by ensuring a stable system is capable to adjust to load changes and performs to specifications before handing the asset to the operating team.

9.1.4 Lube Oil System

Refer to Chapter 5 to find procedures and checklists for the commissioning of lube oil systems. Pay particular attention to cleanliness of the lube oil system.

The HMI will communicate between the core gas turbine and the lube oil system. Use the specific functional check procedure provided by the gas turbine manufacturer to complete this phase of commissioning. At this point, we would like to remind the reader that the check sheets provided in Chapter 5 – covering oil systems – are generic and therefore, suitable for use in commissioning other types of process equipment. However, these lists are not detailed enough to cover complex gas turbines and procedures; specific checklists to the machine at hand must be produced.

9.1.5 Starting System

Like all machines using an internal combustion process, a gas turbine requires a reliable starting or cranking device and a clutch for disengagement once the machine starts to accelerate. Starting systems can be electric, pneumatic, steam or gas driven. Often, a variable speed drive (VSD) electric motor is used for single-shaft gas turbines in mechanical drive applications because the gas turbine cannot provide sufficient torque to startup the whole train.

In two-shaft applications, where the section comprising the combustion air compressor, the combustors and the high-pressure (HP) turbine is frequently referred to as the "gas generator," the starting process is simpler. In mid-stream operations,[2] gas expansion starter motors are used to crank the gas generator until firing and subsequent acceleration occurs. Usually there is enough energy available to break away the power turbine and its unloaded driven machine. Safety critical controls ensure that a shutdown or aborted start occurs in case of a stalled power turbine. A stalled power turbine upon failure of the gas generator to shut down has disastrous consequences in that the energy produced by the gas generator will destroy its blades and rotor.

9.2 Risk Assessment

Gas turbines are thermal machines. Many important parts, such as combustor liners, are therefore, subject to operating time dependent deterioration and predictable end of life. Reliability is achieved by carefully monitoring parameters such as:
- Damage to the inlet guide vanes to the HP turbine
- Combustion chamber damage
- Leakage in the HP combustion air compressor
- Damage to the compressor bearing
- General foreign object damage – FOD

Figure 9.5 illustrates gas turbine vulnerabilities.

Figure 9.6 shows a high-risk number for, as an example, the gas turbine–driven process compressor train sketched in Figure 9.7, compelling startup leaders to pay attention to careful planning, preparation and execution of pre-commissioning and commissioning tasks associated with it. See also Figure 9.8 representing a simplified flow diagram for a similar train in a gas drying process in midstream operations.

Commissioning and startup-related risks are described in Table 9.1.

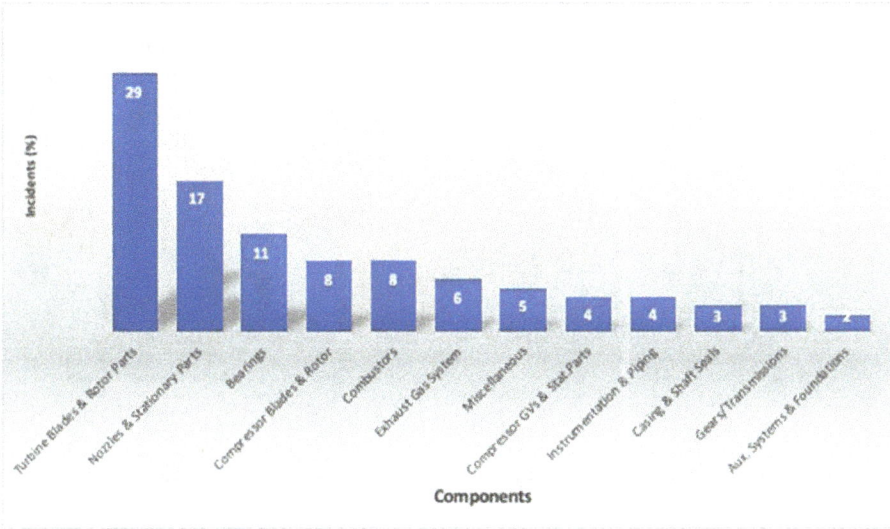

Figure 9.5: Typical mechanical drive gas turbine risk profile.

	1	2
		33,500hp/25 MW
		Set of Gas Turbine/
		CentCmp at 5,000pm
Rel. Risk #		840
Rel.Risk Level:		HIGH
Complexity		30
1.8		10
2.1		10
4.4		4
5.1		1
6.5		5
Severity		28
Table 2.4.1		8
Table 2.4.2		8
Table 2.4.3		8
Table 2.4.4		4

Figure 9.6: Relative intrinsic-risk assessment of a gas turbine–driven compressor train.

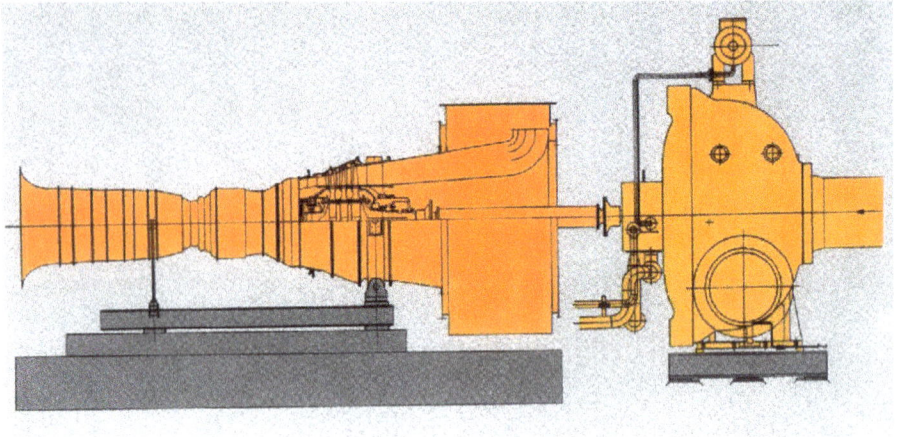

Figure 9.7: Typical gas turbine–driven pipeline compressor set (Cooper-Bessemer).

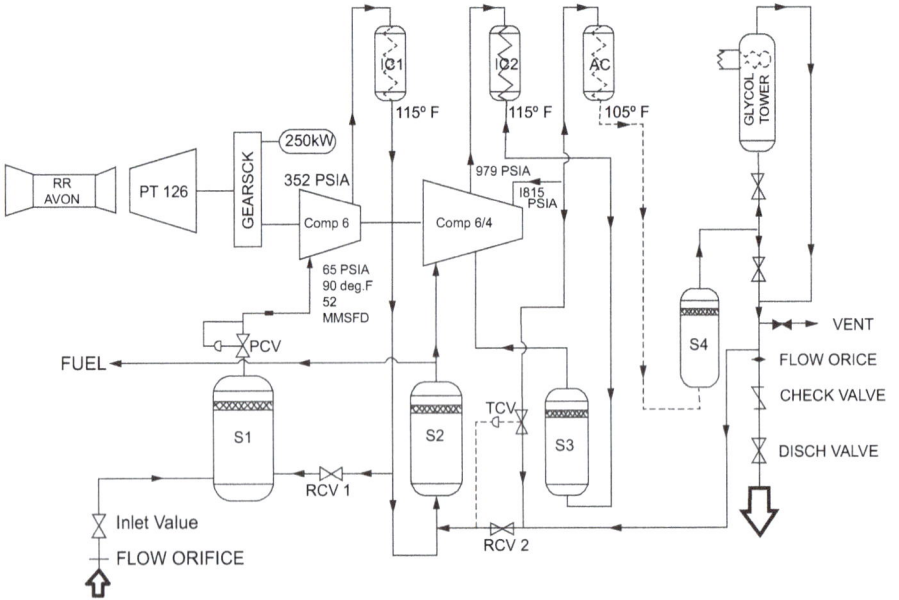

Figure 9.8: Simplified flow diagram of a natural gas drying process using a gas turbine–driven compressor train.

Table 9.1: Project-related risks and their mitigation.

Phase	Risk	Most probable cause[1]	Suggested actions/mitigating measures
Rigging and lifting	**Low** – Inappropriate rigging	QP	– GT packager will have well engineered lifting provisions for main engine and associated auxiliaries.
Handling, staging and storage protection	**Low** – Inadequate long-term storage	OR, HK	– The process starts with preparation for shipment by the OEM. – Package is covered against the elements; if not: – Clean dry storage required
Foundation and grouting	**Medium** – Insufficient skid stiffness – Poor grouting job	QP	– GT skid design review required. – Proper planning by the installer can avoid grouting problems. – Use wooden or metal template for foundation layout – Establishing the unit centerline is key. Most auxiliaries are referenced with respect to the unit's centerline
Piping	**High** – Ingestion of FO after cleaning – Fuel line leakage	OR, HK	– Cleanliness and assembly to the machine should follow practices established as part of the project execution agreement between OEM and the owner organization. – Make sure fuel line is hydro-tested and checked for leaks
Shaft alignment	**Medium** – Misalignment	QP	– Shaft alignment starts with leveling the unit per OEM guidelines.

(continued)

Table 9.1 (continued)

Phase	Risk	Most probable cause[1]	Suggested actions/mitigating measures
Lubrication system	**Low** – Contamination – Wrong Oil Specification – Mix-up	QP, DO, CO	– Cleanliness of the lubrications system is of utmost importance. However, GTs have small oil containments – frequently factory sealed. – Use OEM guidelines for flushing and cleanliness acceptance
Startup and initial operation	**Medium** – Too frequent start attempts and adjustments – Unforeseen environmental events	QP, EN	– Reduce the number of start attempts by planning proper unfired runs. – Awareness of weather conditions (hoar frost – Inlet icing). – Awareness if surrounding conditions (high winds – dust; season for fertilizer application).

[1] Refer to Figure 2.12

9.3 Commissioning and Startup

It is particularly important to pre-commission the gas turbine control system before the gas turbine itself is commissioned. It is strongly recommended that the original equipment manufacturer (OEM) field service representative (FSR) be fully engaged during this critical phase of the project.

Pre-commissioning and commissioning will follow a schedule as shown in Figure 9.9.

9.3.1 Final Check-Out Before Starting

It would be well to perform a final inspection of the unit before the initial startup. Here are our recommendations:
- Visual inspection and or borescopy of all accessible components for potential defects, FOD, OOD or DOD – Foreign Object Damage, Own Object Damage, Domestic Object Damage – or any kind of damage from shipping and assembly activities.
- Check for leakage such as lube oil, fuel gas, sealing compounds – include auxiliaries and peripheral items.
- If necessary, a compressor wash should be scheduled – heed instructions of OEM FSR.

9.3.2 Initial Startup

The initial startup of a mechanical drive gas turbine for a solo run is a part of the project's pre-commissioning plan. The reader might want to refer to the pre-startup procedure recorded in Table 9.2.

The on-site solo run and subsequent coupled load run are to demonstrate that the technical requirements of the gas turbine have been fulfilled. The startup team must assure that the unit is capable to run at the specified operating conditions.

During startup, actual run and shutdown, it is recommended to execute the following tasks:
- If the OEM agrees, run gas turbine up to maximum rpm without ignition using the starter, then measure rundown time for future reference.
- Inspect for leakages.
- If accessible, make note of signs of potential spark development in the exhaust.
- Check exhaust temperature thermocouple array.
- Ascertain vibration data is recorded and collected for evaluation.
- Ascertain other signals, such as temperatures, pressures, flow quantities are treated similarly.
- Measure rundown time after shutdown with agreement of the OEM FSR.

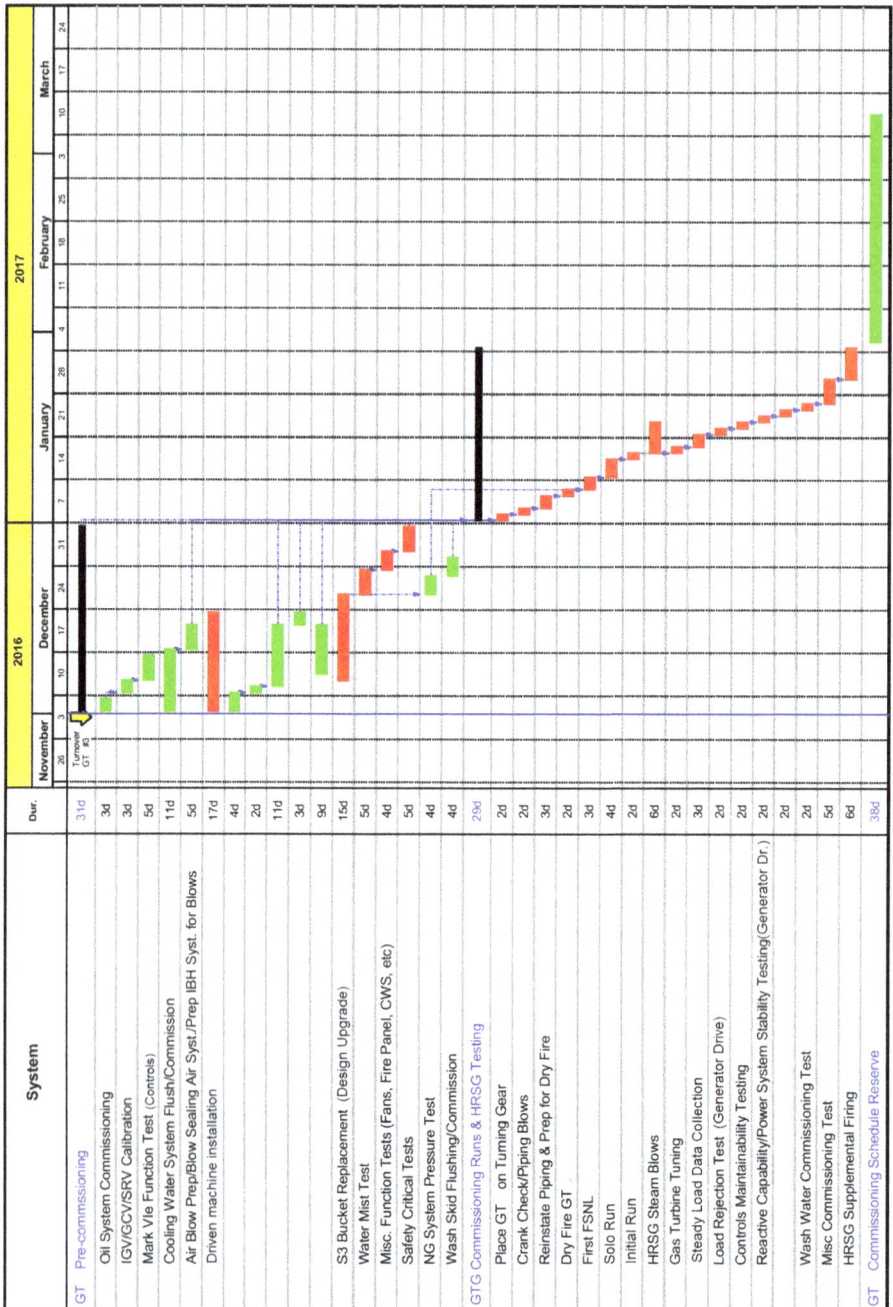

Figure 9.9: Commissioning schedule for a gas turbine–driven compressor train (with HRSG).

Table 9.2: Commissioning outline for a single shaft mechanical drive gas turbine.

Commissioning a Single Shaft Mechanical Drive Gas Turbine
Purpose of Document
This document is meant to be a day-by-day outline of key items that must be done to commission the XYZ GTGs. The machinery planner, with input from electrical and instrument planners, will use this outline to create a detailed commissioning/resources schedule.

Pre-commissioning of auxiliary systems will be per the OEM commissioning manual which has a semi-detailed section for each of the systems.

This outline is based on the OEM commissioning manual. The OEM commissioning manual is segregated into separate mechanical and electrical sections. In addition, some sections of the manual are not applicable to our installation. This outline is an attempt to combine as many mechanical, instrumentation and electrical tests as possible to minimize the total commissioning duration. The OEM commissioning manual will be "rearranged" in day 1, day 2, day 3, and so on tabs so that all pertinent OEM procedures/checklist and driven machine OEM required procedures/checklist that need to be completed on each day are located under the correct tab/day.

Auxiliaries pre-commissioning – Compressor Water Wash (WW) System
Goals:
-   Commission and test WW system to prepare for off-line wash prior to first fire

Preparations:
-   Air blow nozzles and piping downstream of MOVs to confirmed they are not plugged
-   Function test WW skid
-   Install WW ROs
-   Fill detergent tank with water for testing (400 L)

Run procedure
-   Commission WW system per OEM commissioning manual (remove MOVs and flush to ground)
-   Initiate off-line wash
-   Confirm 6 barg downstream of orifice (need temp pressure gauge connected manifold)
    -   *GTG2 original 15.93 mm RO /8 bar at pump discharge/4 bar at manifold*
    -   *GTG2 & 3 Increased to 21.5 mm RO/7.9 bar at pump discharge/6 bar at manifold*
-   Observe spray pattern
-   Initiate on line wash
-   Confirm 7 barg downstream of orifice (need temp pressure gauge connected to manifold)
    -   *GTG3 original _____ mm RO /8.2 bar at pump discharge/7 bar at manifold*
-   Observe spray pattern
-   Inspect inlet plenum, clean IGVs, verify IGV calibration
-   Hand clean inlet duct
-   Final inlet inspection and closure – *GTG2 complete 3/15, GTG3 completed 8/15.*

Table 9.2 (continued)

Post run:
– Correct any deficiencies found during commissioning
– Fill detergent tank with detergent

Auxiliaries pre-commissioning – Water Mist Fire Suppression System
Goals:
– Verify water mist fire suppression system works per design (water flow, duration, nozzle locations)
– Commissioning according to OEM procedures.

Post run procedure immediately prior to first "fail to fire" test:
– Line up water mist fire suppression WM in accordance with OEM procedure.
– Execute pre-startup safety critical test procedures, as required.

Auxiliaries pre-commissioning – oil system
– Commission LO system per OEM manual.
– Flush lube oil system with required temporary screens and jumpers.
– Re-establish permanent lube oil connections without temporary screens.
– Install new lube oil filters.
– Start main lube oil pump. Confirm all associated pressures and flow rates.
– Slowly close the block valve on the lube oil pressure transmitter and then bleed the pressure. Confirm that the auxiliary lube oil pump starts at the required set point.
– Adjust lube oil mist eliminator vacuum in accordance with OEM requirements.
– Confirm all operating parameters of pressure, level and flow rate.
– Collect oil sample for analysis and confirmation of suitable cleanliness and lubricant physical characteristics.

Auxiliaries pre-commissioning – cooling water system
– Flush in accordance with OEM procedures
– Fill with demineralized water circulated and dump a couple of times until water is clean.
– Circulate system and check iron content, fill/drain the cooling water as required.

Auxiliaries pre-commissioning – fuel gas system
– According to OEM requirements blow clear all fuel gas lines.
– Confirm cleanliness of fuel gas filters. If required install new fuel gas filters.
– Leak test of fuel gas system.
Perform tightness check of all fuel gas connections inside compartment that were not pressure tested (critical flanges).

Auxiliaries pre-commissioning – control, monitoring and HMI systems
– This is a complex and critical process. Modern gas turbines cannot be operated until these critical systems have been pre-commissioned. Follow the pre-commissioning procedures of the OEM.

Table 9.2 (continued)

Gas turbine commissioning

Goals:
- Verify unit turns.
- Test emergency trip circuits (local emergency stop, low-lube oil trip).
- Check for any rubs in turbine and driven equipment.
- Slow roll gas turbine per OEM recommended time to relieve any rotor bow due to storage.
- Confirm that instrumentation, monitoring, controls and HMI pre-commissioning requirements complete:
    - All software upgrades complete
    - Safety critical checks completed
    - Verify instrumentation lined up
    - Functional checks complete
    - All communication links from gas turbine systems to the facility DCS are verified.
    - All Auxiliaries pre-commission tests complete
    - Cooling water system
    - Lube oil system
    - Cooling and sealing system
    - Compartment ventilation
    - Compressor wash system
    - Water mist fire suppression system
    - Starting system
- Rack in starting motor

Initial run "fail to fire" run test

Goals:
- Verify function of fuel gas valves proper sequence of operation.
- Check for fuel gas leaks in piping downstream of the fuel gas stop valves.
- Execute fire drill to ensure proper emergency response.

Preparations:
- Ensure fuel gas BLOCK VALVES are closed so that no fuel gas will be introduced during this "fail to fire test" of the complete system.
- Data acquisitions files configured for startup.

Run procedure:
- Select auto/start and accelerate unit up to firing speed
- Confirm gas turbine will complete the fuel gas cycle checks.
- Gas turbine will abort the startup cycle on "failure to fire" and shutdown.
- Confirm monitoring, control, HMI and DCS system pertinent data.

First full speed no load operation

Goals:
- Test startup sequence
- Verify unit critical parameters at FSNL (vibrations, temperatures, etc.)
- Test proper changeover of ventilation fans and gather DP pressure data
- Test overspeed trip function

Table 9.2 (continued)

Preparations:
- Ensure fuel gas BLOCK VALVES are open.
- All SHP blow down facilities available

Run procedure:
- Start unit to FSNL following OEM startup procedure (select auto and give a start order)
- Confirm successive control sequences: purge, fire, warm up, then acceleration
- Confirm presence and stability of flame from all flame detectors
- Verify proper: IGV angle, vibration level, exh temps, oil pressure, fuel flow, etc.
- Trend last chance filter DP
- Maintain at FSNL for 90 min or until unit stabilizes
- Check for exhaust system leaks
- General check of machine
- Overspeed trip function check in accordance with OEM procedures. The gas turbine will trip on overspeed.
- Leave unit on automatic cool down.

Post run:
- Correct any deficiencies found during run
- Check filter house for loose filters, etc.
- Inspect transfer line Last Chance Filter. Remove if clean, reinstall if debris found.

TUNING DRY LOW NO$_x$ COMBUSTION SYSTEM
Lean–lean test and removal of last chance filters
Goals:
- Tune DLN combustion system

Run procedure:
- This is a critical and complex process. Follow the OEM procedure.

Post run:
- Correct any deficiencies found during run

Steady load data collection
Goals:
- Collect steady load data.

Run procedure:
- Start gas turbine in automatic control from the facility DCS.
- Initial assessment of GT performance.

Post run:
- Correct any deficiencies found during run.
- Collect lube oil sample and send to the lab to confirm cleanliness and all typical oil physical properties.

9.3.3 Startup Control System Revisited

The startup system is used to crank the gas turbine without any combustion initiated. In fact, the fuel gas stop valves will be closed during a significant part of this startup cycle. The startup cycle is controlled by the HMI.

The startup system will initiate the turning or cranking of the gas turbine train. This portion of the cycle is known as the purging cycle. It ensures the gas turbine has been purged of any unwanted gas. It is also a time when early checks of the lube oil system will be performed. The two-fuel gas-stop-control valve integrity check will be automatically performed at this time. Once all of these automatic system checks have been performed, including all associated permissive logic requirements have been met, the HMI system will communicate an "all clear" – that the unit is prepared to start, the "*Ready to Start*" display will appear.

With local control, pushing one of the following buttons will initiate a start:
1. Idle
2. Load

For example, a master contactor function will accomplish:
- Secondary auxiliary loop pump starter energized
- Instrument air solenoid valve energized
- Combustor shell pressure transducer line drain solenoid valve energized.
- Instrument air supply established

When the auxiliary lube pump builds up sufficient pressure, the circuit to engage and energize the turbine starting gear starter will be completed. Thirty seconds are usually allowed for the lube oil pressure to build up or the unit will shut down. Upon signalling that the cranking gear has been engaged, the sequence will continue. The turning gear motor will be turned off at usually about 15% of turbine rotor speed. The gas turbine must achieve the minimum speed required for the axial compressor to provide the required combustion air flow to ensure the gas turbine will have a stable combustion. Once the machine has achieved the minimum speed the ignition circuit is energized by activating ignitors in the combustion chambers. A feedback signal triggers the fuel control system to take the lead. The fuel gas-stop -control valves will open. Ignition proceeds through crossfire tubes connecting the combustion chambers. Fire eyes fitted into then combustor chambers serve the purpose of ensuring that all combustion chambers are now fully functioning.

The HMI system will ramp up the machinery train to its designed minimum operating speed. As the machinery train accelerates the starting system will disengage and shut down. All critical instrumentation and control algorithms will be automatically confirmed to be performing according to the OEM's design.

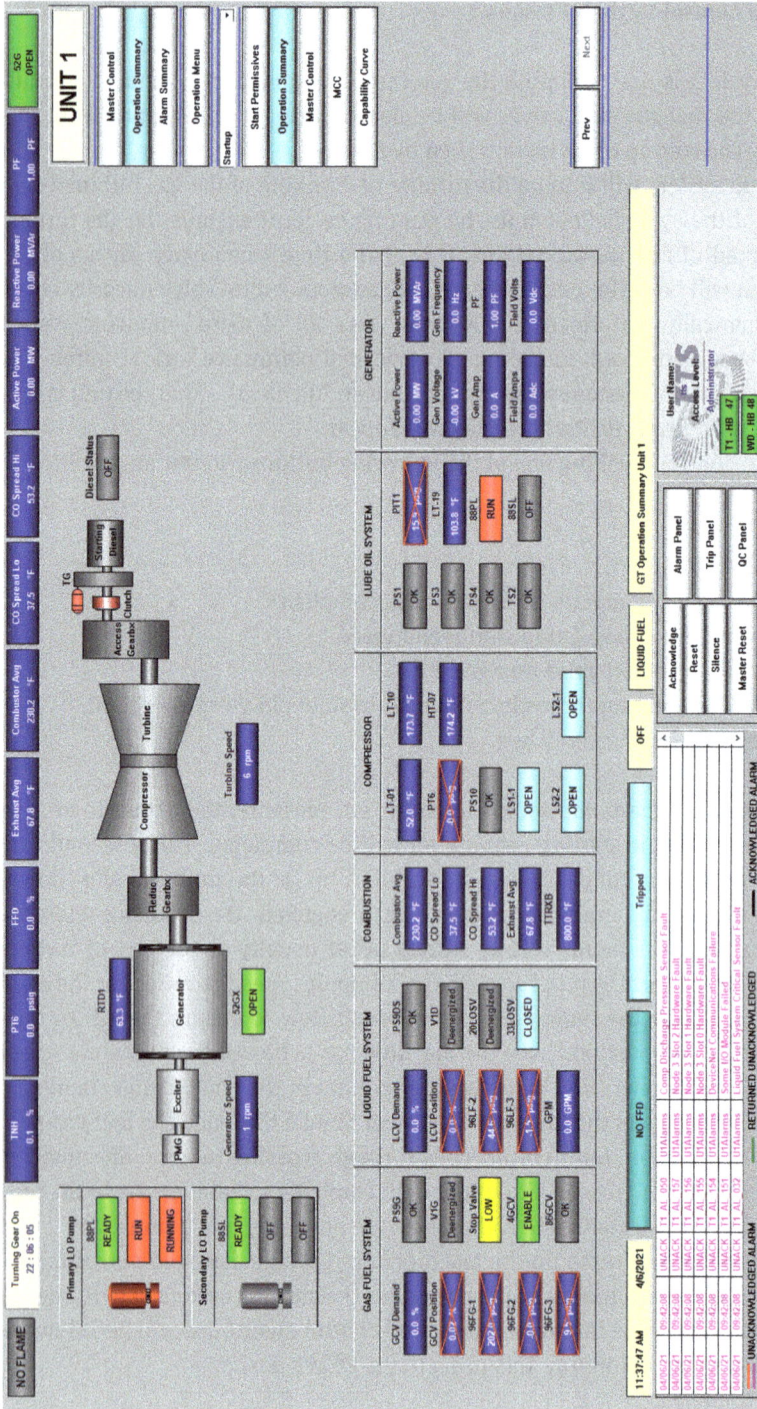

Figure 9.10: HMI display for a gas turbine–driven electrical generator set. (Courtesy of TTS).

PROCESS VARIABLE	DATE:							
	TIME:							
Gas Turbine RMP LP/HP (Max.5950/7250)								
Axial Comp. Discharge Pressure								
Control Oil Pressures: VCO/NCO								
AMI/MAMI								
AM2/MAM2								
Exhaust Temp. Control (EP 65)								
Exhaust Temp. Protection Left/Right								
GT Cooling Water In/Out								
Lube Oil Filter Pressure In/Out								
Lube Oil Header Pres/Temp								
Seal Oil to Gas ΔP/Oil Inlet Temp								
Oil Drain Temperatures:								
No. 1 Lube Oil Gear Drain								
No. 2 Lube Oil/Seal oil								
No. 3 Lube Oil/Seal Oil								
Shaft Position								
Vibration Comp.								
Vibration Gear								

	PROCESS VARIABLE								
	1. Spare								
	2. Turb. Bearing # 1-Temp								
	3. Turb. Bearing # 2-Temp								
	4. Turb. Bearing # 3-Temp								
	5. Turb. Bearing # 4-Temp								
	6. Turb. Bearing # 5-Temp								
	7. Load Gear Brg-Temp								
	8. Load Gear Brg. Temp								
	9. Load Gear Brg. Temp								
TEMPERATURE RECORDER	10. Load Gear Brg. Temp								
	11. Gas Comp. Inlet No. 1								
	12. Gas Comp. Radial Drain No.1								
	13. Gas Comp. Radial Drain No.2								
	14. Gas Comp. Discharge No. 2								
	15. Gas Comp. Thrust Brg.-Drain								
	16. Lube System Temp.Turbine Header.								
	17. Gas Turbine Comp. Inlet								
	18. Gas Turbine Comp. Discharge								
	19. Gas Comp. Seal Oil Drain No. 1								
	20. Gas Comp. Seal Oil Drain No. 2								
	21. Fuel Gas Manifold								
	22. Turb. Temp. Wheel Spare 1st Stg. Forward Outer.								
	23. " " " " "								
	24. Turbine Exhaust Temp.Selected								
	Gas Turbine Fuel Gas Flow								
	Comp. Pressures Suct/Disch.								
	Comp. Temps Suct/Disch.								
	Compressor Gas Flow								

NOTES: Deviations from below should be reported to the instrument depit:

OIL DRAIN TEMPERATURE POINTS

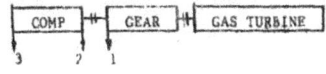

COMP	GEAR	GAS TURBINE
3	2	1

Exhaust Temp. Signal Output		Function
Exhaust Temp. Protection -		
950°F	3.0 psig	
990°F	7.8 psig	Alarm
1010°F	10.2 psig	Trip

Figure 9.11: Generalized log sheet for gas turbines.

The gas turbine is now operating and ready to accept the attached load; operators at this time depend upon the gas turbine HMI display information – see Figure 9.10.

A comprehensive compilation of data requirement for performance evaluation is shown in Figure 9.11., a copy of a log sheet for a heavy-duty industrial gas turbine train stemming from a time where machine-man-interfaces were just on the horizon.

9.3.4 After the Initial Run

The following tasks should be performed:
- Inspection of the combustion air compressor for FOD.
- Inspection of the exhaust path for unusual deposits such as potential metal spray.
- Documentation and review of typical borescope images of important components before and after the initial run.
- Examination and analysis of the lube oil for temperature related changes, such as caused by an oil fire, and for unusual suspended material like metal shavings or contaminants, for example, detrimental labyrinth abrasion particles.
- Examination of filter residue from, for example, machining chips, sealant remnants and particle from cleaning materials.
- Examination of magnetic plug deposits such as magnetic materials of construction originating from antifriction bearings, gears and shafts.
- Inspection for indication of leakage from lube oil, air and hot gases.

References

1 Peter König / Axel Rossmann, RATGEBER FÜR GASTURBINEN-BETREIBER, ASUE-Schriftenreihe, VULKAN-VERLAG ESSEN, 1999.
2 Mid-stream is the part of the hydrocarbon processing business concerned with gathering and transporting oil and/or gas.

Bibliography

[1] H.P. Bloch and F.K. Geitner, Series *Practical Machinery Management for Process Plants*, Volume IV, *Major Process Equipment Maintenance and Repair*, Second Edition, an imprint of Gulf Professional Publishing., at www.elsevier.com 1997, and the preceding edition, Hardcover ISBN: 9780884156635, eBook ISBN: 9780080479002, 700 pages.
[2] Claire Soares, Gas Turbines: A Handbook of Air, Land and Sea Applications, 2nd Edition, 2015, Butterworth-Heinemann, an imprint of Elsevier, Oxford, OX5 1AJ, UK; Waltham, MA 02451, USA, ISBN 978-0-12-410461-7

Chapter 10
C&S of Internal Combustion Engines

10.1 Introduction

Internal combustion engines in the processing industries are characterized by the type of fuel used and the method of fuel ignition. Many other designations are used grouping engines according to their speed, cycle arrangements, mechanical configuration and other design characteristics. Power output ranges between 500 kW or 670 hp and 7.5 MW or 10,000 hp with speeds varying from 300 to 1,800 rpm. The most prevalent power range is 1.0–2.0 MW with speeds of 1,000–1,800 rpm. Engines in this power range also use predominantly natural gas as fuel. We shall therefore limit our discussion of internal combustion engines in the processing industry to gas-fired machines.

Natural gas engines (NGE) are commonly used to power compressors in the upstream and midstream business of the petrochemical industry such as gas gathering and transmission. Other applications are standby electric generators, fire water and irrigation pumps. They are increasingly being used to power primary cogeneration electrical power plants. The main advantages of a natural gas engine over a diesel engine are the lower exhaust emissions of nitrogen oxides (NO_x), carbon monoxide (CO), particulates and in some cases, lower fuel costs. They have matured to a high level of reliability combined, however, with high maintenance intensity.

Gas engines are also extensively regulated having to follow a complicated environmental protection regime – almost everywhere in the world. The member of the startup team assigned to tracking the commissioning of a gas engine train must be familiar with the rules[1] in order to avoid any kind of project delaying surprises.

10.1.1 Speed Ranges

Most process plant engines are used to drive equipment with a limited range of speed requirements, and which can be selected to operate near the point of highest efficiency. Speed increasing or decreasing gears may be used to match an engine with a particular service. Internal combustion engines are classified according to speed in the following broad categories:
- High speed – above 1,500 rpm
- Medium speed – 700 to 1,500 rpm
- Low speed – below 700 rpm

High-speed engines can offer weight and space advantages but will usually require more maintenance than a medium or low speed engine. High-speed engines are

https://doi.org/10.1515/9783110701074-010

often selected for standby or intermittent applications. As a general rule, the lower the speed the longer the service life. Although internal combustion engines are usually selected to run over a limited speed range, they will operate well over large speed ranges just as an automobile engine does.

10.1.2 Engine Types and Characteristics

Spark ignition and compression ignition are the two methods of initiating combustion used in reciprocating internal combustion engines. In practice the ignition method also defines the fuels or range of fuels used. Liquid fuel is typically used in smaller engines of 500 kW up to 1 MW, whereas natural gas is predominantly used by machines with power ratings above 1 MW.

10.1.2.1 Spark Ignition
Natural gas, liquefied petroleum gas (LPG) or gasoline are the fuels used in spark ignition engines. They are often referred to as gas engines or gasoline engines and resemble in appearance – except perhaps for size – and operation the engines used in automobiles. High voltage electrical energy fires one or more spark plugs per cylinder to ignite the air/fuel mixture. Most spark ignition engines can be easily modified to burn any of gaseous fuels. In that case, the fuel delivery system is the only part of the engine requiring significant changes. Figure 10.1 shows a typical large gas fueled engine.

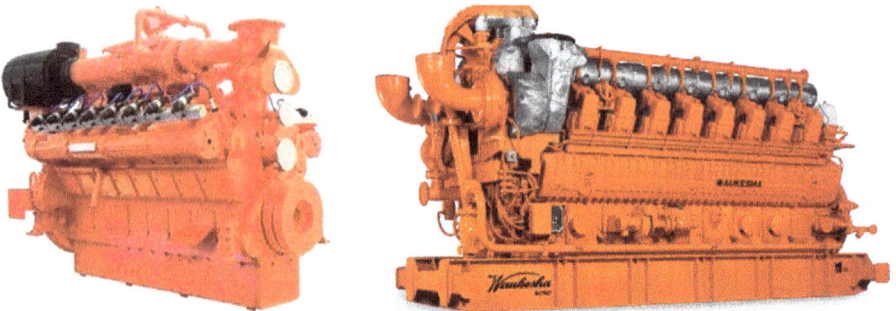

Figure 10.1: Large gas engine.

10.1.2.2 Compression Ignition (Diesel)
Engines that use heat of compression as the ignition source are almost always referred to as diesel engines. A broad range of liquid fuels can be burned in a diesel engine provided proper attention is paid to the handling and preparation of the fuel

as well as to the design of the engine. The type and quality of the fuel can have a significant effect on the service life of the engine.

10.1.2.3 Dual Fuel
Engines may operate in one of two modes. One mode is as an ordinary diesel engine. It may also operate on a gaseous fuel with a pilot injection of liquid diesel fuel for ignition. The pilot fuel provides less than 10% of the total fuel energy at full load.

10.1.2.4 The Four-Stroke Cycle Engine
Is used by most spark ignition engines. It means that the power stroke is achieved in two crankshaft revolutions and consists of the following piston strokes:
1. An intake stroke to draw the fuel/air mixture into the engine cylinder.
2. A compression stroke which raises the pressure and temperature of the mixture inside the cylinder.
3. The expansion or power stroke from the ignition and combustion of the fuel mixture.
4. An exhaust stroke to free the cylinder of combustion products.

The four-cycle diesel engine operates in a similar fashion. During the intake stroke, only air is introduced into the cylinder. Compression of air alone causes a higher temperature to be reached in the cylinder. The fuel is injected into the cylinder at the very beginning of the expansion stroke and spontaneously ignites.

10.1.2.5 The Two-Stroke Cycle Engine
was developed to get a higher output from the same size engine. This cycle is applicable to both compression ignition and spark ignition engines. The two-stroke cycle is completed in one revolution of the crankshaft and consists of two piston strokes:
–   The compression stroke
–   The expansion stroke

Combustion air intake occurs at the end of the expansion and the beginning of the compression stroke. Ignition and combustion take place at the end of the compression and beginning of the expansion stroke.

A particular well proven design deserves mention here: It was the integral gas engine compressor with a V-shaped power piston arrangement as shown in Appendix 2.A, Figure A.1. The key feature was the achievement of compactness by an articulated connecting rod arrangement which allowed two power piston connecting rods to drive onto one master compressor rod for each throw of the crankshaft.

10.1.2.6 Supercharged Engines

utilize a compressor to increase the density of the combustion air before it is inducted into the cylinder. Supercharging increases the power output from a given cylinder size by increasing the engine mean effective pressure.

Two types of supercharging are common: mechanical compressors driven by an engine auxiliary output shaft or a separate prime mover and exhaust turbine driven compressor which obtains its power from expansion of the engine exhaust. This later type of supercharger is commonly called a turbocharger.

The power delivered by an internal combustion engine is directly related to atmospheric conditions. Operation in areas of low atmospheric pressure (high altitudes) will reduce the power output. High inlet air temperature will also reduce the power output.

10.1.2.7 Engine Energy Balance

A gas engine converts the combustion energy in the fuel to mechanical power and heat. The combustion energy is usually distributed as follows:

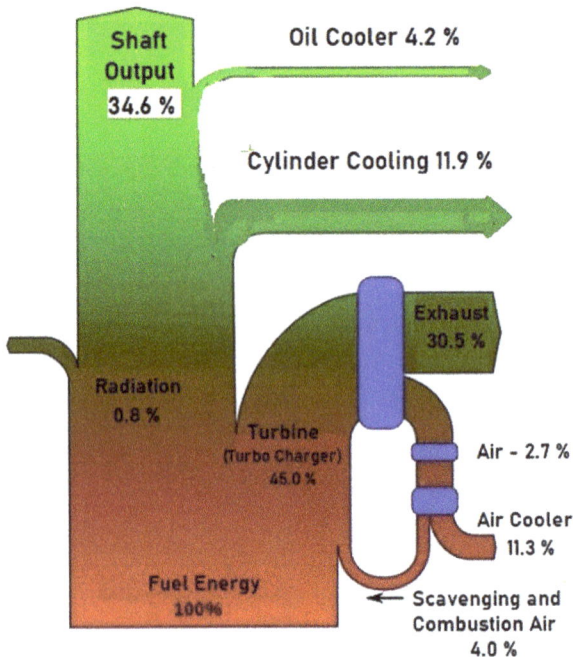

Figure 10.2: Sankey diagram for a typical gas engine.

10.2 Risk Profile

Internal combustion engines exhibit predominantly wear related failure patterns and very few random defects. Figure 10.3 shows bearings as the predominant failure source. However, with cleanliness of the lube oil system and a suitable foundation, the risk of bearing failures can be mitigated or virtually eliminated. There are examples of gas engines that, with prescribed maintenance interventions, have lasted through some 80,000 operating hours without a forced outage. This would be equivalent to 10 years of operation taking maintenance related downtime into account.

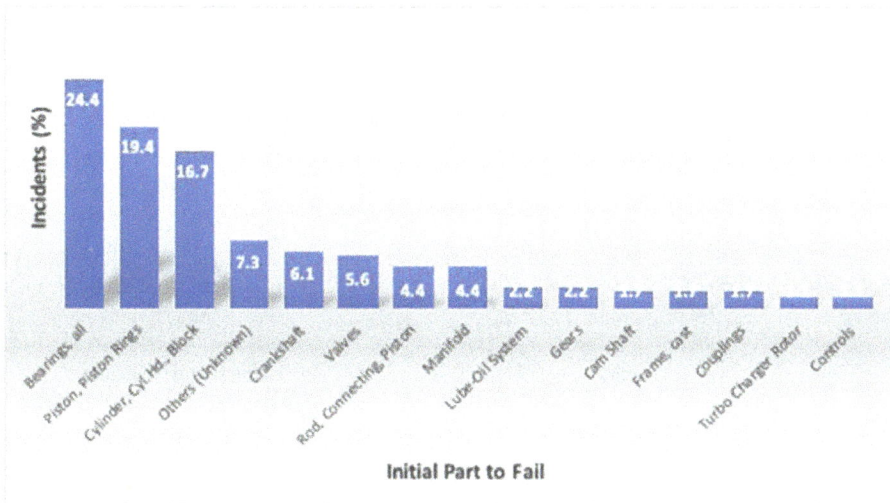

Figure 10.3: Risk profile of large internal combustion engines.

On the other hand, there are project related risks that should be recognized and addressed. Table 10.1 is the result of the startup team's continuous risk assessment.

10.3 Installation, Commissioning and Startup

10.3.1 Installation

The OEM field service representative is usually responsible for preparing the on-engine components and systems. Together with the startup team including the owner's machinery specialist, he should be held to stay abreast of the five project phases:

Table 10.1: Large engine potential risk analysis record.

Phase	Risk	Most probable cause[1]	Suggested actions/mitigating measures
Rigging and lifting	**High – large engine** – Improper lift	QP	– Review lifting plan. If none was provided, insist on one. – Review R&L company's record.
Handling, staging and storage protection	**High** – Overseas destination – Not prepared for long term storage	OR, HK	– Provide proper storage space – dry cool environment.
Foundation and grouting	**High** – Improper foundation – No or improper grouting – Inadequate skid design	QP, EN	– Have grout manufacturer rep at site during pour – Use grout with the proper exotherm formulation suited for the climate conditions at site. – See also Figure 10.6.
Piping and tubing	**High** – No or improper cleaning – Ingestion of foreign objects after cleaning	OR, HK	– Cleanliness and assembly to the machineshould follow practices established as part of the project execution agreement between OEM and the owner organization.
Shaft alignment	**High** – Unusual coupling – installer is unfamiliar with cplg.	QP	– OEM FSR interview
Lubrication system	**High** – Wrong oil chosen – Mix-up – Contamination	QP, DO, CO	– Cleanliness of the lubrications system is of utmost importance. Review recorded work done in the OEM shops.
Startup and initial operation	**High** – Ignition not adjusted – Backfire	FC, OR, QP	– Dry trial runs – Review pre-start checklists

[1] Refer to Figure 2.12

10.3.1.1 Rigging and Lifting

Again, rigging and lifting of heavy loads such as large engines is subject to state or local government inspections and permits. Reputable engine manufacturers would have their field service representative on site to help develop the lifting plan and supervise its execution. Figure 10.4 illustrates a critical lift of an internal combustion engine onto a base plate to be bolted down and coupled to the driven machine, a high-speed reciprocating compressor to be used in the midstream business of the HCPI.

Figure 10.4: Safe lift of a major gas engine onto its base (courtesy of CES L.P., www.ceslp.net).

10.3.1.2 Receiving, Inspection and Storage Protection

This step is for obvious reasons well understood and is most of the time successfully taken care of. By convention, engines received from the factory are internally protected for up to six months for indoor storage. Usually, the vendor or OEM and purchaser will consider environmental conditions. Consequently, engines stored outdoors or in a humid environment may require re-preservation. The purchasing documents should clearly spell out storage protection requirements – usually in increments of six months. However, a significant risk arises when suppliers and owners are forced to extend the storage time of engines due to project scheduling difficulties. If the storage period exceeds six months, the engine should receive additional storage preservation. In this case, the following procedure should be followed:

A. Remove engine from storage and set on temporary supports and restraints.
B. Begin commissioning procedures to ready it for startup.
C. Startup engine and operate until it is hot.

D. Mix an inhibitive-type preservative oil with the engine lubricating oil in the proportions recommended by the manufacturer of the preservative oil. Cooling water used in this run should have an inhibitor added in accordance to the OEM's instructions.
E. Remove air cleaner, then using a manually operated sprayer, squirt can or other means, inject preservative oil of a suitable type into the air intake while the engine is running. Approximately 1 min is ordinarily adequate. If possible, stop engine by slugging enough oil through intake to stall. Continue injecting oil until the engine stops turning.
F. Drain oil and water while hot. If extra protection is desired, the rocker arm covers may be removed and a quantity of preservative oil poured over the rocker arm and valve mechanisms.
G. For gas engines not stopped by slugging, remove spark plugs and squirt or spray several teaspoons of preservative oil into each combustion chamber. Coat spark plugs and reinstall.
H. Wipe engine clean and dry. Apply wax type tape or like material to all openings such as intake, air cleaners, exhaust outlets, breathers and open line fittings.
I. Relieve tension on belts. This is important because continual tension on belts without the working action that occurs in normal operation causes deterioration of the rubber.
J. Apply a coating of heavy preservative compound with brush to all exposed machined surfaces such as flywheels.

While this seems excessive and time consuming, it is well worth the effort. The writers have lived through several situations where vacillation in the case of planned engine storage time extension has resulted in costly rebuilds.

10.3.1.3 Foundation, Base Plate and Grouting

An important step is placing an engine on its foundation. While this sounds straightforward, this step is fraught with hidden risks. As we will see in the following narrative, there are many intricate details to wrestle with. In Chapter 5, we discussed the importance of foundation and machine base grouting in general terms. We continue the discussion here because reciprocating machinery such as internal combustion engines and displacement compressors are particularly vulnerable to inadequate foundation and grouting situations. The finished foundation may pass all possible inspections, but if it is not designed and constructed properly, its flaws will appear years later causing failure and unscheduled downtime.

The writers have collected and partially encountered the following risks in connection with foundation and grouting of reciprocating equipment:
– Improperly designed skids and steel support structures
– Bad crankshaft webs

- High bearing oil temperatures
- Excessive forces from rod run-out/rod drop on reciprocating compressors
- Excessive frame movement
- High vibrations
- Loose or cracked chocks
- Inability to maintain anchor bolt tightness
- Inability to hold alignment

On packaged engine and compressor sets – see Figure 10.5 – used in the upstream and midstream hydrocarbon processing industry, there is no such challenge because machined mounting pads are provided on package skids,[2] even though one has to still find a good place for the skid. Yet, the writers know of an incident involving 60 packaged high-speed engine-detachable-compressor units that could not go on line due to high vibrations. Poor quality of epoxy[3] grout or installation was the suspected cause even though initially placed strengths were acceptable. Skid foundations and grouting would follow the principles illustrated in Figure 10.6.

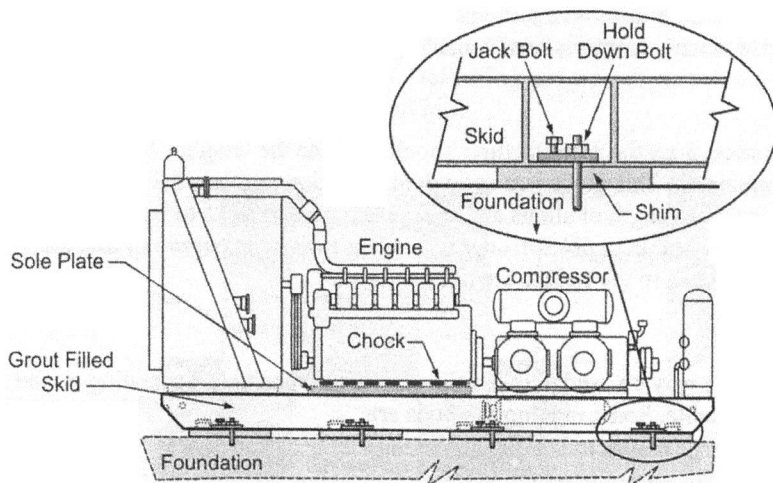

Figure 10.5: Grouting an engine-compressor package in the gas field (adapted from Reference[4]).

Frequently, in process plants and particularly in downstream hydrocarbon processing facilities, engines are placed on reinforced concrete foundations. As a rule, the OEM provides the foundation drawing with the engine outline drawing and all necessary dimensions required to assure an adequate foundation and the minimum space required between buildings or other appurtenances to perform maintenance after the unit is installed. Further, the OEM will furnish the information on unbalanced couples, forces and torque reaction for the engine to assist in the design of a proper foundation. However, the manufacturer usually does not accept the responsibility for

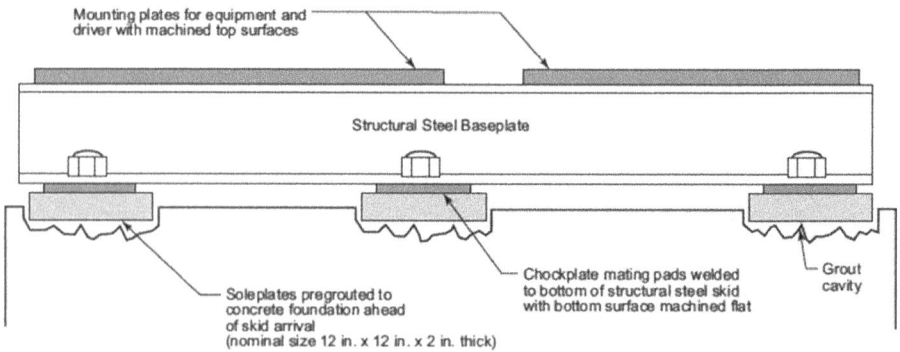

Figure 10.6: Typical mounting plate arrangement for baseplate or skid mounted equipment (adapted from reference[5]).

the design or size of the foundation. It is therefore, at this interface, where the risk of inadequate or incorrect information transmittal with detrimental consequences arises.

The need for excavation work varies considerably with soil condition. If in doubt about the depth required to attain a good footing, soil tests have to be performed. Sometimes, it is advisable to obtain professional service advice regarding geotechnical soil studies, the type, size, depth and reinforcement bar density of the foundation.[6,7] Where more than one engine is to be located in the same area, a mat under all foundations is recommended.

The engine foundation should be isolated from buildings, building foundations and flooring to prevent any vibration being transmitted to the building.

After the excavation has been completed and the foundation forms built, one must refer to the foundation drawing and prepare a template to locate the foundation bolts. Place the template in position and align it for the desired location of the engine. A foundation bold center must not deviate more than 0.125 in. or 3 mm from the drawing.

Before positioning the foundation bolts, slip short pieces of pipe over them as shown in the foundation drawing and wedged them so that each bolt is centered in the pipe. The space between the bolt and pipe will allow for any slight misalignment of the bolts. The template must be secured firmly so it will not move while the concrete is being poured.

The foundation should be poured without interruption to the desired height except in the area of the sole plates. This area is called a "grout pocket" and must be formed to provide the pocket depths – usually 3–3/4 in. or 9.5 cm – as shown on the foundation drawing. Similarly, when the foundation drawing call for rails instead of sole plates, space has to be provided for them.

Seal the foundation after the concrete has cured for at least seven days to prevent oil and water from deteriorating it. Seal the entire top surface of the foundation except areas to be grouted and 1 ft or 30 cm down its sides to provide a cap. Apply sealant according to the manufacturer's instructions. Also seal the sides of the grout pockets to one inch or 25.4 mm below the final finished grout level.

10.3.1.3.1 Grouting Sole Plates

There are two basic approaches to engine sole plate grouting. One, the so-called pre-grouted method where the engine is set on pre-grouted soleplates; and two, setting the engine on sole plates that are not pre-grouted. Appendix 10.A explains these two methods in detail.

10.3.1.4 Piping System

There has to be assurance and evidence that the piping for bearing lube oil and cooling water was cleaned and inspected according to procedures established by the project team. Refer also to Chapter 5.

10.3.1.5 Shaft Alignment

At this stage the engine has been set according to the outline drawings and the driven equipment has been aligned to it. The team must assure correct relative growth data pertaining to the transmission equipment and load has been provided by the OEM and accepted by the commissioning team. Further, the owner's engineer, together with the OEM field service representative should have agreed on the method of turning or barring over the engine while doing the alignment. Similarly, the alignment procedure accepted by the project team will be followed and records will be established accordingly.

10.3.1.5.1 Coupling

The choice of coupling on internal combustion engine driven trains is usually made by the driven machine OEM. Reciprocating compressor drives are often equipped with metal-disk couplings even though the writers consider couplings with elastomeric elements more suitable for this particular application.

10.3.1.6 Lubrication System

Lubrication systems on large internal combustion engines are frequently part of the vendor furnished package. Most engines have received a factory acceptance test (FAT) and the owner can hope in most cases for a clean lubrication system. Nevertheless, the startup team would be well advised to demand oil flushing of the system, especially when there is interconnecting field piping is involved.

10.3.2 Commissioning

When scheduling engine commissioning and startup it is often not obvious that the OEM FSR is in charge of assembling engine components that ship separately from the unit. These items could be the flywheel, electrical components, lubricating piping and other small parts that were removed after the factory test.

The OEM FSR is involved in commissioning the following systems:

10.3.2.1 Lubrication System

Oil sump capacities of 14–6,000 L with an average of 300–800 L or 80–200 US gallons are most common. Typically, a built-in lubricating oil pump driven directly off the forward end of the crankshaft circulates oil through a cooler, filter and strainer and then into the engine pressure system. This is illustrated in Figure 10.7 a simplified flow schematic for a standard engine lubrication system. The full flow oil filter is usually located as close to the engine as possible. Piping between the engine oil inlet and the filter must be stainless steel and is to be flanged to allow accessibility for inspection and cleaning. An electric or gas driven pre-lube pump permits pre- and post-lubrication of the engine and the turbocharger.

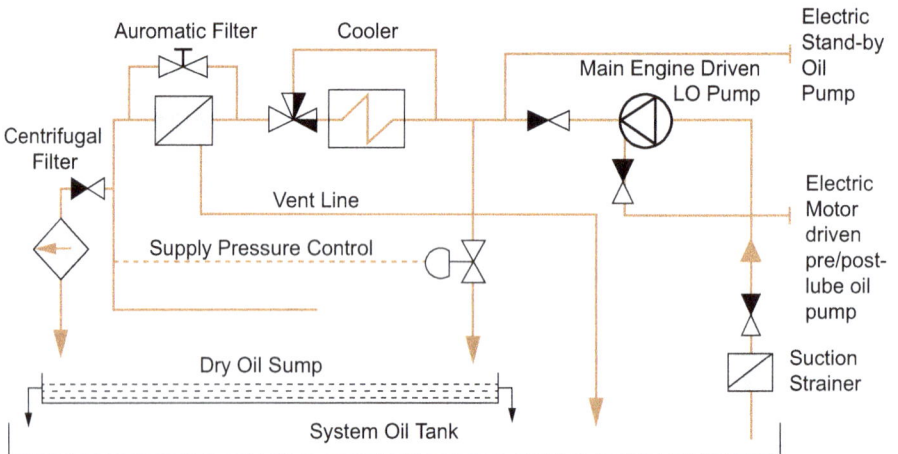

Figure 10.7: Simplified flow schematic of an engine lubrication system.

In four-stroke engines, for example, the clean oil system supplies:
a. Cylinder head.
b. Camshaft including camshaft bearings.
c. Main bearings.
d. Connecting rod.

e. Crankshaft.
f. Valve gear – tappets, push rods, rocker arms and other sliding parts.
g. Turbocharger.

When flushing the engine oil system, a standard flushing procedure should be used. The reader might want to return to Figure 10.7 and examine it more thoroughly: Where would one break the piping connections in order to establish a temporary flushing circuit without contaminating the bearings? It would be well to install a temporary line filter – as shown before in Figure 5.10 – to test system cleanliness, before connecting the oil piping to the machine. The main bearings will be fitted with temporary filters as illustrated in Figure 10.8. They serve the purpose of making sure the oil entering the bearings is clean after the system has been flushed.

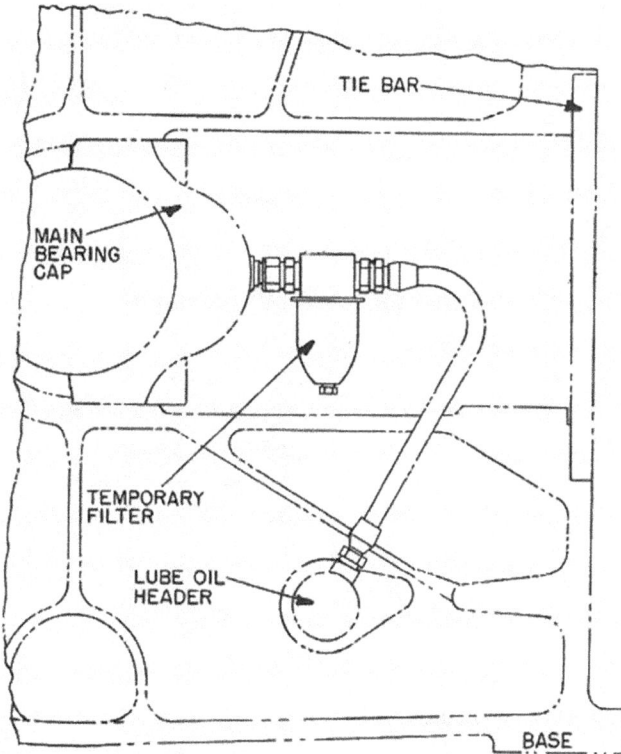

Figure 10.8: Main bearing temporary oil filter (CES/Cooper-Bessemer).

10.3.2.2 Fuel Gas System

The fuel gas piping system consists of:

a. A variable fuel gas pressure regulator
b. Gas receiver
c. Manual gas cock
d. Safety shutoff and vent valve
e. Gas accelerating valve
f. Governor system operated regulating valve
g. Gas injection valves
h. Isolation valves

The variable fuel gas pressure regulator regulates the gas supply pressure. The receiver absorbs pulsation in the gas flow and assures a more uniform gas pressure at the engine. The safety shutoff valve and vent valve will shut off the gas supply and vent the line to the engine if an abnormal operating condition occurs. The gas accelerating valve controls the amount of gas supplied to the engine by the speed governor-operated gas valve during starting.

As an example, the following steps have to be taken:

1. Checking the timing of the gas injection valves
2. Opening of all cylinder isolating or balancing valves
3. Setting of gas injection valve hydraulic lifter
4. Adjusting the engine fuel gas regulator to the correct pressure

10.3.2.3 Cooling Water System

A centrifugal water pump circulates cooling water through the engine jackets. The pump may be built-in, attached to and driven by the engine, or detached electric motor driven. Following the water flow, the cooling water system consists of two pump sets, a thermostatic valve, a twin cooler set configured either as a radiator or as an oil-to-cooling water heat exchanger, and back to the engine cooling water header – a typical cooling water circuit for industrial internal combustion engines.

As an example, the following tasks have to be accomplished:

1. Fill engine and cooling system with water and antifreeze – if required before filling the engine with oil.
2. Apply 30 psi or 2 bar air pressure to the system and inspect the engine for either internal or external leaks.
3. Purge all air from the system.

10.3.2.4 Turbocharger

Combustion or scavenging air is supplied to the cylinders by an exhaust-driven turbocharger located at the flywheel end.

10.3.2.5 Two Intercoolers

through which cooling water is circulated cools the air before it enters the engine cylinders. With the pre-start oil pump running, ascertain oil is delivered to the unit. Turbochargers often have an air-assist feature to start them. This feature has to be tested.

10.3.2.6 Exhaust System

The exhaust system accepts the spent exhaust gases coming from the turbocharger. It can be equipped with environmental controls.

10.3.2.7 Engine Starting Facility

Cranking the engine for starting is accomplished by one of two possible systems: Gas expansion motors or compressed air starting. The motor system consists of an air- or gas-driven reduction gear motors equipped with a retractable pinion. The pinion engages a ring gear on the engine flywheel to crank the engine. A compressed starting air system consists of a distributor that opens air starting valves in the cylinder heads to admit 250 psi or 17 bar air pressure to force the piston down thus cranking the engine.

10.3.2.8 Instrumentation and Control System

Modern engines have sophisticated control systems referred to as engine management systems, or EMS. Figure 10.9 is a graphic representation of the information flow and the resultant machine human interface commonly abbreviated as HMI.

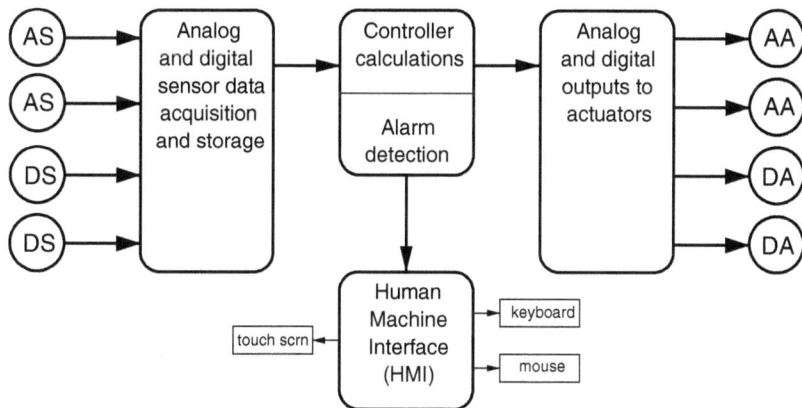

Figure 10.9: EMS functionality.

It is of utmost importance to not consider commissioning and startup of the machine without activating the EMS. Typically, a modern EMS offers functionalities that are necessary for the startup and subsequent monitoring of the machine.

10.3.2.8.1 EMS Functionality

Visual display of system elements combined with central engine and package module control.

Visual display consists of:
1. Color graphics display and eight function keys
2. 10-key numeric keyboard for parameter input
3. Keys for START, STOP
4. Display selection keys and special functions
5. Interfaces
   a. Ethernet for connection to server
   b. Bus connection to intelligent sensors and actuators
   c. Data bus connection to control in- and outputs
   d. SCADA connectivity

A concise and functional graphic compilation of engine parameters is displayed on the screen. User prompts are enabled by direct-acting display selection and function keys.

Main displays:
- Electrical schematic
- Oil and hydraulic schematic
- Gas data
- Engine controllers
- Auxiliaries' controllers
- Cylinder data
- Exhaust gas data
- Spare screen for customer specific purposes
- System display screens
- Parameter manager
- User settings
- Alarm management

Recipe handling enables setting, display and storage of all module parameters. Alarm management consists of efficient diagnostic instrumentation listing and displaying all active fault messages both tabular and chronologically, with the recorded time.

10.3.2.8.2 Central Engine and Module Control

A real-time, modular industrial control system which handles all jobs for module and engine-side sequencing control such as start preparation, start, stop, after-cooling, and control of auxiliaries as well as all control functions.

10.3.2.8.3 Control Functions

- Speed control in no-load
- Power output control
- Exhaust gas emission control
- Anti-knock control: Adjustment of the ignition point, power output and – as site conditions allow – mixture temperature modulation in case knocking is sensed
- Linear reduction of power output in the event of excessive mixture temperature and ignition
- Fully automatic operation, according to remote demand signal
- Remote automatic start
- Fully automatic operation at full load
- Stop with cooling down run for 1 min
- Continuous operation of auxiliaries for 5 min after engine shutdown

10.3.2.8.4 Shutdown Functions

with display:
1. Low-lube oil pressure
2. Low-lube oil level
3. High-lube oil level
4. High-lube oil temperature
5. Low jacket water pressure
6. High jacket water pressure
7. High jacket water temperature
8. Overspeed
9. Emergency stop/safety loop
10. Gas train failure
11. Start failure
12. Stop failure
13. Engine start blocked
14. Engine operation blocked
15. Misfiring
16. High mixture temperature
17. Measuring signal failure
18. Overload/output signal failure
19. Knocking failure
20. Warning functions with display

21. Low jacket water temperature
22. CPU battery failure
23. Operational functions with display:
    a. Ready to start
    b. Operation (engine running)
    c. Off

10.4 Preparing the Initial Startup

The initial startup is supervised by the OEM FSR. It is important to have plant-operators interface with him and be part of the startup process. This is where information and details are transmitted that might not be contained in the OEM supplied operating and maintenance (O&M) manual. If the plant has a crew of captive mechanics,[8] it would be advisable to have at least two mechanics or millwrights work with the FSR. They will, at this time, have familiarized themselves with the contents of the O&M manual and participate actively in the pre-startup steps which lead to the initial startup.

The owner's engineer and the personnel just mentioned will have had part in building suitable checklists, specific for the machine at hand. Such a checklist would resemble what is shown in Table 10.2.

Table 10.2: Gas engine pre-start checklist.

PETRONTA	
*Chemical **Ilderton PVC LE-1 Unit***	
Process:	Machine: (*Describe Function*)
SOP No.:	*Example:* C&S Waukesha 16-Cylinder
Gas Engine Pre-start Checklist	(12V275GL)
Purpose:	To ready the engine for initial startup and solo run.
Safety, Health	– Caution must be exercised when working near rotating equipment.
Environmental	– Hot engine parts
Consideration	
References:	– Posted or shirt pocket / tablet startup operating instructions
	– Punchlist
	– OEM O&M manual
	– VDDR file
	– Work permit manual
	– Task risk assessment document
	– P&D: LE-XXX
	– Receiving, storage protection, foundation, grouting, piping and alignment checklists completed and attached.

Table 10.2 (continued)

Responsibility and Assistance	– Trained operator – OEM FSR – Journeyman machinist – Machinery FLS to coordinate work permit if required
Prerequisites:	– Control loops functionally tested and correct, all setpoints set and verified. – New or calibrated gauges supplied. – P&IDs verified – Engine Management System (EMS) commissioned and tested. – Work permit should be approved prior to startup – Standard PPE – if required
Materials/Tools to be used	– Engine Oil – Description: _____; VG_____ USgal: _____; _____ l

Remarks A. Pre-Start Checks

1. Check the inlet and exhaust system for any obstruction. In some cases, it is easy to remove an expansion joint or pipe piece sufficiently from the turbocharger air inlet to check of the turbocharger spins freely. _____
2. Some engines are not furnished with a pre-lube pump. In that case it is recommended the turbocharger oil supply line be removed and that clean engine oil be poured into the turbo charger bearing cavity. This ensures that the bearings are not starved for oil when the engine is initially started. _____
3. Test jacket water system at 30 psi or 2 bar.
4. Test fuel gas system with 100 psi or 6.8 bar air pressure. _____
5. Gas leak check performed up to fuel shut-off valve performed. _____
6. Check clamping of pipes and tubing; general check for missing pipe plugs, length of studs and other fasteners. _____
6. Lube oil piping stainless steel between filter and engine? _____
7. Ascertain cleanliness of crankcase. _____
8. Ascertain external cleanliness. _____
9. No paint on ignition wiring. _____
6. Temporary main bearing filters installed. _____
6. Fill the crankcase to the correct level with the proper engine oil. _____
10. Starting motor lubricator filled and operative – if applicable. _____
11. Lubricate the governor or any other exposed linkage. _____
12. Treated water in jacket water system and vented. Unplug vent at highest point of the engine cooling system. _____
13. Treated water in air intercooler cooling system. _____
14. Vent thermostat housing so no air is trapped in engine blocks or cylinder heads. _____
15. Remove valve covers and side doors – if furnished – and lubricate valve stems. _____

Table 10.2 (continued)

16. Pre-lube engine – if unit is equipped with a prelube pump. Make sure oil reaches the valve train. Ascertain oil is traveling to the main and connecting rod bearings.	_____
17. Temporary main bearing filters inspected and cleaned. If not sufficiently clean, repeat step 14. Decide to use other flushing methods if unsuccessful. Use milk pad in pump discharge for final cleanliness check.	_____
18. Check gas valve timing.	_____
19. Open decompression valves or indicator cocks – if furnished. Otherwise, remove spark plugs. Bar engine over several times.	_____
20. Blow out starting air or starting motor headers.	_____
21. Coupling half secured for engine solo run.	_____
22. If and when engine turns freely, use the starting system and turn the engine over again; reinstall spark plugs and prepare to start the engine.	_____

10.4.1 Initial Startup and Run-In

Modern control systems, as described earlier, are designed to start large engines from the control panel or console by simply initiating a start command. In the case of a gas engine, no attempt should be made to start the machine manually using the manual gas plug valve as a throttle. No shutdown device or control should be by-passed as an expediency to start the engine. Always check the variable gas pressure regulator to ensure that the correct pressure is available to light off the engine.

In the following paragraphs, an initial automatic starting procedure for a two-cycle gas engine is described:

1. Set required idling speed,
2. Open all cylinder indicator cocks,
3. With the manual fuel gas plug valve closed, initiate a start sequence but stop it when the "fuel on" command signal has occurred. This will permit a complete check of all events up to engine "Light-off."
4. Do not crank engine longer than 20 s if "fuel on" command does not occur.
5. Check ignition timing linkage – if so equipped – and end of cranking to ensure that ignition is fully retarded.
6. Close all indicator cocks. Look for traces of moisture discharged from cylinders through the indicator cocks. If moisture is found, locate the cause and eliminate it before proceeding.
7. Open fuel gas manual plug valve.
8. Make initial start and run at idle speed for 15 min. During this run all instruments and gauges must be monitored constantly particularly main bearing and turbocharger lube oil pressure.
9. Stop engine and wait for completion of post-lube cycle. Always wait at least 15 min after stopping engine to remove crankcase door. Lock flywheel whenever

engine rotation might injure personnel. An unlocked flywheel presents an unsafe condition. Inspect engine inside for possible distress.

10. Inspect main full flow oil filter and clean if necessary. Remove main bearing temporary filter from flywheel end and inspect it. If clean, reassemble and do not disturb others. If the inspected filter is dirty, clean all others. The temporary bearing filters should remain in service until they are perfectly clean after 1 h of operating at 100% speed with the lube oil at normal operating temperature. It is recommended to not exceed 50% of engine rated torque when using temporary main bearing filters.

11. Replace all crankcase doors and unlocked flywheel.

12. Run engine for 2 h at idle speed during this no-load of operation, jacket water water and lube oil temperature will only rise slowly. Be certain to maintain lube oil temperatures slightly below cooling water temperature to avoid possible condensation.

13. Subsequent to a clean main bearing filter 2 h idle run, restart, gradually increase speed to 90% and allow lube oil and cooling water temperatures to settle out at normal running temperatures.

14. Continue to run the engine with main bearing temporary filters installed until the lube oil system is perfectly clean. Only after 1 h of clean operation at 100% speed can the temporary filters be removed. Permanent lines designed to connect the bearing caps to the lube oil header can now be installed.

10.4.2 Engine Startup under Normal Conditions

1. The lengthy procedure previously described does not have to be used to start an engine that has been running properly very recently, but some considerations should be recognized.

2. Bar the engine over to make sure it turns freely because repairs done on a brief shutdown may have created a problem with the engine or a head or a liner could have a water leak into the cylinder. When the engine is operating, this is not noticeable because the pressure in the cylinder is usually higher than the cooling system, and there is not enough time for water to accumulate.

3. If water or any other fluid accumulates into the cylinder, it causes a hydrostatic lock in the cylinder which could bend the connecting rod, break the piston, lift the head off or crack the cylinder liner (if the unit is cranked over by the starter).

4. Turn the pre-lube pump on so that the bearings receive proper lubrication before the engine starts turning.

5. With the ignition in the off position, crank the engine over for about 15 s to purge any gases that may be in the cylinders.

6. With the ignition in the on position, start cranking the engine and then crack the fuel valve so that it is about 1/8 open.

7. Continue to crank and open the fuel valve further to about 1/4 turn open. The engine should have enough fuel to fire and start by this time. If it does not, you may have to check some of the shutdown equipment.
8. As the engine begins to gain speed, shut off the cranking motor and gradually open the fuel valve to full-open position.
9. The engine is now started and running. Run it at low speed with no load for a few minutes until it is warmed up. During this time the engine should be monitored for oil pressure, water pressure and all other engine conditions. Most engine instruments – or EMS – give an indication if operation should continue.
10. When the engine is shut down, the fuel valve must be turned off to stop the engine.

10.4.2.1 Danger
- Never shut the engine off by any other means because the gas valve can be left open. This fills the cylinders, crankcase and possibly the building with fuel gas.
- Engines that operate on other than natural gas fuels have slightly different start-up procedures and precautions. This is where you have to rely on the plant policy manuals, engine manufacturer manuals, engine representatives and the technician's expertise for successful startup and operation of the engine.

At this time, the engine is ready to be permanently connected to the driven machine and handover of the asset can be executed.

Appendix 10.A: Engine grouting procedures

The advantage of approach one is that soleplates can be grouted ahead of engine arrival at the site in case of scheduling or logistics problems. The disadvantage is that more care has to be exercised as pre-grouted sole plates have the tendency of not staying level while waiting for the arrival of the machine. The daily change in ambient temperature and the difference in coefficient of expansion – epoxy and concrete – cause tilting of the sole plate assembly. To avoid this effect, an additional effort has to be made by having to place a hold down pipe sleeve, washer and nut – see Figure 10.A1 – on the foundation bolt. The pipe sleeves must have squared machined ends to ensure that they together with bolts and washers are parallel with and perpendicular to the sole plate surfaces. This assembly permits loading the soleplate simulating the weight of the equipment.

Follow this procedure:
1. An outdoor site would require weather protection over the foundation for protection against wind, sun and rain as these affect the accuracy of the soleplate elevation.
2. Check the foundation bolts for proper location and elevation – refer to the foundation drawing for this information. If a foundation bolt is not vertical and will require pulling to accept the engine base, allow for this by offsetting the soleplate

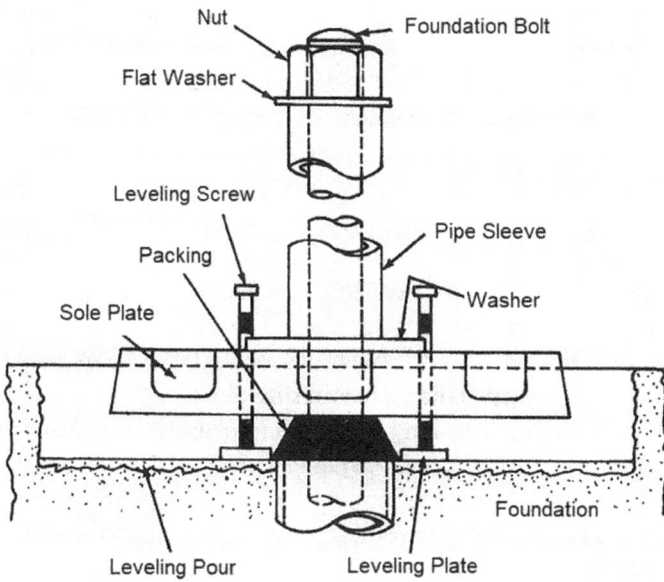

Figure 10.A1: Setting a pre-grouted sole plate (source: CES/Cooper-Bessemer).

in the proper direction. This is to assure that the foundation bolt, once straight, will not touch the soleplate. Figure 10.A1 illustrates the general principle.

3. Pack around the foundation bolts so that grout will not contact the bolts or enter the sleeves – see Figure 10.A2. Likewise, the foundation bolt threads require covering to protect against damaging them.

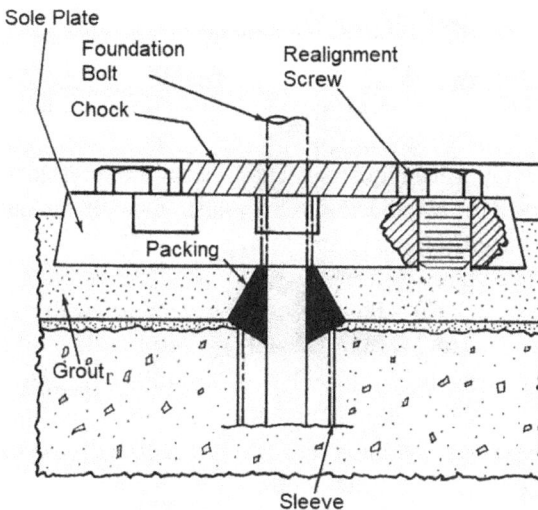

Figure 10.A2: Grouting sole plates.

4. Chip the bottom of grout pockets, or make a grout pour to provide a firm level surface for the leveling screw plates. A minimum of 1 in. or 25.4 mm between the bottom of the sole plates and top of leveling plates should be maintained. Sharp corners in pockets should be avoided.

5. The sole plates, in order to preserve them during shipment, will have been coated with *Tectyl 506*™.[9] Before placing them into the pockets, wash them in MEK[10] or lacquer thinner because of the cored areas two or three washes are required to remove all oily residue. Sandblasting between the second and third wash also provides an excellent bonding surface for the grout. The top machined surface should be protected at all times and all nicks and burrs should be filed smooth. When sandblasting, all tapped holes should be plugged to prevent sand from entering them.

6. Insert three 1/2 in. 20 UNF × 6 in. long leveling screws in the sole plate to support the sole plates at the approximate required elevation. Apply paste wax to the screws, forms and any other items that must be free from grout adherence.

7. Wax or grease the sole plates realignment screws – Figure 10.A2 – and tighten them in the plates so that they are reasonably watertight.

8. Prior to moving the engine over the foundation, place a plastic sleeve or tape over the foundation bolt threads to prevent thread damage.

9. Wash the engine chock, again, in MEK or lacquer thinner. They are located between sole plate and base – refer to Figure 10.A2. Using a file, remove any nicks or burrs on the top and bottom surfaces of the chocks.

10. Clean top of sole plates and set a chock on each sole plate. Check the condition of the engine base and clean, if necessary, the areas where the chocks will rest.

11. Carefully move the engine over the foundation and lower it over the foundation bolts. It is preferable to use jacks when lowering the engine over the bolts as it may be necessary to align certain bolts with the hole in the engine base. Jacks permit this to be done more easily. Ascertain that the engine weight is evenly distributed on all jacks as the engine is lowered and that all foundation bolts enter the base holes.

12. Level the engine by observing a machinist level resting on an accessible machined surface on the base all the by adjusting the leveling screws in the bottom of the sole plates. Again, consult/refer to Figure 10.A2.

13. Assure crankshaft is resting on main bearings.

14. Proceed with initial crankshaft web deflection check[11] to be performed with engine engine flywheel not yet installed and crankcase doors removed.

15. Tighten the foundation nuts to the torque specified on the outline drawing.

16. If deflection is not within limitation listed the OEM's operation and maintenance manual, adjust sole plate realignment screws – Figure 10.A1 – and repeat web deflection measurements until desired alignment is obtained. Due to set taken by the engine base during shipment it may not be possible to obtain the desired alignment by adjust the realignment screws. In this case, it will become

necessary to "pull" the base by tightening some foundation bolts and loosening others while adjusting the realignment screws accordingly.

17. Proceed to grout the engine sole plates.

If our readership wants to find out, follow us through these moves:

1. Grouting should not be attempted before the foundation has cured for at least 15 days or when the ambient temperature is below 50 °F or above 90 °F or below 10 °C or above 32 °C. The best temperature for grouting is between 60 and 80 °F or 15 and 27 °C. At a higher temperature in the acceptable range, the pour thickness should not exceed 3 in. or 76 mm. If a temperature drop of over 40 °F or 22 °C is expected, provisions should be made to maintain a reasonable temperature. The foundation should be held at an acceptable pouring temperature for at least 3 days before pouring to assure flowability.

2. Epoxy resin grout having the following characteristics:
   – A consistency which will allow proper placement.
   – High bonding strength.
   – High dimensional stability at operating temperature, that is, 150 °F or 65 °C.
   – Strength to transmit static and dynamic forces from the engine to the foundation.

3. The surface is to be grouted must be absolutely clean and dry at the time of grouting to assure a good bond.

4. Final level of the grout should be within approximately 1/2 in. or 13 mm off the top of the sole plates. Grout should not be closer than 1/2 in. or 13 mm to any vertical surface of the unit base. Therefore, the foundation trough should be checked and the grout formed where necessary, in order to assure 1/2 in. or 13 mm clearance when the engine is set.

5. Grout the sole plates following correct procedures. Remove any excess grout before it has completely set.

6. Poor grout from one side of the sole plates only to ensure that no air pockets remain under the plates. Epoxy grouting materials present certain health hazards and the grout manufacturer's handling instructions should be strictly adhered to.

7. After the grout has hardened, remove the three leveling screws in each sole plate. It is important that this is done before the foundation bolt nut is removed.

References

1 Powermag.com – Environmental legislation and stationary engines. Visited: Feb.16, 2021.
2 API 11P – Specification for Packaged Reciprocating Compressors for Oil and Gas Production Service. 2nd Ed. 1989.
3 Unlike cementitious grout, epoxy grout is made from Epoxy resins and a filler powder.

4 Learning Module – Alberta Apprenticeship and Training. http://tradesecrets.alberta.ca
5 American Petroleum Institute API RP 686, 2nd Edition, 2009, *Recommended Practice for Machinery Installation and Installation Design.*
6 Robt. L. Rowan & Assoc. Ltd. office@*rirowan.com.*
7 André Eijk, Sven Lentzen, Flavio Galanti, Bruno Coelho, *Compressor Foundation Analysis Tool,* COMPRESSORtech², OCTOBER 2013, Page 24 to 36.
8 As opposed to contract personnel.
9 *tectyleurope.com* Visited: 21/04/08
10 Methyl-Ethyl-Ketone
11 Heinz P. Bloch and John J. Hoefner, *Reciprocating Compressors – Operation and Maintenance,* 1996, Gulf Publishing Company, Houston, TX. ISBN 0-88415-525-0. Pages 231 to 242.

Bibliography

ISO 1204:1990.
Reciprocating internal combustion engines – Designation of the direction of rotation and of cylinders and valves in cylinder heads, and definition of right-hand and left-hand in-line engines and locations on an engine.

Chapter 11
Commissioning of Power Transmission Components

11.1 Drive Couplings

11.1.1 Introduction

Almost all process plant machinery requires a power transmission element between the prime mover and the driven machine. This role is assumed by a coupling. Couplings are used to connect two shafts that turn in the same direction on the same center line. We included gear boxes in this chapter; gear boxes usually require at least two couplings. Three different types of couplings will be identified in this text: rigid, flexible and special purpose. In this text, only flexible couplings will be discussed.

11.1.2 Flexible Couplings

Flexible couplings are designed to dampen vibration, absorb shock loading and accommodate axial movement or end float of the shafts, as well as compensate for minor misalignment.

There are three categories of flexible couplings: "mechanically flexible," such as gear and chain couplings, "material flexible," such as disk, spring, diaphragm, elastomeric and bellows, and "combination," such as metallic grid couplings which combine mechanical with material flexibility.

11.1.2.1 Mechanically Flexible Couplings

Mechanically flexible couplings such as gear and chain couplings provide a flexible connection by allowing coupling components to move or slide over each other.

Some minor misalignment is accommodated by the clearance between gear teeth or chain and sprocket teeth as the case may be, however, shafts should be still aligned as carefully as possible in order to ensure long coupling life.

These coupling types require periodic lubrication using special coupling greases or EP[1] non-channeling[2] greases. The most common form of mechanically flexible couplings, the gear coupling shown in Figure 11.1, even when using a continuously oil lubricated design, has been largely abandoned by particularly the hydrocarbon processing industries (HCPI) wherever feasible due to their unacceptable maintenance load. Approximately 75% of gear or chain coupling failures are caused by misalignment and improper or insufficient lubrication during long operating periods without maintenance shutdowns common in these industries.

https://doi.org/10.1515/9783110701074-011

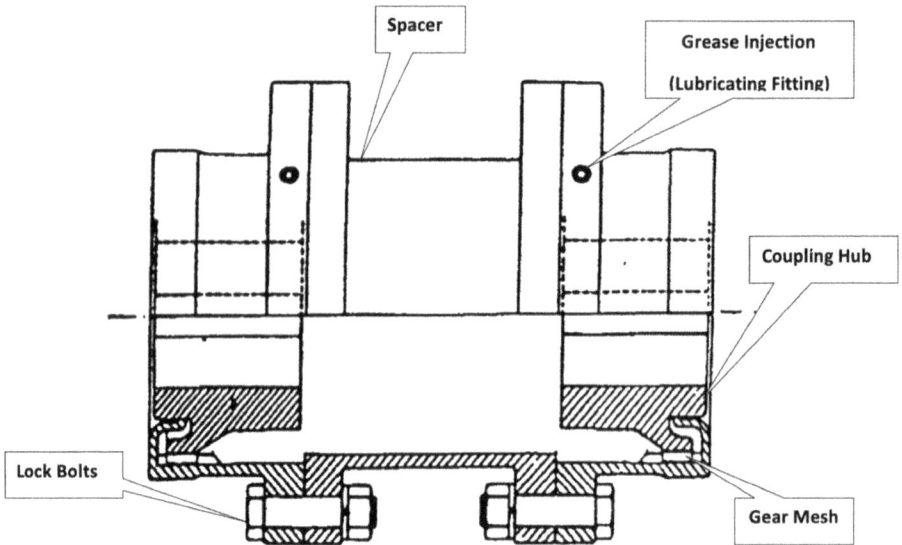

Figure 11.1: Gear coupling.

11.1.2.2 Material Flexible Couplings

As the name suggests, material flexible couplings provide flexibility by incorporating elements that bend and flex. The flexing materials providing the connection between the driver and the driven component include laminated disks, bellows, diaphragms and elastomeric materials that may include rubber plastics, such as neoprene and urethane.

Generally speaking, these coupling types require little maintenance other than alignment checks. Their service life is limited by the fatigue limit of the flexing material itself – see Figures 11.2 and 11.3.

11.1.2.3 Combination of Mechanical/Material Flexible Couplings

These include grid couplings that are compact units capable of transmitting high torques at speeds of up to 6,000 rpm. The construction of this type of coupling consists of two flanged hubs, each with specially designed grooved slots cut axially on the outer edges of the flanges. The flanges are connected by using a serpentine spring grid that fits the grooved slots. The flexibility of this grid provides torsional resilience, can provide periodic lubrication using good quality coupling grease in the areas of the grooved slots and serpentine spring.

While this coupling is still very popular in other industries, North American HCPIs have all but discontinued the use of these couplings, again, due to their maintenance intensity.

Figure 11.2: Disk pack coupling (Thomas).

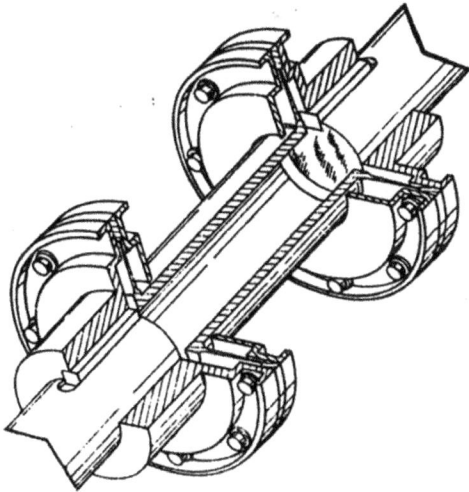

Figure 11.3: Diaphragm coupling (Bendix).

11.1.3 Risk Profile – Why Couplings Still Fail

A risk profile for couplings can be expressed by the experience of a renowned coupling expert. He lists the life expectancy of flexible couplings in order of frequency of occurrence as:
- Human errors in form of:
  a. Improper selection
  b. Incorrect installation
  c. Lack of periodic maintenance

- Corrosion
- Wear
- Fatigue
- Bolt failures

With lubricated couplings, there is the additional cause of improper, inadequate or insufficient lubrication, often brought on by neglect.

11.1.3.1 Application Issues

Table 11.1 compares the salient features of three coupling types most often applied in the HCPI.

Table 11.1: Comparing couplings used in the processing industries.

	Disk Pack	Disk Pack	Gear
Speed Capability	High	High	High
Power to Weight Ratio	Medium	Medium	High
Lubrication Needed	No	No	Yes
Misalignment Capability	Medium	High	Medium
Inherent Balance	Good	Very Good	Good
Overall Diameter	Low	High	Low
Failure Mode	Sudden	Sudden	Progressive
Overhung Moment	Medium	Medium	Very Low
Pulse Generation – Misaligned	Medium	Low	Medium
Capacity for Axial Movement	Low	Medium	High
Resistance to Sudden Axial Movement	Low	Medium	High

11.1.3.2 Alignment Issues

There is a perception that flexible couplings can accommodate a large parallel offset and/or angular shaft misalignment. This is incorrect. Depending upon the coupling type, flexible couplings can only accommodate from 1/4 degree to about 2–1/2 degrees of misalignment. High-speed, high-load applications require much closer tolerances.

The startup leader should be able to tell the indications and causes of misalignment. They include the following:

- Noise at the coupling
- Disintegrated rubber particles in elastomeric couplings, slivers from laminations or lubricant near or directly below the coupling guard – they do not work well in cold climates, depending upon coupling type.
- Process fluid and/or oil leakage from driven or drive shaft – or both shafts
- Premature shaft, shaft key or component failures
- Premature or frequent bearing failures – one or both machines
- "Soft foot" condition at footing of one or both machines
- High operating temperatures at or near the coupling
- Broken or constant loosening (vibration loosening) of bolts on one or both machines
- High vibration conditions, usually at both machines
- Cracked or broken foundation, particularly at or near the foundation bolts
- Continuing or intermittent leaks at pipe joints caused by pipe strain
- High energy consumption
- No compensation for vertical, horizontal and/or axial thermal growth at either the driven machine or the driver during initial alignment procedure
- Settling of the machine or its foundation after installation which would initiate a destructive cycle.

11.1.4 Why Permanently Installed Torque Meters Make Economic Sense

It is no secret that accurate knowledge of the individual performance of gas compression machinery used in the process industries is central to successful plant operation. After all, a large investment has been made. Because of this, and further influenced by prevailing fuel, feedstock and product prices, the success of a plant depends largely on good reliability and optimizing process efficiency and/or output continuously over the life of the plant.

In the petrochemical industry, and particularly in ethylene and fertilizer plants, torque meters, as shown in Figure 11.4, on large turbomachinery compression trains have been recognized for the past decades as providing very valuable plant performance information allowing for optimizing process efficiency, debottlenecking and identifying individual machine performance. They can be an invaluable aid to the startup team when it comes to evaluating the performance, for example, of a steam turbine driven compression train.

With advances in technology the use of torque monitoring couplings continues to become an integral part of predictive maintenance programs wherever large turbo compression trains are being operated. More and more facilities are using permanently installed instrumented torque-measuring couplings in order to understand how their critical equipment is performing so that intervals between scheduled shutdowns can be chosen appropriately. Performance testing and monitoring of

Figure 11.4: Torque meter coupling (illustration of Kop-Flex coupling courtesy of Regal Beloit Corporation).

turbomachinery after commissioning and initial startup is essential in assessing baseline conditions and OEM guarantees. For example, the objective of in-plant field-testing gas turbine driven compressor trains is to verify acceptance criteria such as heat rate, specific fuel consumption, turbine shaft power and compressor gas power.

Thirty years ago, a European gas transmission company recognized this need and decided to use torque meters routinely to test and verify the performance of their gas turbine driven pipeline compressors by installing "drop-in" torque meter couplings on a temporary basis after commissioning and at the end of major inspections and overhauls. Subsequent to a satisfactory field test, they would dismantle the torque meter coupling to be mounted on another gas turbine package for the same purpose. This "wander coupling" concept allowed them to economize on the, at that time, hefty torque meter investment costs for every unit.

Direct measurement of the shaft power between connected machinery enables operators to isolate which machine – driver or driven machine – is responsible for any performance deterioration. Continuous online monitoring of the machinery's output power provides operational trending data.

Torque variations can indicate performance problems such as blade fouling. Over-torque can lead to coupling, shaft or other component failures. When performance declines, more fuel is burned and NO_x emission increases. Torque meters provide a cost-effective method for diagnosing these problems early on so operators can make the necessary adjustments to their system for a proactive maintenance plan. Permanently installed torque meters must be recognized as a machinery health monitoring tool.

Heat balance and energy balance methods rely on measurements of pressures, temperatures, flows, gas compositions and mechanical losses.[3] Each of these measured parameters has its own instrumentation tolerance, which contributes to the overall test result uncertainty. The largest instrumentation tolerance is due to gas composition (up to 5%), with other measurement errors due to pressure (up to 2%), flow (up to 2%), equation of state (up to 2.5%) and temperature – up to 4 °F or 2.2 °C. If the shaft output power is known, the gas turbine heat rate and efficiency can be determined. If a torque meter is used, the total uncertainty for the gas turbine power can be reduced from around 7% to about 1% to 1.5%.

Similarly, the accurate performance measurement of a centrifugal compressor is dependent on the quality of the field data. Again, an important parameter is the shaft horsepower, which can be calculated directly and accurately in real time from indicated rpm and torque if a torque meter is installed. Otherwise, a heat balance method is recommended by ASME PTC 10 (1997). Outright calculation of shaft horsepower absorbed by a compressor is fraught with the attendant measurement tolerances or errors.

All torque meter coupling designs are faced with the task of detecting a physical change due to torsion in the coupling while it is rotating, and getting this information to a stationary output device that is generally located in the control room.

To sum up: Torque monitoring devices have matured and should be considered by bottom-line conscious turbomachinery users for field-based acceptance tests following commissioning and startup.

11.1.5 Commissioning and Startup

Whereas the project phases of rigging and lifting, jobsite receiving and storage protection, foundation and grouting and finally piping do not present significant risks to drive couplings, the alignment phase is critical. Here is where we encounter opportunities for human error, a fact attested to by the abundance of instructive literature dealing with the installation of these important machinery components. The readers are encouraged to refer to this literature if she or he has any involvement with couplings. For example, one of the poorly understood issues around the installation of modern couplings is their hydraulic mounting methods before alignment can begin.[4]

Further, we encounter contradictory references to couplings and alignment of machinery. We hear "coupling alignment" and what is actually meant is shaft alignment. Alignment occurs when two machine center lines are superimposed and form a single line. Misalignment is therefore an error in alignment. It is a measure of how much two lines are away from forming a single line. Evidently, two lines can be at an angle to each other – as they are in the case of angular misalignment – so they can be at a distance from each other in an offset misalignment.

We can move one or both machines in an attempt to bring the centerlines "in line," and this process is called machinery alignment.

While machinery alignment is easy to understand, Figure 11.5 explains coupling misalignment.

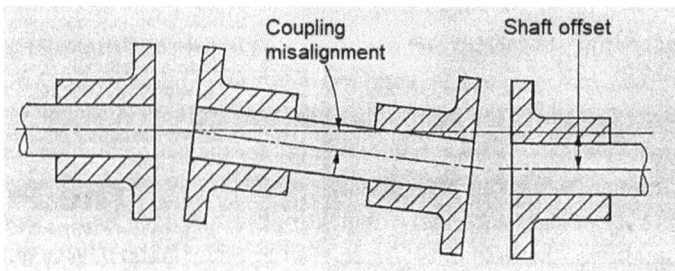

Figure 11.5: Coupling misalignment and shaft offset.

As we see in this figure no couplings are shown. Coupling misalignment is the angle that the spacer's centerline forms with the centerline of either shaft. Therefore, each "half-coupling" can have a different misalignment. What is important is that coupling misalignment is measured in degrees, not in inches or millimeters! Because of this, we should never use the word "coupling alignment." Furthermore, most millwrights align their machines before the couplings are installed. Couplings are used to accommodate misalignment between two shafts. What is not well understood, however, is the fact that all couplings resist being misaligned, and the restoring forces and moments involved in this resistance can damage bearings, seals and even shafts.

Therefore, one must align the machine shafts as well as possible, but try to have zero misalignment is seldom cost effective, and in the case of lubricated couplings it is actually counterproductive.

The question arises then as to how well machines should be should be aligned. How much misalignment is acceptable? One of the writers was assigned to a startup of a major refinery project, and the EPC had negotiated with the millwright union that all machinery were to be straight aligned within zero tolerance without any offset for thermal growth or shrinkage. The final alignment was then to be accomplished upon issuance of the offset values for each machine by the engineering team. Today, we would consider this extreme and not cost effective.

Numerous generic graphs exist, often referred to as "alignment tolerances," that can be consulted but we believe that universal guidelines can be misleading. Cold alignment tolerances must be determined by startup machinery engineers in cooperation with the OEM FSR for each machine. As mentioned earlier, most importantly, machines must be intentionally misaligned, because the goal is to have good alignment when the machines are at operating temperatures.

11.1.5.1 Coupling Guards

One of the components in the context of couplings is coupling guards. The coupling guard's main function is safety. It prevents workers from inadvertently getting too close to the rotating coupling. Even though non-lubricated couplings have been around for many years the associated coupling guards are sometimes a nuisance. They get overheated and can spray oil from bearings and even smoking oil into the surrounding area.

Dry-disk and diaphragm couplings do not have the cooling effect of continuous oil lubrication for gear couplings. Guards for these couplings must be carefully designed so that they do not get hot. The high-speed coupling shears the air inside the guard and imparts energy to the air which gets hot if not cooled or exchanged inside the guard. A coupling standard[5] recommends a limit for the guard temperature.

Any large diameter coupling in close proximity to a bearing housing seal will aspirate oil through the seal into the guard. The reason for this is that the surface speed at the large diameter is less than at the shaft causing a pressure differential and therefore a vacuum at the shaft level. As gear couplings tend to be smaller in diameter for the same torque and run at lower speeds, oil suction becomes less of a problem. It is important to design the guard with as much clearance around the coupling as possible. This is not always possible and owners have resorted to fitting a special expanded metal to protect personnel.[6]

If they are not designed properly and are not fit for purpose, coupling guards can cause a trail of headaches. It is therefore the task of the startup team to assure that guards are free from defects. Deficiencies should preferably be taken care of before handover. If a coupling guard problem is not solved during the commissioning and startup period, it will fester like a slow healing wound for a long time before it gets resolved. Leaking coupling guards are a typical example for an issue that tends to never disappear until, during the operation phase, well after the initial startup, some technical genius takes hold of it and fixes it for good.

Table 11.2 presents a pre-start checklist for drive couplings.

Table 11.2: Pre-start checklist for couplings.

1. Equipment locked and tagged out (LOTO) per plant procedures.	_____
	(Yes or No)
2. Has consideration been given to expected misalignment during startup?	_____
	(Yes or No)
3. Two sets of coupling flange fasteners staged and handy.	_____
	(Yes or No)
4. Remove any solo plates previously installed.	_____
	(Yes or No)
5. Hubs positioned on shaft properly.	_____
	(Yes or No)
6. Key(s) fill keyway and do not protrude beyond the shaft – if Applicable.	_____ (Yes or No)

Table 11.2 (continued)

7. Verify coupling alignment and pre-stretch data (diaphragm coupling).	
	(Yes or No)
8. Distance between shaft ends (DBSE): _____ in./mm	
	(Yes or No)
9. Coupling spacer placed – bolts not binding and torqued to Specifications.	
	(Yes or No)
10. OEM instructions followed for lubricated couplings – if applicable.	
	(Yes or No)
11. Coupling guard fitted – O-rings placed.	
	(Yes or No)
12. Coupled equipment rotates freely with no tight spot.	
	(Yes or No)
13. Cartridge seal assemblies checked for proper installation – pumps.	
	(Yes or No)
14. Cartridge seal assembly tabs removed – pumps.	
	(Yes or No)
15. Coupling guard(s) installed – pumps.	
	(Yes or No)
16. All alignment jackscrews backed off equipment feet.	
	(Yes or No)

11.2 Commissioning and Startup of Special-Purpose Gearboxes

11.2.1 Introduction

The dictionary of rotating machinery[7] defines a gearbox as a device that transfers input rotational motion to an altered output motion using a combination of gears; or a mechanical device using gears that receive rotational mechanical energy at some input speed and converts it to a different output speed and/or changes the direction of rotation. Gearboxes can be either speed increasers or speed reducers depending on the requirement. Figure 11.6 represents a typical industrial gearbox. An estimated 6,000 special-purpose gearboxes are, for instance, in operation worldwide in gas turbine compressor drive applications.[8]

There are many types of gearboxes. The most commonly used in the process industries have parallel shafts, epicyclic or planetary gears. Other classifications are related to the shape of the individual gear teeth, such as spur, helical or herringbone or herringbone, also known as double helical.

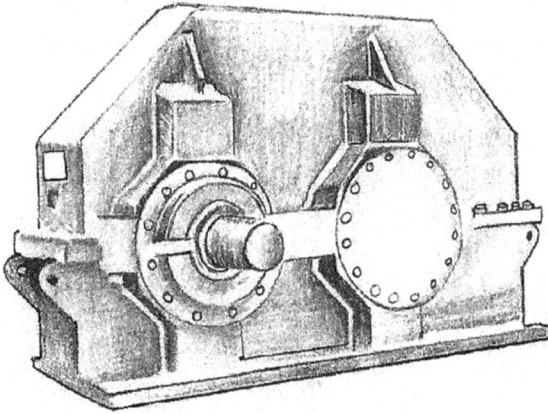

Figure 11.6: Typical industrial gearbox with parallel shafts.[9]

Parallel gearboxes have two shafts, each carrying a single gear. Epicyclic gearboxes split the power transmission between several "planet" gears that move around a central "sun" gear while also meshing with a surrounding "ring" gear. Epicyclic gears have a small footprint and provide co-axial input and output but are limited to a power rating of some 45 MW for gear ratios exceeding six.

Power ratings are within mechanical drive prime mover ratings as shown earlier in Figure 1.2.

In order to understand the limiting factor in gear design, we must translate external operating characteristics such as power ratings and speeds into gear-specific design parameters.

The main factor controlling power and speed limits are:

– Pitch line velocity (PLV) – see Figure 11.7 – leading to the determination of gear tooth sliding velocity which determines gear box heat load
– Elastic deflection produced by torques and bending moments on all parts of the gearbox especially as it applies to the pinion in the smaller of two gears in a parallel shaft gearbox
– Desired input and outputs
– Operating limit of the bearings
– Factor of safety chosen for the application

PLV is a better index than rotational speed because a large gear operating at relatively low rpm may see the same velocity effects as a small gear running at high rpm. PLV can be calculated by using the following formula:

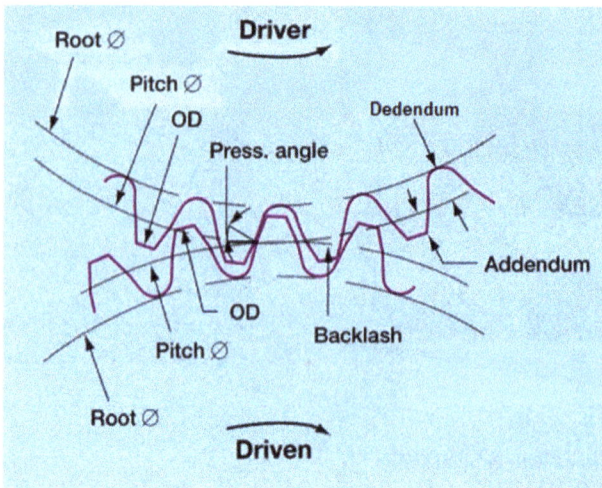

Figure 11.7: Gear pitch line explained.

$$PLV = \pi \times PD(in.) \times \frac{rpm}{12} \ [fpm] \tag{11.1}$$

or

$$PLV = \pi \times PD(mm) \times \frac{rpm}{60,000} \ [m/s] \tag{11.2}$$

Pitch line velocities (PLV) are classified as:
- Low – below 250 fpm or 1.27 m/s
- Moderate – 250–5,000 fpm or 1.3–25.4 m/s
- High – above 5,000 fpm or 25.4 m/s

11.2.2 What Startup Leaders Must Know About High-Speed, Special-Purpose Gear Boxes

The American Petroleum Institute Standard 613[10] defines gearboxes predominantly used in the petrochemical industry as "high-speed special-purpose" type. These are precision gears of parallel shaft design with pinion speeds above 3,600 rpm and PLV over 4,000 fpm or 20.3 m/s. Figure 11.8. shows a typical special-purpose gearbox used as a speed increaser in a motor-driven compressor train. API 613 covers gear rating, related lubrication systems, controls and instrumentation. Accordingly, standard high-speed gear units operate at PLV of up to around 20,000 fpm or 100 m/s and there are units operating with PLV in excess of 35,000 ft/min or 178 m/s and transmitting 30,000 hp or 22.4 MW.

Figure 11.8: Typical high-speed, special-purpose gear box
(courtesy of J.M. Voith SE & Co. KG | VTBS).

Attention to detail and care during commissioning and startup of gear boxes will vary with PLV and power transmission which can be divided into four ranges:
1. 1–100 hp or 75 kW
2. 100–1,000 hp or 750 kW
3. 1,000–10,000 hp or 7.5 MW
4. Above 10,000 hp or 7.5 MW

There are many different gearbox configurations. The extensions of parallel shaft units can be defined and organized according to Figure 11.A1 in Appendix 11.A.

This convention should be used when describing a gearbox in commissioning and startup documents.

11.2.3 Instruments and Controls

High-speed gears are usually used on critical process machinery trains where down time is costly and frequently disastrous. For these installations a vibration monitoring system is cheap protection. There is no question that failures risks can be minimized and forced downtime avoided by using vibration monitoring equipment. We favor using acceleration probes as they are more easily applied to the outside of the case on existing units. Vibration signatures – base lines – should be recorded during factory tests for later comparison to field installations. All bearing housings should be drilled and tapped to permit installation of horizontal and vertical non-contact vibration probes. Better monitoring can be achieved by connecting all sensing devices to a central display such as a SCADA system to assure multiple user involvement.

Monitoring equipment often specified by users include:
- Vibration proximity probes to measure shaft vibration relative to the housing.
- Key phasors that provide timing and phase reference.

- Accelerometers measure vibration in form of absolute casing acceleration.
- Direct reading dial-type thermometers in stainless thermowells to measure bearing temperature.
- Resistance temperature detectors (RTDs) and thermocouples to measure bearing temperatures.
- Temperature and pressure switches as part of alarm and shutdown functions (ESD).

Figure 11.9 illustrates such a monitoring system.

There should be no debate as to whether or not these instruments ought to be activated or armed and ready to function for the initial startup of any gear box if they were specified.

11.2.4 Commissioning and Startup

11.2.4.1 Risk Profile

Many factors such as manufacturing, handling during shipment and receiving at the project site, storage protection, supporting structures, piping loads, alignment at operating conditions, cleanliness of the lubrication and assembly at the site may adversely affect the successful startup of gearboxes.

Most users of critical gears see to it that their units are being tracked through the manufacturing phases in order to mitigate the risks of having to suffer the consequences of manufacturing problems. Qualified inspectors are periodically sent to OEM's facilities to witness and check the following aspects of gear manufacturing on behalf of the future owners:

- Material checks
- Physical inspection of gear parts
- Running test
- Internal examination
- Fit of spare gears
- Reports and attachments

It is noteworthy that during the last quarter of last century failures of industrial high-speed gear boxes became a significant cause of new plant startup delays. Reference[11] lists the following gear problem categories in that period:

- Bearing failures mostly caused by incorrect lubrication practices.
- Pour oil drainage.
- Weld failures.
- General quality control defects.

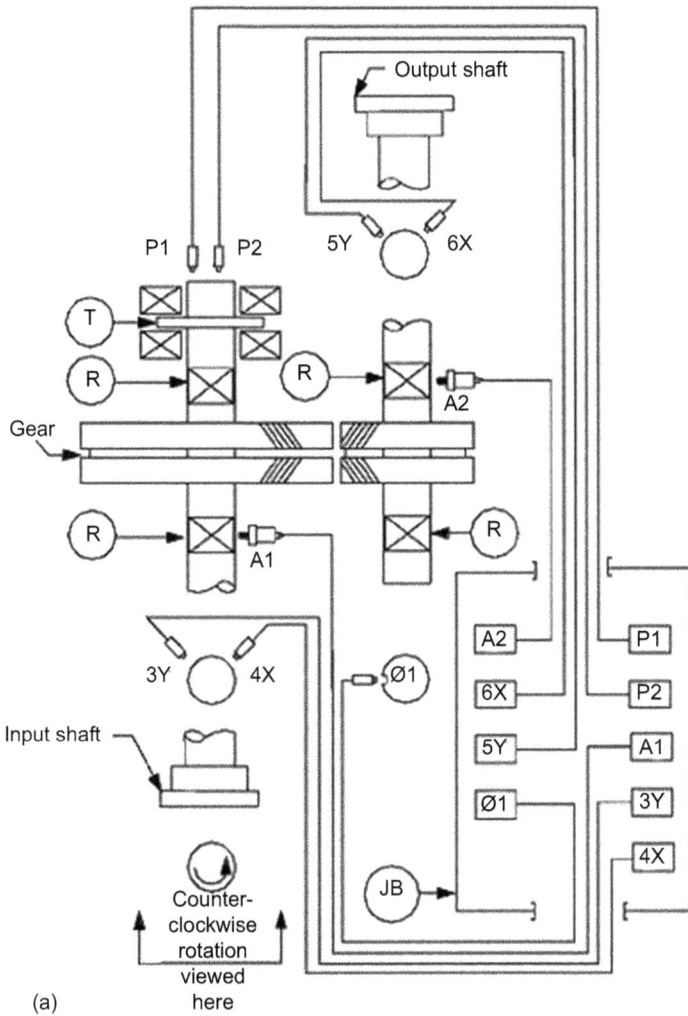

(a)

Radial shaft vibration (input shaft)	Bearing cap vibration (input shaft)	Axial shaft position	Radial shaft vibration (output shaft)	Bearing cap vibration (output shaft)
3Y 4X	A1	P1 P2	5Y 6X	A2

(b)

Figure 11.9: Typical monitoring and protection system arrangement for a double-helical gearbox (adapted from reference[12]).

With the development of high-speed integral compressors covered by API 617[13] and general progress in gear design, gearboxes have become highly reliable provided they are installed, commissioned and started up correctly.

Someone once coined the phrase "Good gears wear in and then wear forever." It suggests that gearboxes generally have a long life and make a reliable component of a machinery drive train. The term "good gears," however, implies that they are well designed in the first place and fit for purpose. Such a gearbox will have been designed according to AGMA[14] and API 613 standards and there should be always a torsional rotodynamic analysis included before its fabrication and assembly.

Life limits of well-designed gearboxes are primarily predicated by mechanical effects such as but not limited to:

– Gear tooth high-cycle fatigue due to torsional vibration or misalignment
– Overload by excessive power application
– Wear caused by rubbing and elastic deformation
– Overheating due to inadequate lubrication.

A risk profile, for high-speed gearboxes after the initial startup, albeit dated, according to published data by a major insurance company is presented in Figure 11.10.

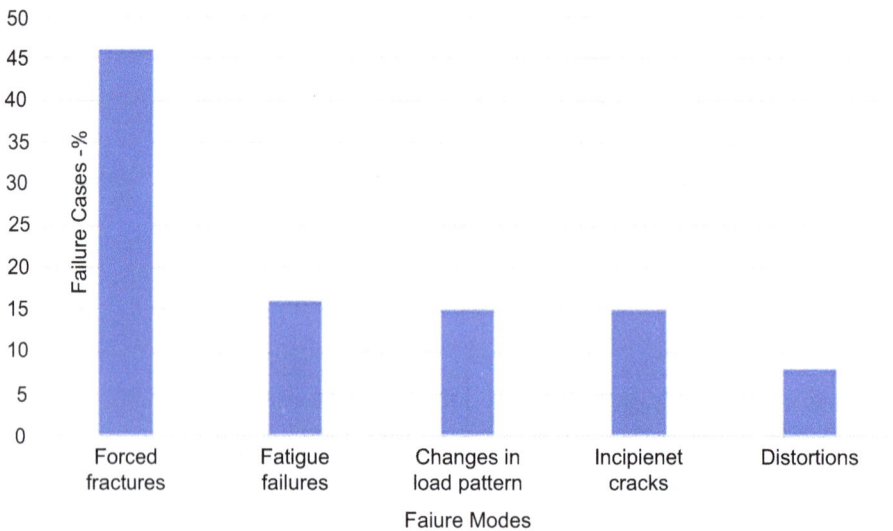

Figure 11.10: Failure causes of high-speed (turbo) gearboxes (source: reference[15]).

11.2.4.2 Rigging and Lifting

Before we investigate this project phase for any risk mitigation, we ought to pay attention to a concern often voiced by gear box manufacturers, namely, treatment of the gearbox for and during shipment. Preparation and actual shipment should be considered as two separate problems. OEM and purchaser should ask with the attendant risks in mind: Should the machine be shipped assembled or dismantled? What sort of shock loading is likely to be encountered? How can all the components be protected from the elements? Second, how should it be crated or boxed for the common carrier's requirements? While the "manufacturer standard" is always readily available, it is not prudent to solely rely on this best guess to avoid potential field storage and installation environmental problems.

If at all possible, the gearbox should be shipped completely assembled to minimize field installation problems. Shafts should be blocked to prevent their axial and rotary motion within the bearings. All openings must be tightly sealed. All internal parts, including bearing areas, should be coated with a light, oil-soluble rust-preventive compound. The casing must be drained of excessive preservative prior to sealing the drain openings. The unit should be tagged with all potential handling and storage instructions.

While the foregoing seems commonplace, self-evident and trivial, one has had to stand in front of a split open shipping box with a high-speed gear box exposed to the Russian winter weather to appreciate the significance of this topic. There are only a few machinery components that are more sensitive to corrosion attacks by inclement weather than gears.

Covered storage is rarely possible and coastal environments, for instance, are extremely hostile during long storage periods which often exceed 12 months. We favor preservatives which are compatible with the lube oil to be used and do not require disassembly of the gear before placing the unit in service.

Export boxing and shipping poses special problems, depending upon the type of vessel. Some operators have used sea-land containers, converted LSTs and ocean-going barges as well as conventional freighters. Equipment is usually stowed in cargo holds but frequently it is lashed on decks and is subjected to ocean spray. Companies specializing in export boxing are available in most major ports and are recommended.

When rigging and lifting of a gearbox is imminent, the machinery startup engineer should concern himself with the lifting plan. It is here where obvious risks arise as stories of unsuccessful lifts of heavy equipment such as a gear box are not exactly rare. Questions that should be asked are: Who made the plan? Is it reasonable? Who will perform the lift? What can go wrong?

Finally, when handling the gear unit for installation care must be taken not to stress parts which are not meant to support the gearbox weight. Gearboxes should be lifted only by the features provided by the OEM such as lifting holes in the casing.

11.2.4.3 Receiving and Storage Protection

After the gear has arrived in the field, a routine inspection procedure should be established to protect the equipment from atmospheric breathing, dust and mechanical damage. All temporary covers and screens should be carefully tagged to be sure that they are removed before startup. One particular gear failure occurred when temporary screens in the oil supply lines adjacent to the bearings were inadvertently left in place when the unit was placed in service.

For extended storage, the following procedure[16] is recommended; see Appendix 5.A for a description of preservation products.

1. Fill the gearbox and piping with oil containing Product B.
2. Plug all vents and allow space for thermal expansion.
3. Apply Product C to the shafts and couplings.
4. Tape the location where the shaft protrudes through the casing or housing.
5. Install a valved pipe on the casing that serves as a filler pipe for adding oil to the fill the casing.
6. On speed reducers, manual turn the input shaft a sufficient number of turns to allow the output shaft to make at least three full turns. Do this every three months and also retape all locations where the shafts protrude.
7. On speed increasers manually turn the output shaft a sufficient number of turns to allow the input shaft to make at least three full turns. Schedule activity every three months and also retape all locations where the shafts protrude.

Receiving inspection is also the point in time where we get intimately familiar with the history of our gearbox. We should find out whether or not there are any outstanding issues reported by our shop inspector who had examined the gear box as components were being fabricated and assembled.[17] These issues should be settled now before they start lingering throughout the project. The startup engineers should examine Table 11.3 representing the record of the mechanical run in the vendor's facilities for any issues impacting the upcoming commissioning and startup of the unit.

Table 11.3: Mechanical run test – high-speed gearbox.

GBX Shop Test (FAT)
Remove headline
Designation: _____
Location: _____
Service: _____
Gears should be given a 4-h mechanical run at maximum speed. This is primarily a balance check. The basic procedure should be to run up to speed, and allow oil temperature to stabilize. The unit is then run unloaded 4 h, and there should be no further undue rise in bearing and lube oil temperatures.

Table 11.3 (continued)

	CONDITIONS	
	DESIGN	TEST
1. Speed (H.S.)	____	____
2. Oil Inlet Temp.	____	____
3. Oil Inlet Press.	____	____
4. Max. Vibration (mils p-p or μm)	____	____
5. Max. Noise Level (db)	____	____
6. Vertical hot rise of the horizontal joint adjacent To each shaft extension	____	____

Inspection

1. Is the high-speed half coupling fitted for the test?

____
(Yes or No)

2. Bearing temperatures: Is the temperature rise across each bearing < 60°C or 33°C?

____
(Yes or No)

3. Vibration Readings; mils, μm, g:

	A Shaft	B Housing	A/B Attenuation
H.S. driving bearing	____	____	____
H.S. blind bearing	____	____	____
L.S. driven bearing	____	____	____
L.S. blind bearing	____	____	____

4. Is the attenuation less than 4?
(If no, mention in your report.)

____
(Yes or No)

5. Bearing Inspection: Are all bearing surfaces without significant damage?
Demand replacement otherwise.

____
(Yes or No)

6. Gear and pinion teeth examined for signs of distress, high spots, etc.? Satisfactory?

____
(Yes or No)

7. After complete cooling (say overnight, so that gear and pinion are the same temperature) a contact check should be made.
a. Is the contact obtained at least 90% of the length of working face?

____
(Yes or No)

b. Is it at least 40% of trailing face?

____
(Yes or No)

Table 11.3 (continued)

Procedure: Contact Check.
A. Gear wheel is to be located in operating bearing shells lubricated by SAE 40 to 60 oil.
B. The pinion is to be located in the operating bearing shells with the bearing internal clearance taken up with paper of suitable thickness. The top half of the bearing should be strapped down.
C. A band of the pinion teeth is to be given a light coating of Machinist's marking blue.
D. A band of gear teeth is to be given a light coat of special gear marking compound.
E. The gear is to be centered in the axial backlash and rotated through the mesh and back again with a steady motion.
F. The contact marking shall be lifted from the gear tooth using matt surface Scotch Brand® tape and placed on heavy bond paper.

7. Copies obtained of OEM's test log sheet, gear contacts _____
    and assembly inspection sheets?   (Yes or No)

8. Complete reassembly witnessed?   _____
                                                            (Yes or No)

    a. Bearings properly located   _____
                                                            (Yes or No)

    b. No bumps during lowering of gears   _____
                                                            (Yes or No)

    c. No bumps during lowering of top cover   _____
                                                            (Yes or No)

    d. Unit weatherproofed   _____
                                                            (Yes or No)

11.2.4.4 Foundations and Grouting

There are several excellent gear installation manuals available from the manufacturers. They cover detailed instructions on:
– Handling
– Mounting of couplings
– Foundation and alignment
– Cold and hot check alignment
– Axial positioning
– Doweling
– Tooth contact pattern checks

In addition, the manuals contain excellent details concerning maintenance, troubleshooting guides and overhaul. They are usually well illustrated and include pictures of damaged bearings and teeth to guide the novice.

Instructions and recommended practices for foundation and base plate configuration and grouting abound. Reference[18] and the OEM's operating and maintenance manual all have ample instructions giving rise to confusion and ambiguity if not harmonized and coordinated. The above standards cover details, like, for instance, "the OEM is to furnish mounting plate or plates with horizontal (axial and lateral) jackscrews, the same size or larger than the vertical jackscrews. The lugs holding these jackscrews shall be attached to the mounting plates in such a manner that they do not interfere with the installation of the equipment, jackscrews or shims."[19] The writers know of occasions where these details were overlooked by the OEM and not noticed by the commissioning crew causing anger and delays in the project schedule.

If the gearbox is to be mounted on a pedestal or baseplate the structure must be carefully analyzed to determine if it will withstand operating loads without excessive deflection. When mounting the unit on steel beams a baseplate should be used. The baseplate should extend under the entire gearbox and be at least as thick as the gearbox base. Both the unit and baseplate should be rigidly bolted to the steel supports with proper shimming to achieve a level surface. If the gearbox is to be mounted on a concrete foundation, grout steel mounting pads into the concrete base rather than grouting the box directly into the concrete. This will facilitate any shimming and alignment required. The concrete should be set and cured before bolting down the gearbox and load is applied.

It is our contention that a gearbox should be set on an OEM furnished sole plate using no shims so that corresponding equipment has to be shimmed to the gear. If shims are required, the owner's engineer should assure that they be solid ground plates, of equal thickness – minimum 0.250 in. or 6.3 mm. All gearboxes should be doweled after the hot alignment check.

11.2.4.5 Shaft Alignment

There has to be a clear understanding as to how to consider the gear unit. In the more complicated case, where the gear box is part of a string of machines, it is advisable to make the gear box the fixed body. This is also predicated by the fact that the gear box has at least two couplings as mentioned before.

If the gearbox has to be brought into alignment place shims under all mounting pads. A feeler gauge is used to determine that the thickness of shims is correct under all mounting pads. The gearbox input and output shafts must be correctly aligned with the driving and driven machine shafts. Even if the total system is delivered on a permanent mounting already aligned at the factory, couplings should be disconnected and alignment checked again. There is always the risk that the alignment may have been disturbed during shipment.

The startup engineer must verify that reliable cold shaft alignment data have been provided considering the thermal growth of all equipment in the string or train. Again, coordination becomes very important, and it is at this stage of the project where risks arise. There has to be a documented alignment plan stating the sequence of activities and their end result.

11.2.4.6 Lubrication Systems

Lubrication and cooling are critical to the operation of high-speed gears. As PLV and gear tooth loading have increased over the years to match speed requirement of driven equipment with prime movers and to reduce equipment costs, so have the lubrication and cooling requirements. Primary purposes of lubrication are to prevent wear of the gear to surface and to reject the heat generated by friction of the mating gear teeth. AGMA have published standards that provide guidance on how to perform critical calculations used to evaluate lubrication and cooling of high-speed gears. Many technical papers have also been published on the subject.[20]

For high-speed gear units, less than 30% of the lubricating oil is used for actual lubrication. Around 70% is used for cooling. In many cases gear boxes receive their lubrication oil from a central system they share with prime movers and the driven equipment. The startup team must assure these systems are according to specifications and, above all, clean – see Chapter 5.

These systems are normally designed to operate with a high-grade turbine oil with a minimum viscosity of 150 SSU at 100 °F or 38 °C or VG 32 and sometimes VG 46 AW with an anti-wear package. A good operating pressure range for the oil is 25–50 psi or 172–345 kPa, with 25 μm filtration.

AGMA 9005-E02 (2002) is a guide to selecting an oil for a gear application. The key lubricant properties are:
-   the class of oil,
-   the viscosity,
-   the pressure–viscosity coefficient.

Oils used in high-speed gearboxes are generally classified as inhibited or anti-scuff/anti-wear. Inhibited oils are formulated with highly refined mineral or synthetic base oils and contain additives that enhance oxidation stability provide corrosion resistance and inhibit phone. Anti-scuff/anti-wear oils contain additional additives that provide protection against wear and scuffing. Viscosity is the ability to resist shear but also flow. Oil viscosity decreases with increasing temperature. The rate of change of viscosity with temperature is generally less with synthetic oils, an advantage in some cases. The pressure–viscosity coefficient is used to assess how viscosity changes with pressure.

We are convinced that certain risks in connection with the above project phases could be avoided if, during the pre-award meeting with the vendor or OEM with train responsibility, she or he would be held to commit to furnishing complete installation instructions, such as but not limited to:

1. Soleplate, base- and bedplate, machinery position and leveling details
2. Foundation bolting and grouting details
3. Cold alignment data – including measuring methods, relative positions and alignment sequence
4. Keying, pinning, dowelling and torqueing details on foundation bolts
5. Pipe support and flange connection details
6. Other relevant details that would prevent calling on the judgement of mechanics and technicians at the project site

11.2.4.7 Startup Checks

Installation manuals discuss the routine checkouts required immediately prior to startup. These include as a minimum:

1. Check the oil level and type in reservoir, if applicable.
2. Ensure the specified oil is used.
3. Tighten all pipe connections.
4. Check the direction of rotation since many gears are unidirectional.
5. All shafts must turn freely.
6. Check mounting of all gauges, switches and other instrumentation.
7. Check all electrical connections.
8. Assure that bearings are pre-oiled (journal bearings).

11.2.4.8 During Startup

The following checks and observations should be made by operators and technicians:
If possible, operate at half load for the first 10 h to allow for breaking in of the gear tooth surfaces.

1. Check for unusual noises and record vibrations from minimum speed to maximum operating speed.
2. Monitor pressure drop across filters. New filters should be available during startup for changeout.
3. Oil temperatures and drain sight glasses (FIs) should be watched closely.
4. After initial run, shutdown and open the case to inspect gear tooth contact.
5. A final shutdown for gear tooth inspection should be made after full load operation and stabilized oil temperatures have been established.
6. After an initial 50 h of operation check coupling alignments and retorque all bolts. Check all piping connections and tighten if necessary.

11.2.5 Gear Inspection

Three spots around the gear should be cleaned and coated with Prussian Blue to check tooth contact after initial loading.

In case there is any doubt about the integrity of the gearbox, the internal gear assembly should be thoroughly examined. This would require a temporary shutdown of lube and seal oil pumps. Any decision to shut down will have to he made jointly with responsible process people.

For external examination, check that all casing feet are in contact with base when hold-down bolts are loose. For internal examination, proceed as follows:

1. Disconnect auxiliary piping and conduit in preparation for lifting of gear cover.
2. Lift off gear cover and carefully place on protective wooden beams or a pallet.
3. Examine upper bearing halves for signs of scoring.
4. Make precision-level check of gear as shown on Figure 11.11, below.

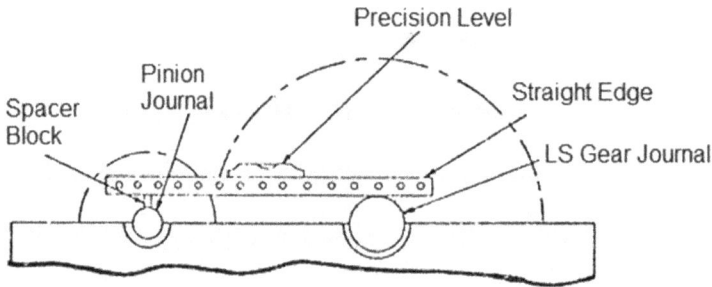

Figure 11.11: Precision-level check of gear setting.

5. Measure and record journal diameters. Then prepare a spacer block.
6. Adjust the level to center the bubble, then move the spacer block and level to the journals at the other end of the gears. The bubble reading will immediately indicate any error in the setting of the gear. Note that this checking procedure works equally well whether or not the machine is setting truly level. It is only necessary to prove that the four casing bores are in a common plane and this check does just that. If one corner is low, the exact amount of error in setting is easily proven by determining the thickness of shim stock that must be inserted between the straight edge and the top of the journal to center the bubble. Any indicated correction in gear casing setting should be made before proceeding with further alignment checks. Allowable limits for out of plane are 0.0002 inches per foot or 17 µm/m.
7. Secure journals in bearings as shown in Figure 11.12. Straps should not be so tight as to prevent the gears from rotating. Apply oil to journals.

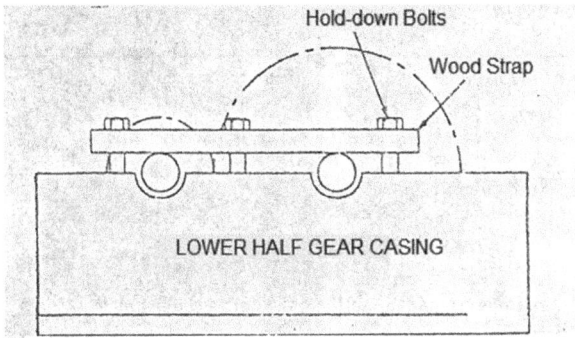

Figure 11.12: Securing gearbox journals in bearings for tooth contact check.

8.   Apply a thin code of red lead to six pinion teeth on both helices.
9.   Identify the six pinion teeth with center punch marks.
     Note: In nearly every case, failure to obtain adequate tooth contact is caused by the bores not lying in the same plane. This condition can be corrected by lifting the corner which is too low. This is in turn is corrected by adding shims underneath the soleplate or fabricated base so that if the gear casing is ever removed loss of shims will not occur.
10. Rotate pinion shaft using a strap wrench until pinion teeth have revolved one full turn.
11. "Scotch Tape" the pattern record and preserve it for future reference. Face contact should be about 90%.
12. Match mark final gear/pinion assembly for future reference.
13. Blue pinion totally for wear-in check.
14. Replace bearing top halves.
15. Replace gear cover after carefully cleaning housing joint surfaces.
16. Reconnect all piping and conduit.
17. Dowel gear in place after the hot alignment check.

Table 11.4 is a checklist that helps prepare the initial startup of the gear and indicates the follow-up the startup team and operators have to conduct after the initial startup.

Table 11.4: Commissioning and startup checklist for gearboxes.

Gearbox Commissioning and Startup Checklist
Remove header
Note: Cross out items not applicable
1.0   Preliminaries
1.1   Mechanical assistance present?

_____
(Yes or No)

1.2   I&E assistance present?

_____
(Yes or No)

1.3   Applicable operation and maintenance manual parts book special tools and
        spares available?

_____
(Yes or No)

1.4   Have the vendor drawings and data requirements (VDDR) been fulfilled?

_____
(Yes or No)

1.5   Are the items/documents listed in the project VDDR form available?

_____
(Yes or No)

1.6   Are there any outstanding items/issues from shop inspector's report?

_____
(Yes or No)

1.7   Receiving, storage protection, foundation, grouting, piping cleanliness and
        alignment checklists completed and attached?

_____
(Yes or No)

1.8   Control loops functionally tested and correct – all setpoints set and verified.

_____
(Yes or No)

1.9   New or calibrated gauges supplied and installed.

_____
(Yes or No)

1.10   All OEM/vendor and site requirements read and understood.

_____
(Yes or No)

2.0   Lubrication System
2.1   GBX breather vent installed and unobstructed.

_____
(Yes or No)

2.2   Piping traced and tagged.

_____
(Yes or No)

2.3   Orifice(s) installed in supply line.

_____
(Yes or No)

2.4   Is drain line vented?

_____
(Yes or No)

2.5   Drain pipes inspected; sight flow indicators clear and oil flow visible.

_____
(Yes or No)

2.6   Oil temperature to specifications: °F or °C _____

_____
(Yes or No)

3.0   Initial Startup and Operation
3.1.   Is lube oil drain functioning properly?

_____
(Yes or No)

3.2   Is temperature rise across bearing as designed?

_____
(Yes or No)

Table 11.4 (continued)

3.3	Are vibration levels at operating speed satisfactory?	_____ (Yes or No)
3.4	Is vibration level at gear passing frequency (GPF) satisfactory?	_____ (Yes or No)
3.5	Check noise level – is it within specified tolerance?	_____ (Yes or No)
3.6	After a 24-h loaded running, inspect lubrication filters for cleanliness.	_____ (Yes or No)

Appendix 11.A

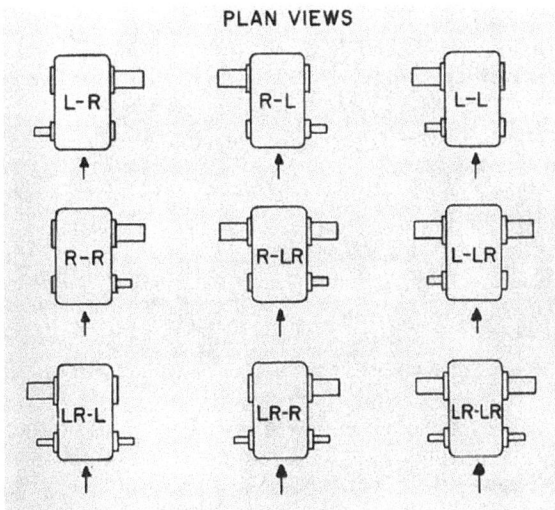

Figure 11.A1: Definition of possible parallel shaft extensions. Code: L = left; R = right. Arrows indicate line of sight to determine direction of shaft extensions; letters preceding the hyphen refer to number and direction of high-speed shaft extension; letters following the hyphen refer to number and direction of low-speed shaft extensions (modified from reference[21]).

References

1 Extreme pressure.

2 A well-known coupling expert, Michael Calistrat, used to explain the difference between "channel-ing" and "non-channeling" greases (used in outboard motors) by using the example of the Red Sea splitting during the Exodus. The sea was channeling to let the Israelites pass, but non-channeling when the Egyptian troops tried to follow them.

3 Kurz, R., K. Brun and D. Legrand, "*Field performance testing of gas turbine driven centrifugal com-pressors*", Proceedings of the 28th Turbomachinery Symposium, Turbomachinery Laboratories, Texas A&M University, College Station, College Station, TX USA, pp. 216–220, 1999. Texas A&M.

4 Heinz P. Bloch, "*Pay attention to puller holes in coupling hubs*", HYDROCARBON PROCESSING, February 1094, p.23.

5 API standard 671.

6 J. Corcoran, "Mission-Critical Couplings", Turbomachinery International, January/February 2005, pp.17–20.

7 Robert X. Perez, Technical Editor and Joe Kane, *ROTATING MACHINERY Diction ary*, Third Coast Publishing Group LLC, Houston TX, 2019, ISBN 978-1-7330413-0-0.

8 Klaus Brun, Rainer Kurz, "ARE YOU AFRAID OF GEARBOXES?", Turbomachinery International, November/December, 2015, p.44.

9 Reference 7.

10 American Petroleum Institute Standard 613, "*Special Purpose Gear Units for Petroleum, Chemical and Gas Industry Services*", fifth edition, February 2003.

11 J. Hans Arndt, Donald B. Kiddoo, "*Special purpose gearing – how to avoid startup problems*", HYDROCARBON PROECSSSING, March 1975, pp.113–118.

12 API 670, "MACHINERY PROTECTION SYSTEMS", 4th Edition, 2000.

13 API 617, "Axial and Centrifugal Compressors and Expander-compressors"; Part 3, "*Integrally Geared Centrifugal Compressors*", 8th Edition, September 2014.

14 American Gear Manufacturers' Association.

15 Allianz-Handbuch der Schadenverhütung, Allianz Versicherungs – AG, 3rd Edition, VDI Verlag, Munich 1984, pp.738–740.

16 Heinz P. Bloch, Don Ehlert, Fred K. Geitner, "*Optimized Equipment Lubrication, Oil Mist Technol-ogy and Storage Protection*", Reliabilityweb.com, Ft. Myers, FL, USA, 2020, ISBN 978-1-941872-98-7.

17 See Reference 8.

18 American Petroleum Institute API RP 686, 2nd Edition, 2009, *Recommended Practice for Ma-chinery Installation and Installation Design*.

19 See Reference 11.

20 Turbomachinery Laboratory, http://turbolab.tamu.edu
John M. Rinaldo, "*GEAR BOX SPECS – GETTING IT RIGHT*", TMI HANDBOOK · 2017, pp. 24–26.

21 AGMA Standard 420.04

Bibliography

P Lynwander, GEAR DRIVE SYSTEMS – Design and Application, Marcel Dekker, Inc., New York and Basel, 1983, ISBN 0-8247-1896-8.

Chapter 12
C&S Large Multistage Reciprocating Process Compressors

12.1 Introduction

Reciprocating compressors are the most common type of process machinery used for compression of air and particularly for light gases. They are essential for generating high pressure (see Figures 12.1 and 12.2). They are flexible and can work under variable operating conditions while maintaining high efficiency.

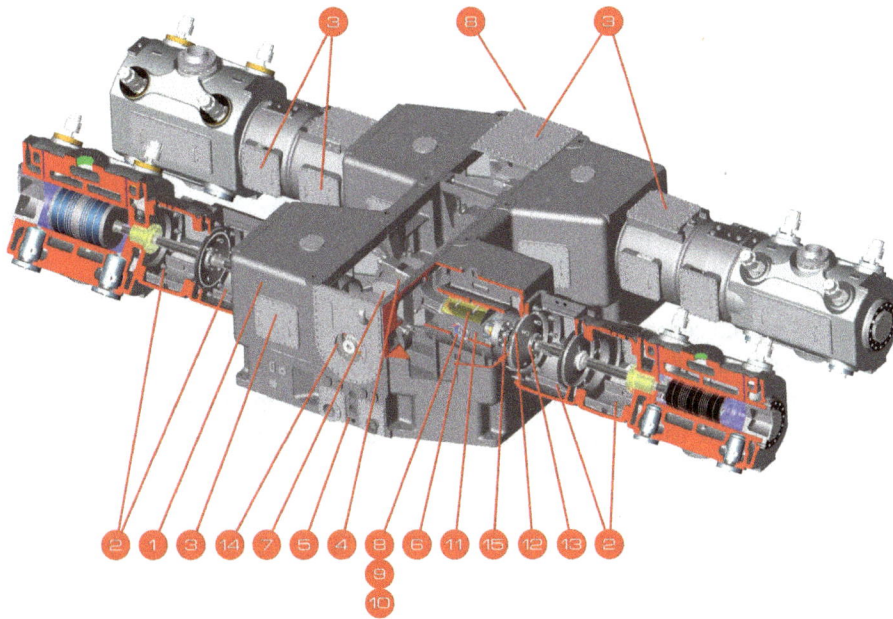

Figure 12.1: Multistage reciprocating process compressor: 1, frame; 2, distance piece; 3, inspection door; 4, main bearing bolts; 5, crankcase; 6, forced oil feed from crosshead; 7, identical, interchangeable main and crankpin bearing sleeves; 8, axial crankshaft bearing; 9, crosshead pin bearing; 10, crosshead pin; 11, crosshead; 12, space between crosshead and oil wiper packing; 13, oil wiper hoods; 14, sealed crankshaft and crankcase; 15, crosshead coupling (courtesy of LMF, Leobersdorf, Austria).

However, they can have high maintenance costs and consequently low availability if they are improperly commissioned, started up, operated and maintained. Recent investigations showed maintenance costs of around US $45.00 per horsepower per year or ~ 60.00 $/kW/a for reciprocating compression equipment as opposed to some

https://doi.org/10.1515/9783110701074-012

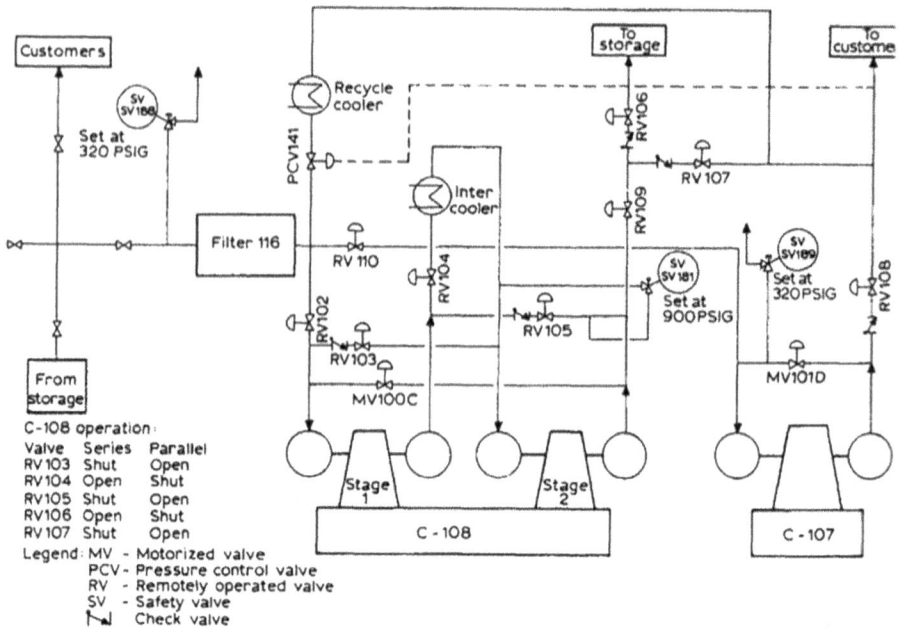

Figure 12.2: Simplified piping and instrumentation diagram showing two reciprocating process gas compressors.

10.00 US$ per horsepower per year or ~ 13.00$/kW/a for turbocompressors. This information should indicate to our readers that these machines have to be treated with caution and care during commissioning and initial startup. Additionally, because these machines are usually applied in high-pressure processes special safety, health and environmental issues can arise.

12.2 Risk Profile

Despite the application of advanced technologies in the development and manufacture of wearing parts, such as valves, piston rings and packings, these components continue to be the main sources of reciprocating compressor failure risk. Figure 12.3 shows the current risk profile of this type of machine. Valves are still the most vulnerable part of reciprocating compressors. Valves fail due to several well recognized reasons listed in order of likelihood:

- Foreign object damage (FOD)
- Liquid carryover
- Spring fracture due to fatigue
- Failure of moving components such as channels, plates, rings and poppets
- Installation oversights

- For hypercompressors used in LDPE, service alignment of the plunger (pistons are not use in these machines, they are replaced with solid tungsten carbide plungers) is critical.

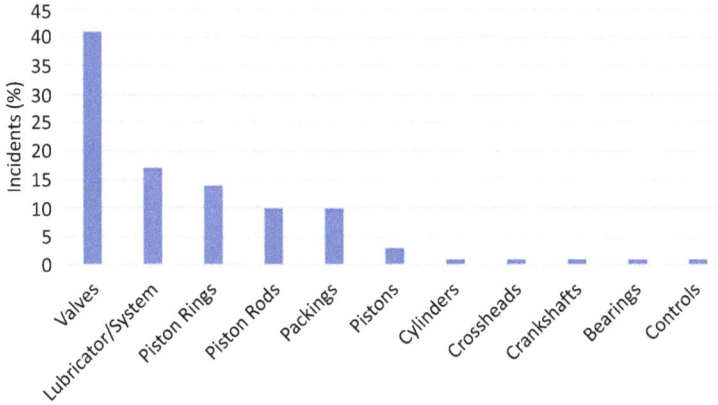

Figure 12.3: Reciprocating compressor risk profile.

Figure 12.4 represents the relative RIN calculation result for a six-cylinder reciprocating compressor driven in hydrogen service driven by a 7,500 hp or 5.6 MW synchronous 50 Hz motor operating at 330 rpm – see Figure 12.5. Figure 12.5 also shows the cylinder numbering convention. This is important for communication purposes.

1	2
	1,000hp/750KW
	Motor driven
	4-Throw Rcp.Cmp.
Rel. Risk #	**600**
Rel.Risk Level:	MEDIUM
Complexity	24
1.3	8
2.3	2
4.5	2
5.1	4
6.3	8
7.0	0
Severity	25
Table 2.4.1	5
Table 2.4.2	8
Table 2.4.3	8
Table 2.4.4	4

Figure 12.4: Relative inherent-risk assessment of a motor driven six-cylinder reciprocating process compressor unit.

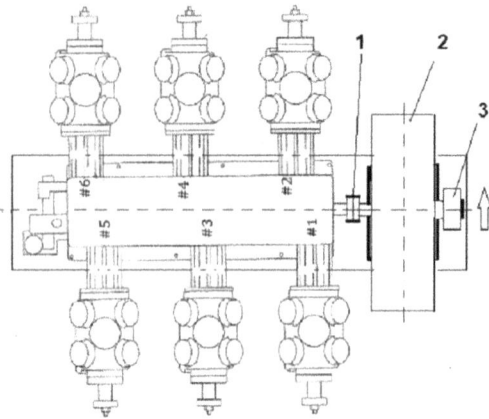

Figure 12.5: Sketch of a reciprocating process compressor unit – top view: 1, flanged connection; 2, electric motor; 3, motor outboard bearing.

Commissioning and startup-related risks are covered in Table 12.1.

Table 12.1: Project-related risk mitigation – reciprocating compressors.

Phase	Risk	Most probable cause[1]	Suggested actions/mitigating measures
Rigging and lifting	**High** – Rigging error	QP	– Qualify R&L contractor – Review Lifting Plan
Handling, staging and storage protection	**High** – Inappropriate storage protection due to unforeseen project delay resulting in rework – Failure to revise plans	OR, HK	– Follow API 618: "The equipment shall be packed for domestic or export shipment as specified. Lifting, load-out and handling instructions shall be securely attached to the exterior of the largest package in a well-marked weatherproof container. Where special lifting devices, such as spreader bars, are required, the supply of these shall be subject to agreement. Upright position, lifting points, weight and dimensions shall be clearly marked on each package." – Review plans for handling and storage protection.

Table 12.1 (continued)

Phase	Risk	Most probable cause[1]	Suggested actions/mitigating measures
Foundation and grouting	**Medium** – Flawed design – Incorrect installation procedures	QP	– Employ third-party foundation expert/consultant or have grout manufacturer rep present at site during pour – Use grout with the proper exotherm formulation suited for the climate conditions at site.
Piping	**High** – Faulty flange bolting – Ingestion of foreign objects after cleaning	OR, HK	– Flange bolting plan – Guarantee cleanliness. Assembly to the machine should follow practices established as part of the project execution agreement between OEM and the owner organization. – Piping configuration must stick to the plan used per pulsation analysis report (Analogue or Hybrid piping analysis)
Shaft alignment	**Medium** – Installers not familiar with special coupling(s)	QP	– Follow OEM instructions and project best practices[2]
Lubrication system	**High** – Contamination in the OEM shop during manufacturing (foundry sand) – Contamination during installation	QP, DO, CO	– Use OEM QA plan when available. – Because external piping must be fabricated and installed as part of the lubrication system, the cleanliness obtained during FAT helps but should be discounted, therefore the system should be flushed again to achieve cleanliness standard per ISO 4406:99.
C & S	– OEM field tool kit unsuitable – Procedure: Incorrect torquing of valve cover fasteners	FC, OR, DO	– Interview to check out OEM FSR – Include in inspection plan – Arrange for owner's witness for torquing procedure including tool calibration

Table 12.1 (continued)

Phase	Risk	Most probable cause[1]	Suggested actions/mitigating measures
Startup and Initial Operation	– By-pass line not opened – Load applied too fast	QP	– Operator training – Review operating SOP

[1]Refer to Figure 2.12
[2]Based on: API RP 686, 2nd Edition, 2006, *Recommended Practice for Machinery Installation and Installation Design and / or API RP 1FSC, 1st Edition, 2013, Facilities Systems Completion Planning and Execution.*

We would like to point out another, albeit not immediately obvious, experience-based risk mitigation that precedes commissioning activities. As we mentioned before in connection with other process machinery, this is the technical tracking of a major machine through its project phases after the purchase order is placed. It is part of a process we referred to previously as MQA. The reader might recall, distinct project phases are design, manufacturing, testing, shipment and delivery. These phases should be followed by a competent and experienced owner's engineer. As a minimum, the startup leader should be familiar with the outcome of the compressor's mechanical shop test that precedes the machine's shipment and delivery to the plant site. Yet, there is general agreement among experts that however well a reciprocating compressor system has been designed, fabricated and tested, its ultimate success depends on proper installation and commissioning. In recognition of this fact, many different organizations[1] and last but not least the engine and compressor manufacturers or original equipment manufacturer's (OEM) have issued guidelines for these last phases of the project.

12.3 The Factory Acceptance Test (FAT)

The purpose of an in-shop running test of reciprocating compressors is to determine and correct any flaws or errors in the machinery manufacture which would delay commissioning of the machine and ultimately impact its reliability. The test consists of a flush check of frame lubrication cleanliness before running, using a 100-mesh or smaller screen before filter, a mechanical run of approximately 8 h and a physical examination of machine internals after the run. This will ascertain that all working parts will operate satisfactorily in the field. Usually, because of power limits of shop drivers, the test will be run with the machine unloaded. Screens should be inserted in suction and discharge, but valves should be left in. The surge bottles are mounted and liquid knock-out vessels, if possible. The manufacturer must have taken all necessary steps to satisfy himself that the machine is ready to run before any test to be witnessed by the owner's representative. Again, the site commissioning leader should be abreast of potential problems or outstanding items discovered during that very important test. One of our colleagues created the adage that this careful behavior should result in the new machine being greeted as an old acquaintance when it arrives at the construction site. The shop run record – see Table 12.2 – should become a part of the commissioning file.

Table 12.2: Mechanical run test in original equipment manufacturer (OEM) shop.

Designation: _____

Location: _____

Service _____

1. Is lube screen free of foundry sand and weld slag? If not, crankcaše and oil cooler should be reopened to find where it came _____

2. Does the lube screen show sufficiently clean that the machine maybe safely run? Minor amounts of matter on the screen that will be removed by frame filter are acceptable _____
   (Yes or no)

3. Is auxiliary lube pump capable of supplying pressure? _____
   (Yes or no)

4. Is the pressure drop across frame oil filter normal? _____
   (Yes or no)

5. During the run is there sufficient oil on rod to indicate satisfactory operation of the lubricator? _____
   (Yes or no)

6. Are oil drops showing in all lubricator pumps and is lubricator developing full pressure? _____
   (Yes or no)

7. Are all valve covers cool and do all valves appear to be operating properly? _____
   (Yes or no)

8. Is the machine free of knocks or undue vibrations? If not, machine must be stopped immediately and problems corrected. _____
   (Yes or no)

9. Does the lube oil screen show sufficiently clean that the machine may be safely run? Minor amounts of matter on the screen that will be removed by frame filter are acceptable. _____
   (Yes or no)

10. Is auxiliary lube pump capable of supplying pressure? _____
    (Yes or no)

11. Is the pressure drop across frame oil filter normal? _____
    (Yes or no)

12. During the run is there sufficient oil on rod to indicate satisfactory operation of the lubricator? _____
    (Yes or no)

Table 12.2 (continued)

13. Are oil drops showing in all lubricator pumps and is lubricator developing full pressure?	_____
	(Yes or no)
14. Are all valve covers cool and do all valves appear to be operating properly?	_____
	(Yes or no)
15. Is the machine free of knocks or undue vibrations? If not, machine must be stopped immediately and problems corrected.	_____
16. On completion of run request a contact, infrared or laser thermometer, dial indicators, suitable micrometers, and feeler gauges.	_____
17. Have the valve covers, valves, cylinder heads, and crankcase covers been removed and have all packing cases been pulled out of bores? Record measurements.	_____
18. Have all measurements been recorded?	_____
	(Yes or no)
19. Are rod clearances in head and partition bores large enough to prevent scraping of rod after normal piston and crosshead wear?	_____
	(Yes or no)
20. Are all gasket seats free of paint, casting defects, or tool chatter marks? (Use a flashlight oriented on its side and look for radial shadows.)	_____
	(Yes or no)
21. Are valve gaskets solid metal type?	_____
	(Yes or no)
22. Is the bore for stuffing box in frame end head free of casting defects?	_____
	(Yes or no)
23. Is the rod securely fastened to piston and crosshead such that it cannot back off?	_____
	(Yes or no)
24. Is there sufficient space in distance piece to readily change packing in cylinder, intermediate diaphragm, and oil scraper rings?	_____
	(Yes or no)
25. Are separate covers supplied for each space in distance pieces for packing access?	_____
	(Yes or no)
26. Are vent and drain connections from packing and distance pieces securely piped and labeled?	_____
	(Yes or no)
27. Is the cylinder head gasket face free of any open plugs to cored water passages in either cylinder or head?	_____
	(Yes or no)
28. Are the gas passages cored such that there are no cavities or depressions for a liquid trap?	_____
	(Yes or no)
29. Are the cylinders firmly supported on the distance piece or cylinder body and not on the head?	_____
	(Yes or no)

Table 12.2 (continued)

---

30. Is the frame oil circulation so connected such that the filter is last in the stream before injection to bearings?

_____
(Yes or no)

31. Is there an easily removable plug in the piston?

_____
(Yes or no)

32. Is there any indication of penetration of stud drill or tap in the drilling for cover studs?

_____
(Yes or no)

33. Are the cylinder analyzer holes drilled and located properly? They should be open to bore through Liners and not covered by rings when piston at end of stroke.

_____
(Yes or no)

34. Are the suction and discharge valves truly non-reversible?

_____
(Yes or no)

35. Are the supports for the lubricator and piping, and frame oil cooler and piping firm enough to prevent vibration in operation and sturdy enough to withstand shipping without damage?

_____
(Yes or no)

36. Are the liners, pistons, rods and crossheads free of any score mark deep enough to catch a finger nail? If not, item must be washed clean and score must be smoothed off.

_____
(Yes or no)

37. Is the crankcase nameplate securely fastened? Does it provide machine description and serial number and will it be sufficiently viewable when on-site?

_____
(Yes or no)

38. Does each cylinder carry a nameplate securely fastened and does it indicate cylinder description and serial number and piston end clearances?

_____
(Yes or no)

39. Are the rails or sole plates for the machine precoated with epoxy paint for proper adherence to epoxy grout?

_____
(Yes or no)

40. Are laminated shim packs of stainless steel available for shipment with machine?

_____
(Yes or no)

41. Is rust protection being applied to machine before shipment and is it equal to that specified?

_____
(Yes or no)

42. Is the shipping crate sturdy enough to protect the machine, especially the external piping and gauges? All integral pipes and lines must remain installed and connected for shipment.

_____
(Yes or no)

43. Question the supplier if the shipper has adequate experience and if shipping method is completely satisfactory to prevent any damage to the machinery in transit.

_____
(Yes or no)

12.4 What Reciprocating Compressor C&S Files Should Contain

The commissioning and startup files should also contain Table 12.3, a very important aid in conjunction with API recommended practices for machinery installations[2] to be used by field inspectors and startup engineers.

Table 12.3: Checklist for contractor (EPC) drawing review.

1. Have KO facilities been installed on all suction and interstages?	_____ (Yes or no)
2. Do recycle lines re-enter upstream of the KO facilities?	_____ (Yes or no)
3. Are KO facilities equipped with a gauge glass, LLA and shutdown switches?	_____ (Yes or no)
4. Are the suction lines traced between the KO drums and the machine flanges, including pulsation bottles? Alternatively, is cylinder cooling medium (preferably water/glycol mixture) designed to be warmer than incoming gas?	_____ (Yes or no)
5. Do KO facilities have automatic drains?	_____ (Yes or no)
6. Is there a safety valve on each compression stage and on the coolant header?	_____ (Yes or no)
7. Has a remote shutdown switch been provided?	_____ (Yes or no)
8. Have all the alarm and shutdown devices specified been provided?	_____ (Yes or no)
(a) Low-lube pressure alarm	_____ (Yes or no)
(b) Low-lube pressure trip	_____ (Yes or no)
(c) High-temperature alarm on each cylinder, discharge	_____ (Yes or no)
(d) Low-cylinder lube flow alarm	_____ (Yes or no)
(e) Others (detail)	_____ (Yes or no)
9. Are TI's and PI's specified for suction and discharge of each stage and TI's on discharge and coolant outlet of each cylinder	_____ (Yes or no)
10. On machines which are specified to be reaccelerated, have controls been provided to unload the cylinders during reacceleration?	_____ (Yes or no)
11. If these controls are of the bypass type, is there a check valve on the compressor discharge?	_____ (Yes or no)

Table 12.3 (continued)

12. Have unloading facilities been provided for startup?	_____
	(Yes or no)
13. Is there a coolant block valve on each cylinder with a low point drain and a high-point vent?	_____
	(Yes or no)
14. Does each machine have double block valves and a vent to avoid blinding for valve repairs?	_____
	(Yes or no)
15. Have pressure taps been provided around the spool which will contain the temporary suction screen?	_____
	(Yes or no)
16. Do the oil coolers have provision for back-flushing the water side?	_____
	(Yes or no)
17. Will all instruments be changeable on the run?	_____
	(Yes or no)
18. Are all lines to remote pressure gauges valved at the tie-in to the main line?	_____
	(Yes or no)
19. Is there an isolatable closed circuit at the machine so that the pre-startup run-in can be performed?	_____
	(Yes or no)
20. Are there purge connections which allow gas-freeing the compressor in preparation for maintenance?	_____
	(Yes or no)

Layout

1. Are the main and interstage suction lines from the KO drums or filters cleanable by the method proposed? If the lines are quite long it may be cheaper to put filters adjacent to the machine to reduce cleaning cost.	_____
	(Yes or no)
2. Are all the cylinder, snubber and gas cooler supports from the machine foundation?	_____
Note: This requirement does not apply to remote coolers.	(Yes or no)
3. Are all the piping supports either on the machine foundation or on separate footings going down below the frost line?	_____
	(Yes or no)
4. The pulsation study will indicate which gas lines have high shaking forces and an estimated magnitude of these forces. Are these lines suitably supported and clamped?	_____
	(Yes or no)
5. Is there sufficient clearance to pull all pistons and cooler bundles?	_____
	(Yes or no)
6. Is all contractor's piping such that it does not interfere with access to any valve, distance piece, crosshead or crankcase cover?	_____
	(Yes or no)
7. Can all piping to each cylinder be removed to permit cylinder removal?	_____
	(Yes or no)

Table 12.3 (continued)

8. Where the contractor (EPC) is doing some of the oil piping, is the piping between the filters and the machine stainless steel?	_____ (Yes or no)
9. Are packing vents run separately to a safe location?	_____ (Yes or no)
10. If these vents run into a disposal header under pressure, is there a block valve and check valve at the connection to the header?	_____ (Yes or no)
11. Has each packing vent line a means of individually monitoring for leakage?	_____ (Yes or no)
12. Have distance piece vents and drains been provided?	_____ (Yes or no)
13. Has all small bore piping and connections been gusseted?	_____ (Yes or no)
14. Has piping been modified as required per acoustic study results?	_____ (Yes or no)
Process piping layout: Suction piping	
15. Are the main and interstage suctions steam traced and insulated?	_____ (Yes or no)
16. Are the machine laterals taken off the top of the header?	_____ (Yes or no)
17. Is there a manual drain on the header?	_____ (Yes or no)
18. If the compressor is handling a flammable gas, are the isolation valves 25 ft or 7.6 m from the machine?	_____ (Yes or no)
19. Is the spool piece for the temporary strainer readily accessible for both inspection and removal?	_____ (Yes or no)
20. Normally, it should be mounted immediately adjacent to the suction bottle. A strainer is required at each stage unless frame type intercoolers are used.	_____ (Yes or no)
21. Are blind flanges available at each end of the suction headers to facilitate cleaning, inspection, etc.?	_____ (Yes or no)
22. If the contractor's piping ties directly onto CI cylinders, are the flanges flat faced?	_____ (Yes or no)
23. Are all valves supported?	_____ (Yes or no)
24. Vertical unbraced lines can result in excessive vibration. Has unnecessary flexibility been avoided wherever possible?	_____ (Yes or no)
25. Is the suction KO drum within 50 ft or 15.2 m of the machine, or if not, has a proven separator been provided within this distance?	_____ (Yes or no)

Table 12.3 (continued)

Process piping layout: Discharge piping system	
26. Are all fittings such as oil separators adequately supported?	_____
	(Yes or no)
27. Is the piping system flexible enough to keep thermal load stresses on cylinders within acceptable limits?	_____
	(Yes or no)
28. Liquid condensate in intercoolers and aftercoolers will not run uphill into KO facilities. Assume that it will mist. Therefore, on intercoolers and on aftercoolers where condensate removal is desired, have liquid separation facilities been provided at the low point in the line between the cooler and the KO drum?	_____
	(Yes or no)
29. On machines mounted above grade with mezzanine floors, has grating been provided for access to all machines, valves, distance pieces and block and bypass valves?	_____
	(Yes or no)
30. Cylinder ends sometimes blow out. Is all equipment requiring operator attention such as instrument panels, instruments, and block and bypass valves out of direct line with the cylinder ends?	_____
	(Yes or no)
31. Consider maintenance on the machines. Is there a way of removing all sections of the machines for maintenance? Suitable laydown available for cylinders? If on mezzanine is deck strong enough?	_____
	(Yes or no)
32. **Consider operability**	
(a) Are all instruments clearly visible?	_____
	(Yes or no)
(b) An operator's primary sense is touch. Can he feel all valve covers?	_____
	(Yes or no)
(c) Run through the starting sequence. Can all the operations required be done by one person?	_____
	(Yes or no)
33. For **integral gas engine compressors** consider the need for re-grouting. During the life of the machine it will likely have to be re-grouted. Is the building design such that heavy lifting equipment can be brought in to lift and move the whole unit?	_____
	(Yes or no)

This document is the last official opportunity for the owner's representatives to influence the EPC or OEM design and deliverables before installation of the equipment commences.

12.5 Installation

12.5.1 Rigging and Lifting

Generally, three crane capacities are recommended
1. Capacity of overhead crane for maintenance, for example, a typical 9,000 hp or 7 MW special-purpose train would require 12 ton.
2. Maximum maintenance weight for mobile crane during IMR&O around 55 ton.
3. Maximum skidded package installation of the above size train using a mobile crane would require around 100 ton. Crane height also should be taken into account. For the above example it would be in the range of 36–42.6 ft. or 11–13 m.

12.5.2 Receiving, Inspection and Storage Protection

This step is for obvious reasons well understood and is most of the time successfully taken care of. The purchaser's and user's expectation are met when the following measures have been taken:
- Internally protected for up to six months for indoor storage by factory.
- Site climatic and environmental conditions understood by OEM.
- Compressors stored outdoors or in a humid environment have been subjected to preservation measures.
- The purchasing documents are clearly spelling out storage protection requirements – usually in increments of six months.
- Unplanned extended storage times trigger appropriate revaluation of current preservation.

We recommend the following longer-term preservation measures for the compressor:
1. Refer to Appendix 5.A for preservation product specification.
2. Purge all cylinders of hydrocarbons.
3. Blank or blind the compressor suction and discharge.
4. Fill the crankcase, cooling water jacket and valves with Product B.
5. Install a valved standpipe, allowing space for thermal expansion.
6. Cover all exposed machine parts with Product C.
7. Top up the level of product B in the cooling water jacket.

Again, suitable project policies must be created timely to assure the machine is protected from deterioration after arrival at the construction site, during field staging and erection as well as during the period prior to commissioning.

12.5.3 Foundation, Base Plate and Grouting

In Chapters 5 and 10, we discussed foundation, baseplate and grouting issues in detail. Very similar steps should be planned and executed when it comes to reciprocating compressors and their drivers. Again, it cannot be stressed enough that finished foundation may pass all possible inspections, but if it is not designed and constructed properly, its flaws will appear years later causing failure and unscheduled downtime.

As in foundation and grouting for other reciprocating equipment, there are many risks in connection with these activities involving reciprocating compressor units. The risks are:
– Bad crankshaft webs
– Excessive frame movement
– High vibrations
– Loose or cracked chocks
– Inability to maintain anchor bolt tension
– Inability to hold alignment

12.5.4 Assembly and Maintenance Tools

Like in connection with other major equipment, an important list in the compressor C&S files should be the vendor or OEM furnished assembly tool list. There should be some proof of the fact that these tools are on site, current and usable, and if they are not, expensive project delays may happen. A typical list for reciprocating process compressors would, for example, contain:
1. Bearing removal tool
2. Piston extractor
3. Valve extractor
4. Piston fit-up tool
5. Hydraulic tightening system
6. Crosshead assembling tool
7. Special lifting tools
8. Partition plate assembling tools
9. Mandrels for wear bands

12.6 Commissioning

A two-stage reciprocating process compressor commissioning procedures can be broken down into five distinct phases:
1. Compressor and system completeness checks also referred to as pre-startup phase.

2. Compressor run-in with valves removed.
3. First-stage valves in. Compressor run-in and piping system blowing in progress.
4. First- and second-stage valves in. Compressor run-in and air load test.
5. Run-in and load test on process gas.

12.6.1 Safety Precautions

Installation, operation and maintenance of a compressor and its auxiliary components present certain hazards that are unique to this type of equipment. All personnel prior to working with or on the compressor and its auxiliary systems should understand well established safety precautions. Pressure safety valves must be pressure relief tested and installed to protect against a blocked discharge. Failure to comply with these warnings can result in an accident, causing property damage, personal injury or a fatality. The following points should be paid attention to in this context:

– Volatile flammable liquids must not be used as a cleaning agent for compressor parts. A safety solvent must be used, and the parts thoroughly dried before assembly. There should be appropriate provisions for ventilation when using any halogenated solvents. Consult national and local safety and health organizations regulations covering cleaning materials and their use.
– Air jets must not be used for cleaning work clothes, due to the risk of air entering the bloodstream via body openings or small skin wounds. When air jets are used for cleaning parts, sandblasting or for spray painting, skin and eyes must be adequately protected against flying debris.
– Piping with temperatures greater than 175 °F or 80 °C which may be touched by personnel must be suitably guarded or insulated before startup.
– A reciprocating compressor must not be placed in operation unless adequate safeguards have been provided to protect operating personnel. Service on a machine shall always start with good housekeeping practices consisting of cleaning the floor and the outside of the machinery to remove oil that could have accumulated and cause maintenance personnel to slip.
– The compressor must be fitted with safety valves or rupture disks shown earlier in Figure 12.2 to limit the discharge pressures to a safe maximum. We should never think of installing a valve between a compressor cylinder and the safety valve or rupture disk. Safety valves must have their settings tested at least once a year, and more often under extreme operating conditions, using an appropriate bench test. Startup personnel should see that a recent safety valve test record is available prior to the initial startup of a reciprocating compressor. If a safety valve or rupture disk releases during operation, stop the unit immediately and determine the cause. Pressure relieving devices vented to the atmosphere must have their outlet connections directed away from operator stations.

12.6.2 Prime Movers

Please refer to Figure 1.3 showing prime mover applications. In the downstream section of the hydrocarbon processing industry (HCPI), for example, the most common prime movers for reciprocating process compressors are electric motors. In the oil and gas transportation, the midstream business of HCPI, high speed compressors driven by internal combustion engines are common.

In the case of an electric motor driver, control wiring must be carefully installed according to National Electrical Codes, Occupational Safety and Health Acts and any other code requirements at the installation site as pointed out in Chapter 7.

12.6.3 Piping Cleanliness

When connecting and bolting up pipe flanges, the premise is that the piping has been cleaned during the pre-commissioning phase. Nevertheless, when first starting, it is advisable to use a temporary line filter in the intake line near the compressor to catch any dirt, chips or other foreign material that may have been left in the pipe.

Even though cleaning procedures[3] have been carefully followed on the compressor piping, a temporary filter – such as Type PT American Filter or equal – should be installed in the line to the suction bottle to remove particles 230 μm or 0.009 in. in diameter or larger. If the compressor is an "NL" – nonlubricated – design, the filter should be designed to remove particles 140 μm or 0.0055 in. in diameter or larger. Provision must be made in the piping to check the pressure drop across the filter and to remove the filter cell for cleaning. If the pressure drop across the filter exceeds 5% of the upstream line pressure, remove the filter, clean thoroughly and reinstall. The filter cell should be removed and left out only when the inlet line is free of welding beads, pipe scale and other foreign objects.

One piece of hardware deserving to be mentioned again at this point are strainers. They are underappreciated and frequently overlooked or forgotten. Their function is to protect close tolerance positive displacement machinery by catching the "things that shouldn't be there and won't go through without damage to the machine."[4] We saw that strainers come in a variety of arrangements. The most common types in the compression industry are cone and basket strainers. These are similar in that a basket strainer is a truncated cone but the basket strainer is shorter for the equivalent flow area or gives more flow area in the same length. Figure 12.6 shows a basket type strainer. Both cone and basket strainers have a metal flange at the large end, which is held in place between mating ANSI flanges with a gasket on each side. This stack up of gasket-flange-gasket requires additional space, typically 0.25 in. or 6.35 mm. Strainers in compression service are installed with the cone or basket pointed upstream. This causes debris to collect in the annulus at the large end of the column. Flow within a pipe is faster in the center and slower close to the pipe wall due to friction. By collecting the debris near the pipe wall, the negative impact on flow and resultant pressure drop are minimized.

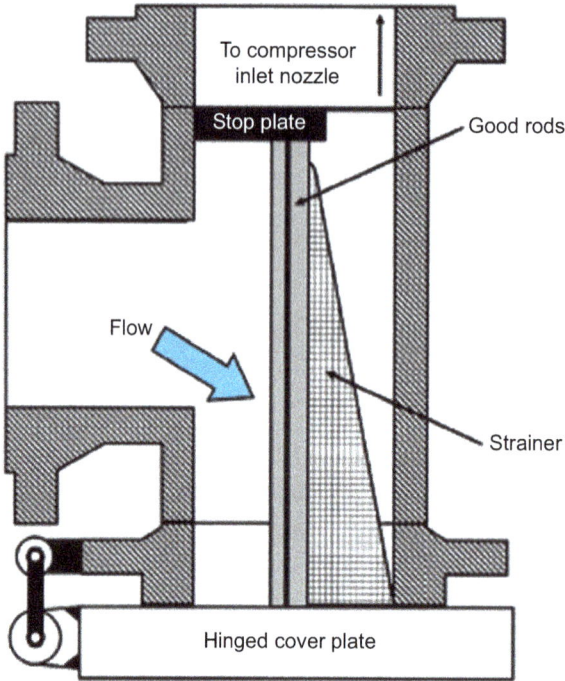

Figure 12.6: Basket strainer for compressor suction pipes.

12.6.4 Pipe Alignment

We should never forget: Piping strain causes high stresses that can lead to failures of piping and the equipment to which it is connected. There are cases where piping strain can be easily avoided by flexible connections. Most reciprocating compressor applications, however, are at high pressures where process piping has to be rigidly connected the compressor. Here, pipe strain can only be controlled by directing careful attention to the fitting and alignment process of mating pipe flanges prior to bolting them together. Adherence to procedures as shown in Chapter 5, Appendix 5.G to 5.H will go a long way. The consequences of not following these and other good procedures[5] may turn out to be revising pipe supports and clamps or more drastic measures such as cutting and re-welding the flanges or connecting piping to ensure that the alignment is within acceptable limits for parallelism and concentricity.

12.6.5 Inlet Gas Separation and Scrubbing

Entrained liquids in the gas stream can cause serious and even catastrophic damage to reciprocating compressors. It is therefore, important to provide for adequate liquid separation upstream of each stage of the compressor. Most compressor packages in midstream operation of the HCPI, for example, include inlet scrubbers for each stage, but they are typically intended as a safety device to protect the cylinder from a small amount of condensation entering the cylinder; they are not furnished to be used as the primary method of liquid removal and cannot handle large amounts of liquids. Therefore, whenever there is a possibility of free liquids in the gas stream an adequately sized separator or slug catcher must be installed immediately upstream of the compressor package.

The last paragraph is valid for all reciprocating compressor installations, as operators are always wary of liquids entering the gas stream upstream of their compressors. Liquid levels in suction vessels or knock-out drums are usually automatically controlled; however, it is very important to be able to manually check for the presence of liquid during commissioning and startup to ascertain the automatic blowdown is working. This activity must be on a checklist.

12.6.6 Shaft Alignment

Shaft alignment of the driver to the compressor is of extreme importance for a reliable operating experience. To paraphrase what we said about alignment in other chapters, proper alignment including allowance for thermal growth is an important task during the installation and commissioning phase. Aligning the two machines sketched in Figure 12.5 would have to be extremely accurate as the inboard side of the synchronous motor rotor is supported by one of the compressor bearings. There should be a clear understanding about acceptable tolerances before the unit can be handed over to the owner.

12.6.7 Lube Oil System

Prior to initial startup, the lube oil system must be flushed as part of the precommissioning activities. The flushing procedure is similar to what we described earlier in this text. However, there are some details that are unique to reciprocating machines. Here the risk of dirt accumulation is heightened because of the fact that reciprocating compressor housings or frames are castings with an inherent tendency to accumulate foreign objects, in short, dirt. The authors are familiar with several examples where undiscovered dirt had wreaked havoc with bearings and other lubricated components in spite of the customary attempts to assure cleanliness during all manufacturing stations ranging from foundry operations to preparations for shipment.

We recommend the following procedure similar to the one described in Chapter 5 to remove debris or foreign objects, which may damage bearings or lube oil system components:

1. Remove all frame inspection covers and verify the compressor frame to be free of dirt, debris or other items inappropriately left in the compressor frame. Disconnect the tubing from the main bearing caps.
2. Connect a temporary hose between the auxiliary lube oil pump suction line and a barrel of clean lube oil. Connect a temporary hose between the compressor frame lube oil drain and a collecting barrel. Place a flanged connection and a special filter carrier[6] between the frame lube oil inlet flanges. This flange can be found at the discharge of the frame lube oil pump.
3. Replace oil filter elements and reassemble the filters. Double check proper assembly against filter cross-sectional drawing as filter designs vary.
4. Refer to Figure 12.7.
5. Circulate oil from the compressor frame to a waste oil collecting tank. Use the electric motor driven auxiliary pump for the flush, as well as both lube oil filters. Monitor the flow out of the main bearing cap tubing and as soon as clean flow is established to the main bearing cap tubing, reconnect this tubing.
   Note: If any debris is noted coming from the main bearing lube oil tubing, the frame lube oil sump should be wiped out with clean lint-free rags.
6. Continue oil circulation until the milk pad and mesh show no signs of contamination. It will probably be necessary to change the gauze pad frequently during this evolution. Adjust lube oil pump relief valves to proper setting (75 psig or 520 kPa(g)).
7. After the system is deemed to be clean, circulate oil and monitor for proper flow to each of the running gear components.
8. Monitor the crossheads, connecting rods, wrist pins and main bearings for flow.
9. When satisfied with the lube oil system flush, shutdown the system and change the lube oil filter elements.
10. Fill the lube oil filter and inspect for leaks when operational.
11. Prior to running the compressor, ensure oil temperature is at least 60 °F or 16 °C.

It is a good practice to have procedure steps such as above accompanied by a marked-up sketch such as Figure 12.7 in order to show points of pipe breaks for flushing.

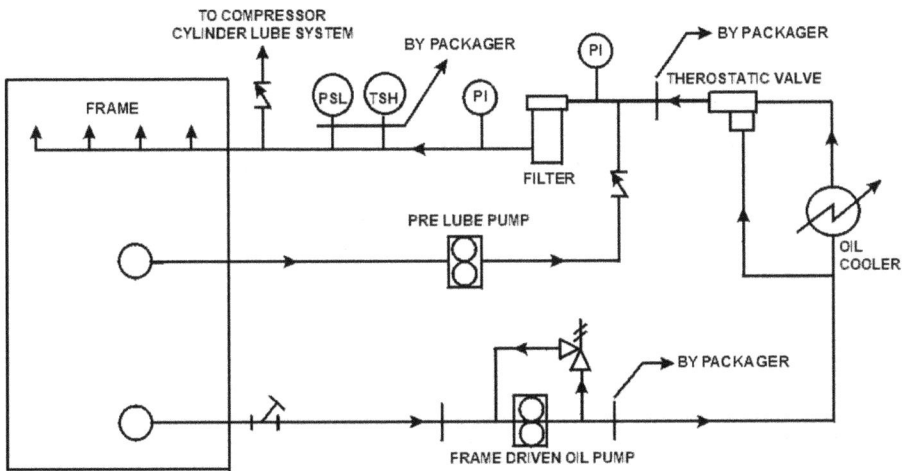

Figure 12.7: Simplified flow diagram – reciprocating compressor lubrication system.

12.6.8 Cooling Water System

The system requires flushing as follows:
1. Install a 100 mesh[7] screen at the cooling water header.
2. Fill the system with water. Ensure all air is removed by opening all high-point vents.
3. Let fresh water circulate through the system and monitor sight flow indicators for proper flow and periodically vent the system to remove trapped air.
4. Leaks, if present, should be corrected at this time.
5. Periodically check the mesh screen and change, as needed.

Once initial flush is complete, drain the system and remove mesh screen.

The cooling water system must be operating prior to compressor startup. If the compressor is started with the cooling system inoperative, it must be shut down. Do not start the cooling system until the cylinder water jackets are cool to the touch or thermal shock may cause damage to the cylinders.

12.6.9 Compressor and Motor Driver

The following procedure is based on the assumption that the motor rotation has been verified, the above auxiliary systems have been prepared for startup and all compressor alarms and shutdowns have been checked – see also Table 12.4. The following steps should be considered:

Table 12.4: An example of a typical reciprocating compressor pre-commissioning checklist.

Compressor number _____

	Cold/Prelim.	Hot	Final Corrections	Comments
Compressor mechanical				
Crankshaft deflection				
STG 1 rod run-out (V/H)				
STG 1 X-head clearance				
STG 1 piston end clear (HE/CE)				
STG 1 rider ring clearance				
STG 1 internal insp. (HE/CE)				
STG 1 packing lube/vent insp.				
STG 1 bolt tightness check				
STG 2 rod run-out (V/H)				
STG 2 X-head clearance				
STG 2 piston end clear (HE/CE)				
STG 2 rider ring clearance				
STG 2 internal insp. (HE/CE)				
STG 2 packing lube/vent insp.				
STG 2 bolt tightness check				
Crankcase thrust end clear				
Crankcase rod end clear				
Lube oil sys. flush clean				
Lube oil distrib. points pumping				
Lube oil connections tight				
Crankcase prestart inspect.				
Lubricator system clean				
Lubricator dist. points pumping				
Lube oil pumps run-in				
Lubricator pumps run-in				
Compressor suction pipe insp.				
Compressor drums insp. (suct./inter.)				

Table 12.4 (continued)

	Cold/Prelim.	Hot	Final Corrections	Comments
Strainers in place (suct./inter.)				
Motor/compressor CPLG bolts tight				

	Checked	Data Logged & Filed		Comments
Motor – elect./mech.				
Phase rotation				
Motor leads megger				
Bearing insulation megger				
Motor air gap check				
Motor mag. ctr. check				
Control wiring and interlocks check				
RTD checkout				
RTD calibration check				
Motor CT checked				
Motor check for debris				
Rotor position relative				
LEF coupling				

	Installed	Calibrated	Simulated	Comments
Instrumentation				
Suction drum high-level alarm				
Suction drum high-level shutdown				
Suction/interstage/discharge flow OR				
Process high temp. shutdown				
STG 1 high discharge temp. alarm				
STG 2 high discharge temp. alarm				
Interstage drum high-level alarm				
Interstage drum high levels shutdown				
STG 1 high water temp. alarm				
STG 2 high water temp. alarm				

Table 12.4 (continued)

	Installed	Calibrated	Simulated	Comments
OB motor bearing temp. alarm				
IB motor bearing temp. alarm				
Motor winding. (RTD) high temp. alarm				
Lubricator-level alarm				
Low-lube press. alarm/aux. pump start				
Low-lube press. shutdown				
Low suction press. alarm				
Low suction press shutdown				
Barring rig motor interlock				
Suction safety valve				
Interstage safety valve				
Discharge safety valve				
Interstage pressure control valve				
Discharge pressure control valve				

1. Install startup screens in compressor inlet piping.
2. Visually inspect compressor area to ensure all piping properly installed. A walk through of each subsystem with the appropriate P&ID is recommended.
3. Check all bolting for proper tightening and threaded connections, including "bull plugs" in all pulsation control vessels.
4. Check all small bore[8] piping supports.
5. Ensure the compressor area is cleaned up and all unnecessary equipment or tools are removed and that there is no danger of anything falling from the overhead when the compressor is run.
6. Open the compressor bypass valve. This will ensure that the initial operation in the field is unloaded and allow for a check of a no-load mechanical run. This represents the first step to insure safe operation.
7. Remove one suction valve per end from each cylinder, or unload each cylinder end. Be sure to re-install valve covers once the suction valves are removed, this will prevent foreign material from falling into the cylinder bore.
8. Manually bar the compressor over and listen for any abnormal noises. Bar the machine over several times. This is to ensure the compressor turns freely.

9. The bar over check is extremely important and should also be done anytime the machine has had any extensive intervention. Damage to the compressor could occur if this is not done.

Check lube oil temperature; ensure temperature is above 60 °F or 16 °C. Start the auxiliary lube oil pump. Monitor oil pressure and level. Inspect lube oil system for leaks.

Ensure jacket water cooling system is operational and verify flow with the sight flow indicators.

Generally, one would now proceed as listed in Table 12.5. This checklist was used for a high-speed reciprocating compressor package in the midstream business of the HCPI – see Figure 12.8. The checklist is thorough and would also, with a few alterations, be applicable to large multistage reciprocating machines shown earlier in our Figure 12.5 sketch.

Table 12.5: Reciprocating process compressor pre-startup checklist – package environment.

	Initials
1. Are applicable Operation and Maintenance Manual, parts book, special tools & spares available?	_____
2. Has the 100 psi or 7 bar (or as specified) pressure check been completed and system purged with nitrogen?	_____
3. Has the final flange torqueing been done and documented?	_____
4. Have the Vendor Drawing and Data Requirements (VDDR) been fulfilled?	_____
5. Are the items/documents listed in the project VDDR Form available?	_____
6. Have the application operating conditions been determined? Suction: _____ psig / kPa _____ Temp: _____°F / °C Discharge: _____ psig / kPa _____ Temp: _____ °F / °C Min. RPM _____ Max. RPM _____ (for IC engine driven units)	_____
7. Is a copy of the compressor performance prediction run of expected operating conditions on hand & has valve type/design been checked for this application?	_____
8. Are expected startup & operating conditions within compressor design limits? If in doubt, contact the OEM representative for confirmation of the compressor's operational limitations.	_____
9. Drain any condensed water from crankcase oil sump & lines. Check thoroughly crankcase is clean using lint free rags. Remove rust inhibitor (desiccant) bags from crosshead guides & cylinders.	_____
10. Drain and flush compressor oil make-up tank & supply lines of debris & condensed water.	_____
11. Confirm any discharge bottle supports are loose at ambient temperature.	_____
12. Soft-foot on packaged machines: Confirm the frame leveling jackscrews have been backed off. Have frame & guide supports been properly shimmed & bolts re-torqued? Compressor must not be twisted or bent.	_____

Table 12.5 (continued)

13. Have crosshead guide supports been properly shimmed tight plus + 0.005" or 127µm for small cylinders 7" or 178mm and below and 0.010" or 254µm for large cylinders above 7" or 178mm? Have bolts been re-torqued?	_____
14. Has the compressor to driver cold alignment been checked at site? Are alignment readings within specification and readings recorded?	_____
15. Have coupling / flywheel adapter and drive coupling bolt torque values been verified & recorded?	_____
16. Has compressor crankshaft axial thrust been checked & recorded? Is drive coupling free floating when installed (check coupling design)?	_____
17. Have piston end clearances been checked and recorded?	_____
18. Have variable volume control pockets (VVP) been set to the desired clearance settings & stem jam nuts been re-torqued to specification?	_____
19. Has the oil system been thoroughly cleaned following good oil system flushing and cleaning practices describe earlier in Chapter 5?	_____
20. Has the oil filter been installed and compressor crankcase filled with oil to the proper level?	_____
21. Is the crankcase oil supply isolation valve open?	_____
22. Pre-Lube: Confirm pre-lube pump rotation if electrically or pneumatically driven,	_____
23. Oil system, oil filter & oil piping been primed with oil? **Note:** Electric motor driven units require an automated pre-lube pump system.	_____
24. Is the compressor crankcase oil level controller vented and set to the proper level?	_____
25. Has the crankcase "low oil level" shutdown & alarm been adjusted and functionally checked?	_____
26. Has the low oil pressure shutdown tubing been installed & shutdown setting verified? **Note:** Minimum oil pressure 30 psig or 200 kPa falling pressure.	_____
27. Is oil cooler installed & is oil flow counter to water flow? **Note:** Maximum oil supply temperature to main bearings is 185°F (85°C).	_____
28. Has the force feed lubricator box been filled with oil? **Note:** If frame & cylinder oils are not compatible, assure there will be no mixing. Has the force feed lubrication system been primed and purged of all air?	_____
29. Are all of the packing case drains/vents installed and open?	_____
30. Replace all inspection covers.	_____
31. Are the packing & cylinder (force feed) lubrication pump(s) adjusted to break-in (maximum flow) rates? Normal lube rate: ____seconds/stroke.	_____
32. Has force feed lubrication system "No Flow" switch(es) been installed and functionally checked?	_____
33. Is the force feed rupture disc assembly(ies) installed and checked for correct disc color & pressure rating? (See operator's manual for proper disc selection.)	_____
34. Have the frame & cylinder cooling systems (if applicable) been filled with proper coolant / antifreeze and purged of air? Compressor piping, suction drums and inter-stage drums have been chemically cleaned and inspected.	_____
35. Have the suction lines been blown out to remove water, dirt, slag, and any other foreign objects?	_____
36. Have knock-out drum level controls been commissioned for easy blowdown. Has the manual bypass been identified?	_____

Table 12.5 (continued)

37. Have the suction, inter-stage, & discharge pressure high/low shut downs been set and functionally checked?	_____
38. Have the safety relief valves been installed in the proper locations for each stage of compression to protect piping and cylinder MAWP ratings?	_____
39. Have the discharge gas temperature shutdowns been installed, set & functionally checked?	_____
40. Is there a compressor frame vibration shutdown device installed, set & functionally checked?	_____
41. Have the temporary inlet debris screens (100 mesh) been installed at the suction flange of each cylinder?	_____
42. Has the gas piping been purged of all air for machines compressing a combustible gas?	_____
43. Is the system settling-out pressure known? It is: _____ psi/kPa	_____
44. Have all critical fastener torque values been checked & recorded? **Caution:** Loose fasteners may result in a safety hazard or equipment failure. Frequently OEMs list critical fasteners in their O & M Manual.	_____
45. Has the driver rotation been verified to match the compressor rotation?	_____
46. Verify the compressor and its driver are free rolling with minimum force. For engine drives, has the unit been rolled with the air starter to ensure it Is free turning? For electric motor drives, has the unit been barred over by hand to ensure it is free turning?	_____
47. Have the compressor frame & cylinder lube oil systems been pre-lubed prior to starting?	_____
48. If applicable, has the driver over-speed shutdown been electronically set & verified? **Note:** Do not exceed compressor's max. unloaded inertia speed, see instructions for operational limitations.	_____
49. Package compressors: Have startup instructions for all package equipment been reviewed & performed?	_____
50. Has the OEM or packager representative reviewed the unit's startup & operating instructions with the site personnel?	_____
51. Has absence of liquids in the suction pipe been ascertained? Has suction KO drum been blown down manually?	_____

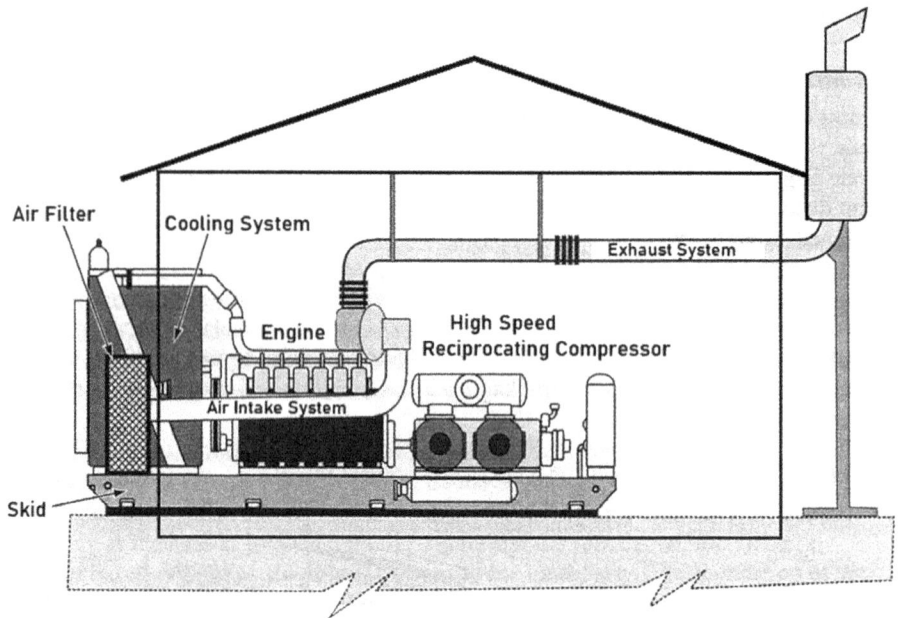

Figure 12.8: Typical gas field reciprocating compressor package installation (adapted from reference[9]).

12.7 Startup

12.7.1 Initial Unloaded Startup

The compressor is now ready for initial startup. Table 12.6 contains the recommended procedural steps. During the initial startup and any subsequent startups after maintenance, for example, it is imperative to closely monitor the unit for abnormal noise, vibration or temperature. The equipment must be shutdown immediately if an abnormal condition exists. It is important to station personnel around the installation to monitor the compressor and motor driver for abnormalities. The next steps are:

– Verify proper rotation of the motor driver.
– Start the compressor drive motor. Let the motor come up to operating speed then turn the motor off. Monitor and time the compressor coasts down.
– Note the amperage drawn by the compressor motor upon startup.
– The amount of current drawn by the motor is a good indication of free running gear parts. Amperage drawn in each subsequent start (for the same 0% load condition) should be the same or slightly less.

Table 12.6: Typical run-in instructions for large multistage reciprocating compressors.

A. PREPARATORY CHECKS

1. Has pre-startup, Table 12.5 (Pre-Start Checklist), been completed?	_____
2. Motor drivers are electrically checked-out.	_____
3. Is the motor suitable to sustain total rod load? (FINAL CHECK)	_____
4. Instruments of the compressor system have been set, checked, and simulated.	_____
5. Mechanical checks of compressors are complete.	_____
6. Lube oil flushing of compressors is completed as outlined in Chapter 5.	_____
7. Cooling water system to lube oil coolers, compressor cylinders, gas suction coolers and gas inter-stage coolers has been flushed and commissioned.	_____

B. INITIAL RUN-IN (UNLOADED)

1. Remove suction valves, blind cylinder suction or install baffles (as required).	_____
2. Main oil pump running with auxiliary oil pump on standby auto start. Check auxiliary pump start and low oil pressure shutdown.	_____
3. Check shutdowns or trips to be bypassed for this run-in.	_____
4. Start lubricator motor. Bleed and check oil flow (per OEM instructions) at each lubricator connection.	_____
5. Bar compressor over for one complete revolution to ensure it turns freely.	_____
6. While observing lube oil pressure, bearing temperatures, and compressor cylinder temperatures, start motor driver. Allow speed to increase to just below synchronous speed then manually shut unit down. Record the time of acceleration and the roll down time. Investigate noises and any suspected malfunctions.	_____
7. Bar compressor over for one complete revolution.	_____
8. Have operating personnel stationed at the control panel to be able to activate ESDs in the event of a malfunction	_____
9. Start motor driver	_____
10. Run compressor for 5 minutes noting bearing temperatures, oil pressure, and noises.	_____
11. Shut unit down and record roll down time.	_____
12. Pull motor breaker and shutdown lubricators and lube oil pump.	_____
13. Open crankcase top covers and feel bearing caps and connecting rods for excessive heat or use infrared temperature gun to scan top cover, crosshead doors etc.	_____
14. Open crosshead side covers and check crosshead shoes and pins for elevated temperature.	_____
15. Open one suction valve cover on each end of each cylinder to check condition of cylinder walls.	_____
16. After closing up the machine, start lube oil pump (auxiliary pump in auto.) and lubricator. Bar over compressor, then start motor driver.	_____
17. Allow machine to run for 15 minutes keeping close watch for problems which may be developing.	_____
18. At conclusion of a 15 minute run, shutdown machine, time roll down, then repeat inspections of steps 13 through 15 above (optional if time permits).	_____
19. Run machine for 1 hour, watching for signs of potential problems. Time roll down at conclusion of run and repeat steps 13 through 15 (optional if time permits).	_____
20. Run machine for 24 hours logging complete data of the run-in log sheet every hour.	_____
21. At conclusion of run, time roll down, hot check crankshaft deflection, measure rod run-out and cross-head clearances, take oil sample from crank-case, plus the checks of steps 13 to 14.	_____

Table 12.6 (continued)

Temperatures	1	2	3	4	5	6	7	8
Oil at pump <180 °F or < 82 °C								
Main bearing lube oil <225 °F or <107 °C								
Main bearing behind throw to cylinder number <225 °F or <107 °C								
Conrod shell to cylinder number <225 °F or <107 °C								
If temperature excessive, rerun until satisfactory and then examine part for wipes								
Rod drop to cylinder (piston high *(H)* low *(L)* with dial indicator) <5 mils or <127 µm								
Rod centerline runs below head bore centerline. Micrometer in stuffing box bore top and bottom of rod								
Rod clearance on bottom-cylinder head bore. Feeler gauge reading								
Clearance at crosshead shoe-top. Long feeler gauge. Ample clearance required, but must be uniform across the whole shoe								

Piston clearance in liner with feeler gauge. (If rider rings
supplied must be ample to allow for wear)
Name of Inspector: _____   Assert No.: _____
Serial No.: _____
Date: _____ Make: _____ Type: _____
Size:_____

*Measurements taken
immediately after run test
(dimensions in thousandths of
an inch or µm) Cylinder numbers*

Frame oil pressure before filter/after filter (<5 psi or <35 kPa across filter)_____
Standby lube pump pressure_____
Cylinder lubricator pressure if available_____

C. OPERATION WITH FIRST-STAGE LOADED	**Comments**

Note: Second-stage valves are removed – Install all first stage valves.
1. Remove blind between first-stage suction bottle and cylinder and install temporary screen. Remove spool piece at inlet to bottle. Blind on process side and install cone screen on compressor side.
2. Remove all blinds in compressor inter-stage except at inlet to second-stage bottle.
3. On second stage, blind cylinder inlet and lift suction bottle about 2 in. or 50 mm. Open discharge bottle outlet flange, blind on compressor side and separate flange about 2 in. or 50 mm.
4. Commission and test the first-stage water temperature and discharge temperature alarm.
5. Blind outlet of suction drum and lift open flange 2 in. or 50 mm. Open first-stage startup bypass line.

Table 12.6 (continued)

C. OPERATION WITH FIRST-STAGE LOADED	Comments
6. After pre-startup checks, start compressor. First stage should be taking suction at open suction spool piece and discharging through the startup bypass to the suction drum then to the atmosphere via an opening provided upstream of drum.	
Note: Be certain the second-stage bypass lines are closed.	
8. Allow suction system to be blown for 1 h. Then, open the second-stage bypass which connects upstream of the inter-stage cooler and slowly close the first-stage startup bypass.	
9. Allow the machine to run in this condition for 1 h. i.e., taking suction through the open spool at the inlet to the suction bottle, discharging to the inter-stage via the bypass upstream of the cooler, and back blowing to the second-stage discharge bottle.	
10. Open the valves in the suction pressure control spill back or recycle from the inter-stage and slowly close the second-stage bypass upstream of the intercooler to blow from the first-stage discharge through the intercooler and inter-stage drum back through the suction pressure control line. Blow for 1 h.	
11. Open second-stage startup bypass and slowly close valve in suction pressure control line to blow the entire inter-stage through the second-stage bypass and out at the opened second-stage discharge bottle flange. Blow for 1 h then shutdown machine.	
12. Remove blind at inlet to second-stage bottle. Restart compressor and allow to run overnight blowing inter-stage through opened flange at inlet to second-stage cylinder.	

D. OPERATION WITH BOTH STAGES LOADED	Comments
1. Install second-stage valves.	
2. Connect second-stage bottle with temporary suction screen installed.	
3. Remove blind on second-stage discharge and connect flanges.	
4. Provide atmospheric release downstream of suction pressure control valve.	
5. Commission second-stage discharge temperature and water temperature alarms.	
6. Commission suction and inter-stage level alarms and trips.	
7. With suction pressure control line and second stage discharge open (all bypasses closed), start compressor. Allow to run without load for one- half hour – clearance pockets or VVPs are open.	
8. Close clearance pockets, then pinch suction pressure control line blow off until the first-stage discharge pressure is _____psig or _____ kPa to limit discharge temperature to 300 °F or 150 °C. Pinch second-stage discharge blow off until the final discharge pressure is _____psigor _____ kPa to limit discharge temperature to 300 °F or 150 °C.	
Note: It may be necessary to readjust the blow-off until equilibrium is established.	

E. RUN-IN ON PROCESS GAS
Note: This run-in will closely resemble normal compressor operation except recycling will be controlled to simulate actual operating conditions. Prior to this run-in, all required instrumentation must be fully commissioned – see Table 12.4.

Table 12.6 (continued)

D. OPERATION WITH BOTH STAGES LOADED	Comments
1. Inert the system per process instructions.	
2. With cylinder bypasses open, pressure system to _____ psig or _____ kPa(g).	
3. Follow compressor operating instructions for preparing the unit for startup.	
4. Start compressor and allow to run unloaded for 15 min.	
5. After assuring the suction pressure control line is wide open, close the first-stage bypass line completely.	
6. After assuring the second-stage bypass upstream of the cooler is wide open, close the second-stage startup bypass.	
7. Check that the inter-stage cooler has water circulation, then pinch the suction pressure control line valve to bring the first-stage discharge pressure to psig or _____ kPa(g). Keep close attention to maintaining the suction pressure at about _____ psig or _____ kPa(g).	
8. Pinch the second-stage bypass upstream of the intercooler to produce a discharge pressure of _____ psig or _____ kPa(g). Adjust recycles as required to maintain this condition of pressures for about 1 h.	
9. Pinch suction pressure control line valve to produce an inter-stage pressure of _____ psig or _____ kPa and pinch second-stage bypass upstream of intercooler to produce a final discharge pressure of _____ psig or _____kPa(g). Allow to run in this condition for 1 h.	
10. Pinch suction pressure control line valve to produce a first-stage discharge pressure of _____ psig or _____ kPa(g) and pinch the second-stage bypass upstream of the intercooler to produce a final discharge pressure of _____ psig or _____ kPa(g). Run in this condition for a minimum of 4 h.	
11. After loaded run-in is complete, perform hot checks on cross-head clearances, piston rod run-outs and crankshaft deflection. Check cylinder walls, rider ring top clearance, and piston end clearance. Check suction screens for cleanliness.	

If during the next run checks, overheating is suspected because of smoke emanating from the crankcase breather or if abnormal heat is detected on the compressor externals, do not remove the frame covers for at least 30 min. This will allow the running gear parts to cool down. If this precaution is not always taken, a crankcase explosion may occur.

– Restart the compressor motor and once oil pressure is verified, place the auxiliary lube oil pump in auto.
– Run the unit for approximately one minute. During this run closely monitor oil pressure, check for excessive heating and listen for any abnormal noises.

12.7.2 Unloaded Shutdown

Shutdown the unit, verify that the auxiliary oil pump starts as the main oil pump slows and pressure decreases. Time the coast down and lock-out the motor controller. Remove all frame inspection covers and inspect the running gear components, main bearings, connecting rod bearings, wrist pins and bushings, and the crosshead shoes.

If no abnormalities are found, re-install the frame covers and repeat the compressor runs at longer intervals; 5 min, 15 min, 30 min, 1 h and 4 h. No additional visual inspections are required during these incremental runs of 5 min, 15 min, 30 min, 1 h and 4 h unless abnormal operation is detected.

Monitor oil pressure continuously, check for excessive heating and listen for abnormal noises. Time each compressor coast down and monitor motor amperage. We must always assure the motor driver is locked out whenever the compressor frame or cylinders are entered. This is to prevent injury or death to personnel inspecting the compressor components.

Run the compressor continuously for at least 8 h. Record lubricating oil pressures and temperatures once stabilized. During this period, closely monitor the unit and shut it down immediately if an abnormal condition develops.

After completing the no-load break-in run of the compressor, reinstall the valves that were previously removed from the cylinders. Caution should be taken to ensure inlet and discharge valves are properly located and installed. Proceed to the following checks:

- Check the alignment of the compressor and motor. If it is satisfactory, dowel the motor base permanently in place.
- Check clearances of all running parts. Generally, all items listed in Table 12.4 should be checked again where applicable.

At this juncture, valves and valve covers will be installed. Incorrect placement of the inlet and discharge valves in the cylinder can cause an extremely hazardous condition.

Installing an inlet valve in a discharge valve location is generally not possible, however there are designs where this error can occur. Installing an inlet valve in a discharge valve hole or installing a discharge valve upside down may cause an explosion. As a rule of thumb, valves must always be placed in the cylinder with the valve center bolt AWAY from the cylinder bore. The valve crab or unloader cage should be fastened to the valve, when possible, to ensure the valve is not reversed at installation. Inlet valve covers and their hold-down bolts are the most highly stressed components and deserve therefore, special attention.

12.7.3 Torqueing of Valve Cover Fasteners

Based on the general failure risk of valve covers, it would be well to review the procedure for proper valve cover assembly. The owner's engineer should assure the valve and valve cover assembly sequences are witnessed and approved.

Best torque practices must be utilized. Torque practices are crucial to the long-term reliability of the valve cover design. Good procedures should address the following issues:

- **The Friction Factor.** There should be a clear understanding about the use of petroleum lubricants on threads before attempting bolt-up. Quantities and type of lubricant should be defined for consistency of results. There is an up to 30% reduction in stud stress level obtained when a nut is torqued with a dry thread.
- **The role of the operator.** Competent field technicians should have a good understanding of the torqueing sequence.
- **Geometry.** It would be difficult to understand why technicians[10] would not be held to make gap measurements of completed valve cover joints once the torqueing is completed. This will assure a square fit of the valve assembly in the cylinder bore.
- **Tool accuracy.** A testing and calibration schedule for torque wrenches must exist. If such a program is not implemented the result of tasks involving the use of torque wrenches will become questionable. The more accurate the tool or stud tightening measuring system, the more reliably a stud can be preloaded.
- **Relaxation.** There is no doubt a relaxation effect in any bolted assembly – it can be overcome by periodic torque checks.

12.7.4 Startup for Break-In

Crack the main block valve in the suction line to the compressor and allow the pressure to build in the unit to the minimum suction pressure required to allow startup. Using a leak detector, check and repair any leaking joints that may be found. After completing the leak check and making necessary repairs or adjustments, blow down the unit.

The compressor is now ready to start, purge and load as required by the operation and service that the unit is intended for. When placing the unit in service, pay particular attention to the cylinder break-in routines that are usually specified by the OEM and vary widely from machine to machine as a function of piston and wear band materials of construction. As a general rule, piston and packing rings made of TFE® or similar materials do not normally require the gradual loading break-in procedure necessary for metallic rings. However, consideration must also be given to the break-in requirements of the compressor running gear, bearing surfaces and to the gradual increase of the compressor load, especially in high pressure applications.

When the main drive and compressor running gear are determined to be operating satisfactorily and with the frame oil temperature at 90 °F or 32 °C, the compressor cylinders can be loaded, and cylinder break-in can be started.

12.7.5 Compressor Cylinder Break-In

The normal procedure for "breaking in" new cylinders and the compressor involves several short runs at gradually increasing loads. All compressor parts and operating systems must be closely monitored for abnormal conditions during the break-in period.

Check to be sure that the cylinders are properly assembled, studs and nuts tightened to specification, piston rods aligned within prescribed run-out limits and all gas end piping properly cleaned and installed.

During the break-in period, whenever an abnormal noise or condition is observed in the main drive, compressor running gear or gas end, immediately shutdown the compressor and determine the cause of the problem.

When the compressor running gear has been satisfactorily broken in as described previously, it is ready to be placed in the process.

12.7.6 Loaded Startup and Operation

Ensure all liquid collection points are drained, open low point drains in piping and in pulsation dampeners, drain as needed.
1. Check that the cooling water system is properly lined up to the cylinders and the lube oil cooler outlet valve is open.
2. Verify proper cooling water temperatures and flows.
3. Assure lube oil system is operating (auxiliary tube oil pump running) and all instrumentation has been checked.
4. Open compressor bypass control valve.
5. Open isolation valves.

Once proper lube oil temperature is reached, start main motor driver as per OEM's instructions.

When the unit comes up to operating speed and stabilizes it may be loaded by gradually closing the bypass valve. The auxiliary lube oil pump should be placed in standby when proper pressure is observed at the main lube oil pump discharge valve.

Upon initial starting, the rate for building up pressure will depend to a large extent on the performance of the piston rod packings and the downstream system volume. As the leakage rate decreases with the gradual wearing-in of the packings, the discharge pressure can be increased.

Watch piston rod packings for excessive leakage, overheating or other signs of distress. If and when packings overheat, reduce the operating pressure. All equipment must be closely monitored for correct operation.

12.8 Condition Monitoring

Modern reciprocating compressor trains are equipped with condition monitoring that is hopefully cost effective. It would include necessary items to identify malfunctions at an early stage to lower the risk of accidents and the severity of failures. The most common method is vibration monitoring. In its best form, it would feature:

1. Continuous monitoring triggering alarms and shutdowns. Velocity transducers are preferred over accelerometers because of their better signal-to-noise ratio. The optimum configuration is the end of the crankcase about halfway up from the base plate in line with main crankshaft.
2. Each crosshead accelerometer (alarm).
3. Electric motor vibration (shutdown).

Proximity probes are located under the piston rods and used to measure the rod position and determine wear of the piston and rider bands. They detect faults, such as cracked piston rod attachment, a broken crosshead shoe or even liquid carryover to a cylinder. They are utilized for alarm and not for shutdown functions. Optimum cold runouts and normal conditions operating runouts are about 50 μm or 2 mils and in the order of 50–150 μm or 2–6 mils peak to peak, respectively.

Temperature and Pressure Monitoring:

1. High gas discharge temperature – each cylinder (alarm and shutdown).
2. Rod Load (monitoring)
3. Pressure packing and piston rod temperature (alarm).
4. High crosshead pin temperature (alarm).
5. High main and motor bearing temperatures (alarm).
6. Valve temperature (monitoring).
7. Low oil pressure (alarm/shutdown)
8. Oil temperature out of frame (alarm).
9. High jacket-water temperature, each cylinder (alarm).

Shutdown functions should be two-out-of-three voting to avoid unnecessary trips. Usually, deviation for compressor frame vibration and machine temperature-related trips can be allowed.

12.8.1 Valve Monitoring

Check inlet valve covers for excessive heating and knocking. Increased valve cover temperature usually indicates a leaking valve or valve seat gasket, while a knocking sound or pounding may indicate a loose valve. Usually, this temperature rise will affect the whole cylinder, but it occurs predominantly at or in the vicinity of the failing or failed valve.

Operators find leaking valves by checking the temperatures of each valve cover as part of their rounds. Another practice is to have technicians periodically check all valves after a sudden increase in overall discharge temperature for a given cylinder. Various instruments have been used to assess cylinder valve condition such as ultrasonic leak detectors and thermographic cameras. However, these methods require much time from skilled analysts. Additionally, these approaches have certain limitations: First, valve temperatures vary with load and process conditions. Second, low ratio cylinders may not exhibit any discernible valve temperature rise until the valve is in significant distress. Gas composition may also have an impact on the degree of valve temperature rise caused by valve failure. Small temperature changes associated with some gases and low-pressure ratios may cause a very small increase in temperature on the valve cover until the valve has major problems.

Numerous torsional vibration problems continue to occur in reciprocating compressor trains. The main reasons are lack of comprehensive torsional vibration analysis at the design stage, improper application and maintenance of couplings – mainly flexible couplings – and lack of monitoring. As a rule of thumb, the electric motor shaft diameter should be equal to or greater than the reciprocating crankshaft diameter.

12.9 Shutdown

The compressor must be gradually unloaded by opening the compressor bypass valve or valve unloaders, as the case may be. Allow the machine to run with the bypass valve open for 3 min. Stop main drive motor and close isolation valves. The lube oil system may be left in operation. There should be a plant site specific policy that defines lube oil pump running times after shutdown. Additionally, there should be a rule that when the compressor is stopped for any extended time, it should be turned about a quarter turn every week using a bar-over device. A manual barring device is used for small compressors. A pneumatic device is furnished – without area classification problems – for compressors rated over 1000hp or 750 kW.

The compressor unit is now ready to be handed over to the owner or operator.

12.9.1 Safety Concerns After Shutdown

Whenever a compressor is shut down for inspection or repair, positive steps must be taken to prevent any accidents. A "lock-out" or LOTO procedure would entail:
1. Lock out the motor controller
2. Attempt to start the motor to confirm the motor lockout is affective.
3. Purge the compressor cylinders
4. Block flywheel or crankshaft before opening and entering the compressor.

5. When the unit is equipped with a barring or flywheel locking device, this device must be used to prevent rollover. Blocking of the crankshaft or crossheads is an alternate method of preventing accidental rollover.

A compressor cylinder or any other part of the compression system should never be opened without first completely relieving all pressure within the unit and taking necessary precautions to prevent pressurizing of the system. This should be a guide to all operating and maintenance personnel:

- Compressors handling toxic or flammable gases must be isolated from the process piping by means of blinds, or DVBs[11] when a major intervention is required. Before opening such compressors, the equipment must be purged or evacuated. In addition, a warning sign bearing a legend such as "WORK IN PROGRESS – DO NOT START" must be attached to the starting equipment.
- Whenever the compressor is shutdown because of an indication of overheating, a minimum period of 30 min must elapse before the crankcase is opened. As already mentioned, premature opening of the crankcase can result in a crankcase explosion.
- Corrective action must be taken when the cylinder vent gas leakage is excessive or when there is a sudden increase in the leakage rate.

References

1 Gas Machinery Research Council, *GMRC Guideline for High-Speed Reciprocating Compressor Packages for Natural Gas Transmission and Storage Applications*, July 1013.
2 API RP 686, *Recommended Practice for Machinery Installation and Installation Design*, 2nd Ed., December 2009.
3 See Appendix 5.F.
4 Steven B. Todaro, "Design strainers into the system and plan for them as part of commissioning", Gas Compressor Magazine, October 2018, Houston TX 77379, info@thirdcoastpublishing.net
5 See Reference 2.
6 See Figure 5.10.
7 A 100-mesh screen has 100 openings per in. or 25.4 mm, and so on. As the number describing the mesh size increases, the size of the particles decreases. Higher numbers equal finer material.
8 Auxiliary piping around the compressor comprised of smaller pipe, frequently less than 1in. or 25.4 mm in diameter. This smaller piping is often ignored by piping designers and therefore, tends to vibrate due to lack of sufficient support. See more in Chapter 5.
9 Learning Module – Alberta Apprenticeship and Training. http://tradesecrets.alberta.ca
10 Not always sure about the proper use of this word. The term should describe a person who uses tools and had a trade education. The name of this tradesman/woman ranges (regionally) from millwright (Britain), to machinist or technician (US, Canada) to artisan (South Africa).
11 DVB – Double Valves and Bleed.

Bibliography

[1]   P.B Heinz, "Compressor and Modern Process Application," John Wiley and Sons, 2006.

[2]   P.B Heinz "A Practical Guide to Compressor Technology, Second Edition," John Wiley and Sons, 2006.

[3]   W.A. Griffith, E.B. Flanagan, "Online Continuous Monitoring of Mechanical Condition and Performance For Critical Reciprocating Compressors," Proceeding of the 30th Turbomachinery Symposium, Texas A&M University, Houston, Texas, U.S.A., 2001.

[4]   P.B Heinz and J.J Hoefner, "Reciprocating Compressors Operation & Maintenance," Gulf Publishing Co., 1996.

[5]   D Hickman, *Specifying Required Performance When Purchasing Reciprocating Compressor – Part I, II*, CompressorTechTwo, August-September, October 2007.

[6]   D. Woollatt, "Reciprocating Compressor Valve Design: Optimizing Valve Life and Reliability," Dresser-Rand Technology Journal, Volume 1, pp. 44–51, 1995.

[7]   M. Steven Schultheis, Charles A. Lickteig, Robert Parchewsky, "Reciprocating Compressor Condition Monitoring," Proceeding of the 36th Turbomachinery Symposium, pp. 107–113, 2007.

[8]   S.W Robin, "Reciprocating Compressor: Reliability Improvement Focusing on Compressor Valves, Piston and Sealing Technology," Compressor Optimization Conference, Aberdeen, Scotland, Jan. 30–31, 2007.

[9]   R Rahnama, K Eberle, S Crocker, *Effects of pipe strain and flange misalignment on vibration in reciprocating compressor systems*, Paper presented at the GMRC Gas Machinery Conference held September 30 to October 3, 2018, in Kansas City, Missouri, USA.

Chapter 13
C&S of Rotary Screw Compressors and Blowers

13.1 Introduction

A screw compressor is a twin-shaft rotary piston machine functioning on the principle of displacement combined with internal compression. The gas handled is conveyed from the suction port to the discharge port, entrapped in steadily diminishing spaces between the convolutions of the two helical rotors, being compressed up to the final internal pressure before it is released into the discharge nozzle – see Figure 13.1.

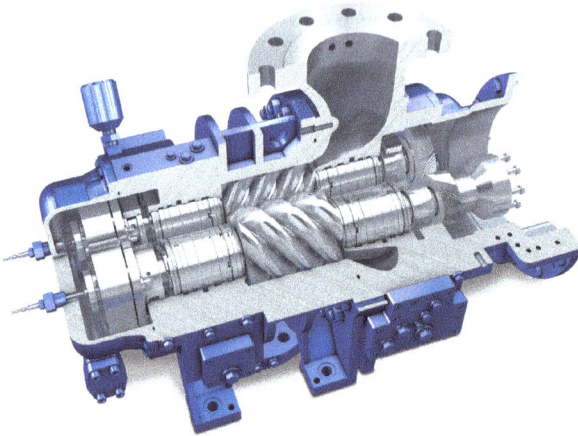

Figure 13.1: Cross section of a typical oil-free rotary screw compressor (courtesy of AERZEN).

Screw compressor pressure limits are normally dictated by temperature, due to the heat of compression. However, rotor deflection and bearing life are also key factors in pressure limits. In an oil-free screw compressor – Figure 13.1 – sometimes referred to as a "dry" screw, the rotors do not contact one another, but are separated by extremely small clearances and driven by timing gears. Seals at the conveying chamber ensure that the conveyed gas does not come into contact with the lubricating oil or outside air. Thus, the gas is never contaminated by oil, and vice versa. Helical screw compressors are capable of producing discharge pressures of up to 1,500 psi or 100 bar, flow ranges from 120 to 50,000 acfm or 200 to 80,000 m^3/h and a power requirement of up to 10 MW. A typical average rating would be 400 psi or 28 bar, a minimum pressure ratio of 1.5, 47.6 Mcfh or 3,250 m^3/h at 3,000 rpm with a maximum shaft power of 1,675 hp or 1,250 kW.[1]

 Another version of the helical screw design is the oil-injected screw compressor, sometimes referred to as a "flooded" or "wet" screw. This type of rotary screw

https://doi.org/10.1515/9783110701074-013

compressor has no timing gears as shown in Figure 13.2. Rather, one rotor drives the other directly. There are also no internal seals. The conveyed gas is mixed with the lubricating oil inside the machine, and gas and oil are discharged together. The oil is then separated from the gas stream before the gas is discharged into the downstream system. The oil, sometimes referred to as *lubricoolant*, is then cooled, filtered and again injected into the compressor conveying chamber, bearings and driveshaft seal. Because the gas is mixing with the oil, care must be taken in oil selection to prevent oil contamination by the gas or gas contamination by the oil.

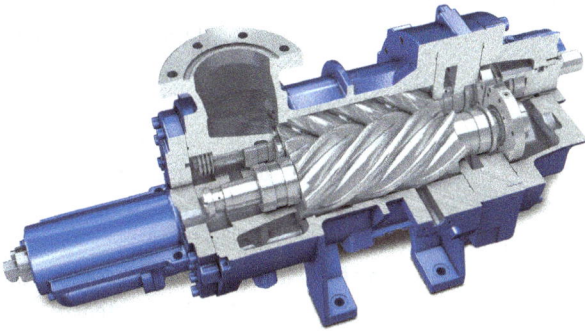

Figure 13.2: Cross section of an oil-injected screw compressor (courtesy of AERZEN.).

In contrast to the dry screw, an oil-injected machine usually features a built-in slide valve that can be utilized to automatically control compressor capacity.

Oil-injected screw compressors require only one seal where the driver shaft exits the casing. Generally, the driver is a direct-coupled electric motor. Optimum coupling selection is one with high torsional stiffness, usually metallic flexible couplings.[2]
Screw compressors have many inherent advantages over other compressor types, namely reciprocating and centrifugal, in the low-to-medium-flow range.[3] Since they are displacement compressors, they will draw a constant inlet volume, can meet the varying differential pressure requirements and are not significantly affected by gas density or composition. Technical and cost advantages of oil-injected screw compressors have made them the compressor of choice for many small- and medium-capacity applications.

Although originally intended for air compression, oil-injected screw compressors, in the twenty-first century, compress a large number of process gases, including solution gas, in oil and gas plants of the upstream business of the petrochemical industry. Further, we find screw compressors in oil refineries, petrochemical units, air separation plants, refrigeration units, gas processing, natural gas systems, boosting of gas turbine fuel supply pressure, flare and recovery gas applications, various cryogenic plants, LNG/LPG units, evaporation plants, mining and metallurgical plants.

Oil-injected screw compressors employ a pressurized lubrication system. They require an oil separation system that is usually combined with the lubrication system. Experience has shown that bearing and seal component reliability is a direct function of auxiliary system component selection and design. Auxiliary systems are major sources of potential cost reduction for screw compressor vendors. Since screw compressors are relatively new, the specifications may not reach the required sophistication in some specific areas.

Figure 13.3 shows a typical simplified process flow diagram of the production of sodium carbonate – also known as soda ash – utilizing two dry or oil-free helical screw compressors. Figure 13.4 illustrates an upstream process application of an oil flooded screw compressor which could be typically applied for flare gas movement in an oil refinery.

One would hope that special care has been exercised in selecting and sizing the oil pumps. Oil pump operating discharge pressure is determined based on compressor discharge pressure plus a margin – usually 44 psi or 3 bar – to ensure proper oil injection. Ample margin, approximately 25–30% is needed for rated flow as compared to normal oil flow. The recommendation for this oil system is:

- Two identical motor-driven oil pumps according to a current industrial standard.[4]
- Two 100% lube oil filters – filter body and internal made of stainless steel.
- Twin lube oil shell and tube-type cooler according to TEMA C or alternatively air cooler API 661 with spare fan, if cooling water is not available.
- An oil purifier or oil conditioner – as either a portable or installed device to decontaminate or purify a circulating oil stream that has been slowly degraded by an influx of contaminants. Such a device will not normally require shutdown of the compressor.

13.2 Blowers

Rotary piston, or lobe-type blowers, derives from the *Roots* compressor principle dating back to 1864. They are positive displacement machines and are used in a large variety of process plant applications including pneumatic conveying of bulk materials, pressurized aeration in water treatment plants, creation of vacuum, and gas movement in the pharmaceutical and metalworking industries. Their power requirement ranges from fractional horsepower or kW to hundreds of kilowatts.[5] A typical application is the transfer of polymers with a pressure increase of 15 psi or 1 bar and power requirements of approximately 300 hp or 224 kW.

Similar to helical screw compressors, they are twin-rotor type machines. Two shafts are axially parallel to one another and located centrally inside the casing. Timing gears ensure that the rotors turn without lobe contact. The rotors are supported on ball and roller bearings. The clearance between the rotors is kept to a

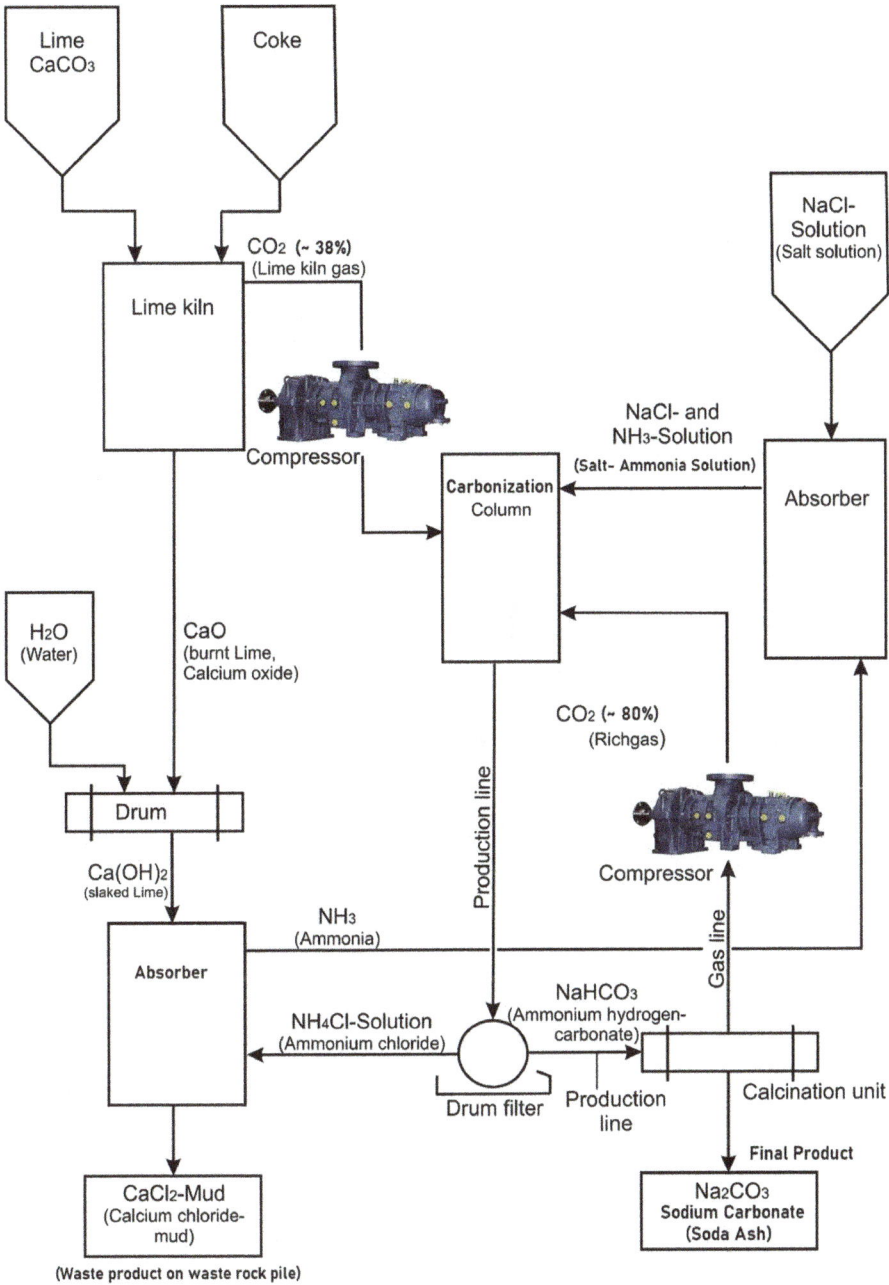

Figure 13.3: Helical screw compressors used in soda ash production (courtesy of AERZEN).

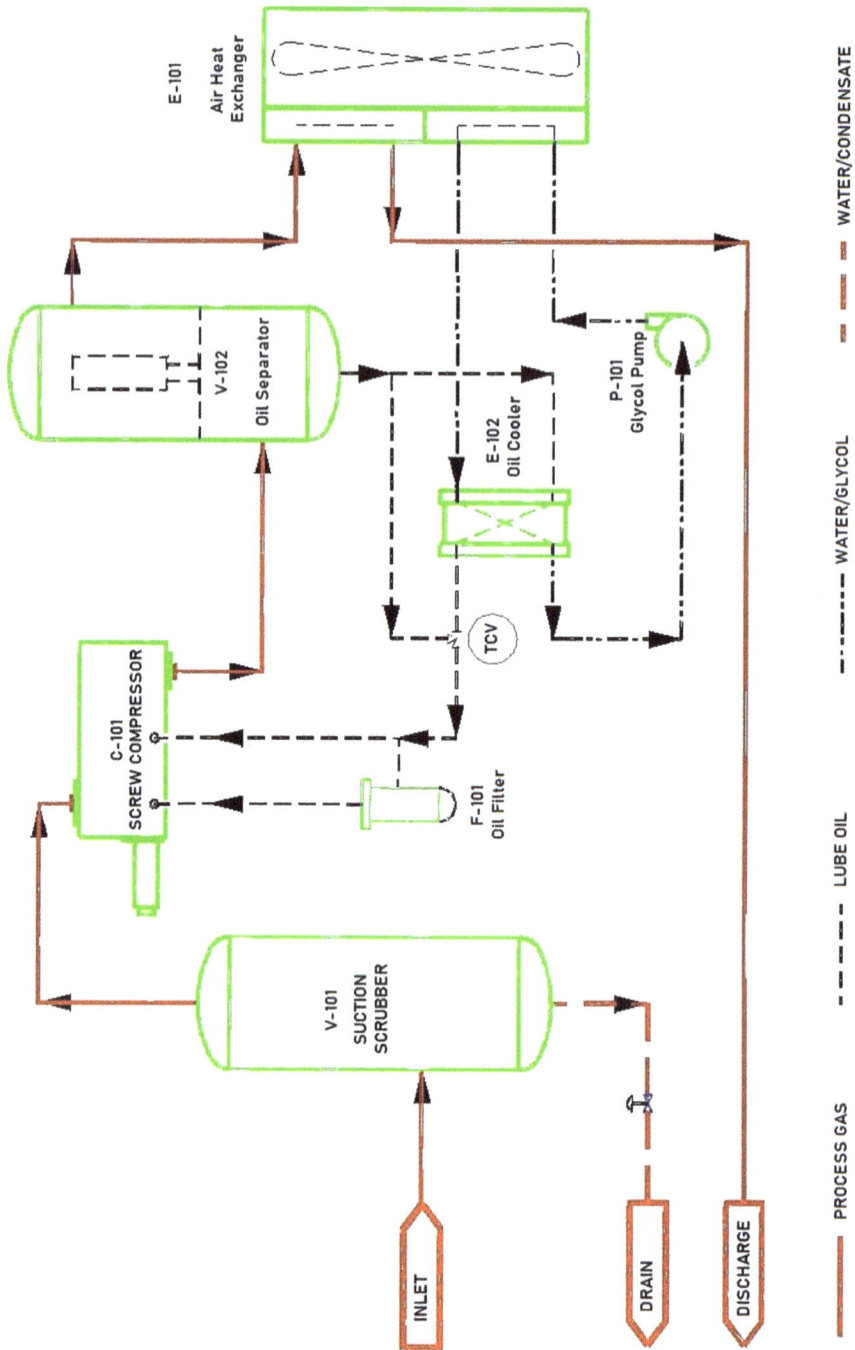

Figure 13.4: Typical simplified process flow diagram for an oil flooded helical screw compressor application.

minimum and selected for the expected pressure differential and thermal load under working conditions. Figure 13.5 illustrates their method of operation.

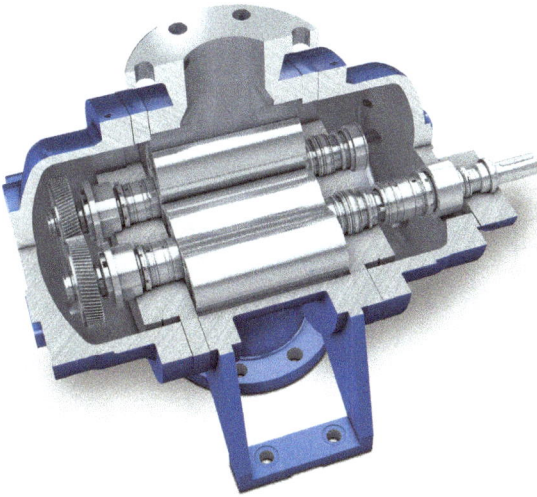

Figure 13.5: Blower – lobe type (courtesy of AERZEN).

13.3 Commissioning

13.3.1 Risk Profile

Helical screw compressors as well as blowers have shown remarkable mechanical reliability. The design is mature and proven. Vulnerabilities must be seen as emanating from ancillaries, such as lubrication and sealing systems but also process related factors such as foreign objects in the gas or air stream and other contaminants. As to wet or oil-flooded machines, there is an additional hazard to be taken into account. There will be a rapid rise in discharge temperature when the oil supply is interrupted. This high temperature can lead to rotor seizures and the destruction of the machine. Protective devices, RTDs, must be redundant and reliable. While 35,000–40,000 operating hours can be expected from bearings, often continuous operation is limited to around 5,000 h because auxiliary components such as filters and separators require cleaning or replacement.

The keys to long-term availability of a screw compressor are modern seal and bearing designs, and proper arrangement and modularization that enhances efficient maintenance. Risk of failure reduction can be achieved by installing self-acting dry gas seals in favor of traditional contact seals prone to wear and random breakage. Helical screw compressors, in many applications, belong to machinery category III as they are often un-spared. Blowers, on the other side, are frequently classed category II depending on size. Their vulnerability is connected to the fact that excessive heat of compression can lead to rubbing of the rotors against the casing internal surfaces. This often results in an expensive wreck. Again, appropriate instrumentation must be furnished and functional prior to startup.

An inherent risk assessment is presented in Figure 13.6. It is based on an evaluation of the process shown in Figure 13.4.

1	2
	550hp/410kW
	Motor driven O.I.
	Hel.ScrewCmp at 1800rpm
Rel. Risk #	288
Rel. Risk Level:	MEDIUM
Complexity	18
1.5	4
2.3	2
4.4	4
5.1	1
6.4	7
Severity	16
Table 2.4.1	5
Table 2.4.2	5
Table 2.4.3	2
Table 2.4.4	4

Figure 13.6: Inherent-risk assessment of an oil-injected helical screw compressor unit.

Commissioning and startup related risks are covered in Table 13.1.

Table 13.1: Project-related risk mitigation.

Phase	Risk	Most probable cause[1]	Suggested actions/mitigating measures
Rigging and lifting	High – R & L complex	QP	– Make Lifting Plan mandatory as an OEM deliverable.
Handling, staging and storage protection	High – Inadequate protection resulting in rework and delay	OR, HK	– The process starts with preparation for shipment by the OEM. API 619 has excellent instructions for preparation. Review plans for handling and storage protection. – Prevent moisture intrusion: Use Vapor Phase Inhibitors (VPI); desiccants; dry instrument air as a purge where possible. Fill bearings with the proper lubricating oil.

Table 13.1 (continued)

Phase	Risk	Most probable cause[1]	Suggested actions/mitigating measures
Foundation and grouting	**Medium** – Neglected due to low priority assessment	QP	– Helical screw compressors are frequently assembled on a package and the rules for skid placements on a foundation should be followed. – Have grout manufacturer rep at site during pour. – Use grout with the proper exotherm formulation suited for the climate conditions at site.
Piping	**High** – Not enough attention given to cleaning – Ingestion of foreign objects after cleaning	OR, HK	– Cleanliness and assembly to the machine should follow precise established procedures as part of the project execution agreement between OEM and the owner organization. – Piping alignment to the compressors must follow guidelines given in current recommended standard.[2]
Shaft alignment	**Medium** – Misalignment – Neglect	QP	– Shaft alignment starts with leveling the unit. Blowers as well as helical screw compressors and their drivers should be accurately leveled. A minimum permissible deviation from the horizontal line would be customarily 0.002in. per foot or 0.2 mm per linear meter. – Usually, blower or screw compressors are considered the fixed machine and the driver is aligned to it.

Table 13.1 (continued)

Phase	Risk	Most probable cause[1]	Suggested actions/mitigating measures
Lubrication system	**High** – Wrong oil specification – Mix up with different lubricants – Contamination	QP, DO, CO	– Because external piping must be fabricated and installed as part of the lubrication system, the cleanliness obtained during FAT helps but should be discounted. Therefore, the system should be flushed again to achieve cleanliness standard per ISO 4406:99. – For oil-flooded screws, accurate gas composition is critical so the correct lubricant that can be selected. If the gas composition changes before or after startup, this must be addressed immediately to minimize risk of lubricant carryover and/or impact on machinery health. Oil-flooded screw compressors in upstream oil and gas service typically utilize a synthetic oil. The inherent risk is a mix-up of lubricoolants.
Startup and initial operation	**High** – Disregarded critical speed range – Seizure due to FO		– Operator training and provisions in automatic startup sequence. – Avoid extended run at critical speed – See "Piping" line items above.

[1] Refer to Figure 2.12
[2] API RP 686, 2nd Edition, 2006, *Recommended Practice for Machinery Installation and Installation Design and/or* API RP 1FSC, 1st Edition, 2013, *Facilities Systems Completion Planning and Execution.*

13.3.2 Preliminaries

As stated in other sections, it is very important to understand what transpired before machinery units arrive at the construction site. The startup team's machinery specialist would be intimately familiar with the proceedings and exchanges between the vendor and owner during the bidding stage and the coordination meeting with the vendor or OEM representatives.

In the case of an oil injected machine – as alluded to before – special care should have been given to oil selection, separator retention time, oil pumps, coolers, oil filters and vendor experience for similar applications. The oil recovery vessel, known as the primary separator, is generally equipped with suitable internals such as demisters, coalescent filters devices or alternative devices.

Startup leaders should also be familiar with mechanical run tests performed in the vendor's shop prior to delivery. Inspectors' reports should have been read carefully in order to find out if any unresolved issues exist. Table 13.1 will serve as a guide to uncovering outstanding issues.

Table 13.2: Checklist – mechanical test run record (FAT) for helical screw compressors.

1.0 Mechanical inspection

1.1 During assembly of the equipment before testing, was each component (including cast-in passages) and all piping and appurtenances cleaned chemically or by another appropriate method to remove foreign materials, corrosion products and mill scale?

(Yes or no)

1.2 Did all furnished portions of the oil system meet the cleanliness requirements in API Standard 614?

(Yes or no)

1.3 If specified, was there an inspection made for cleanliness of the equipment, all piping and appurtenances furnished by or through the vendor before heads were welded to vessels, openings in vessels or exchangers closed, or piping finally assembled?

(Yes or no)

1.4 If specified, were the following verified: Hardness of parts, welds and heat-affected zones as being within the allowable values?

(Yes or no)

2.0 Hydrostatic testing

2.1 Were hydrostatic tests on pressure containing parts performed according to specified standards?

(Yes or no)

2.2 Are there any outstanding issues connected with testing (e.g., repaired castings, etc.)?

(Yes or no)

3.0 Mechanical running test

3.1 Were the contract shaft seals and bearings used in the machine for the mechanical running test?

(Yes or no)

3.2 Was the lubrication system tested as per API 619 specifications?

(Yes or no)

Table 13.2 (continued)

3.2 Were all warning, protective and control devices used during the test, checked, and adjustments shall be made as required?	_____ (Yes or no)
3.3 Was the contract coupling used during the test?	_____ (Yes or no)
3.4 Did the unit show satisfactory vibration behavior?	_____ (Yes or no)
3.5 Was the mechanical running test conducted at maximum continuous speed for a minimum of 4 h?	_____ (Yes or no)
3.6 If the unit has hydrodynamic bearings, were they removed, inspected and reassembled?	_____ (Yes or no)
3.7 Where timing gears' contact areas visually inspected and the results of the inspections recorded?	_____ (Yes or no)
3.8 Was the casing (including end seals) pressurized with an inert gas to the maximum sealing pressure or the maximum seal design pressure, as agreed upon by the owner and the OEM?	_____ (Yes or no)
4.0 Performance test – if specified	
4.1 Was a complete unit test performed?	_____ (Yes or no)
4.2 Any unresolved issues?	_____ (Yes or no)
5.0 Auxiliary-equipment test	
5.1 Were auxiliary equipments such as oil systems, gears and control systems tested in the OEM's shop?	_____ (Yes or no)
5.2 If specified, was the compressor, gear and the driver dismantled, inspected, and reassembled after satisfactory completion of the mechanical running test?	_____ (Yes or no)
5.3 If specified, was there a full-pressure/full-load/full-speed test performed?	_____ (Yes or no)
5.4 If specified, was a spare parts test (such as couplings, gears, bearings and seals) conducted?	_____ (Yes or no)
5.5 If specified and in the case of an oil-free screw machine, was there a heat run performed (at the maximum allowable speed, with the discharge temperature stabilized at the specified maximum operating temperature plus 11 °C or 20 °F for a minimum of 30 min)?	_____ (Yes or no)

13.3.3 Pre-Commissioning and Commissioning

Pre-commissioning activities are described in Table 13.3.

Table 13.3: Pre-commissioning helical screw compressor units.

Initials

1. Are applicable operation and maintenance manual, parts book, special tools & spares available?   _____
2. Have the vendor drawing and data requirements (VDDR) been fulfilled?   _____
3. Are the items/documents listed in the project VDDR Form available   _____
4. Have the application operating conditions been determined?
    a. Suction: _____ psi/kPa _____ Temp: _____°F/°C
    b. Discharge: _____ psig/kPa _____ Temp: _____°F/°C
    c. Min. RPM _____ Max. RPM _____ (for IC engine–driven units)
5. Are expected startup and operating conditions within compressor design limits? If in doubt, contact the OEM representative for confirmation of the compressor's operational limitations.   _____
6. Soft-foot on packaged machines: Confirm the casing leveling jackscrews have been backed off. Has casing been properly shimmed & bolts re-torqued? Compressor must not be twisted or bent.   _____
7. Has the compressor to driver cold alignment been checked on site? Are alignment readings within specification and readings recorded?   _____
8. Have coupling adapter and drive coupling bolt torques been verified and recorded?   _____
9. Has the oil system been thoroughly cleaned following oil system flushing and cleaning practices descrbed earlier in Chapter 5?   _____
10. Oil system, oil filter and oil piping been primed with oil?
    Note: Electric motor driven units require an automated pre-lube pump system.   _____
11. Has the low oil pressure shutdown tubing been installed and shutdown setting verified?
    Note: Minimum oil pressure 30 psig or 200 kPa falling pressure.   _____
12. Is oil cooler installed and is oil flow counter to water flow?
    Note: Maximum oil supply temperature to main bearings is 185 °F (85 °C).   _____
13. Compressor piping, suction drums have been chemically cleaned and inspected.
14. Have the suction lines been blown out to remove water, dirt, slag and any other foreign objects?   _____
15. Have the suction, inter-stage, and discharge pressure high/low shutdowns been set and functionally checked?
16. Have the safety relief valves been installed in the proper locations to protect piping and casing MAWP ratings?   _____
17. Have the discharge gas temperature shutdowns (RTDs) been installed, set and functionally checked?   _____
18. Is there a compressor frame vibration shutdown device installed, set and functionally checked?   _____
19. Have the temporary inlet debris screens (100 mesh) been installed at the suction flange?   _____
20. Has the gas piping been purged of all air for machines compressing a combustible gas?   _____

Table 13.3 (continued)

21. Have all critical fastener torques been checked and recorded? **Caution:** Loose fasteners may result in a safety hazard or equipment failure.Frequently OEMs list critical fasteners in their O&M manual.	
22. Has the driver rotation been verified to match the compressor rotation?	_____
23. Verify the compressor and its driver are free rolling with minimum force.	_____
24. For engine drives, has the unit been rolled with the air starter to ensure it Is free turning? For electric motor drives, has the unit been barred over by hand to ensure it is free turning?	_____
25. If applicable, has the driver over-speed shutdown been electronically set & verified?	_____
26. Package compressors: Have startup instructions for all package equipment been reviewed and performed?	_____
27. Has the OEM or packager representative reviewed the unit's startup and operating instructions with the site personnel?	_____

The first order of commissioning is to attend to the lube and seal oil system of the screw compressor unit. The procedure shown in Appendix 13.A will serve as an example for the procedure to be followed.

All owner's connections to the unit have been made. As in other compression equipment, it is good practice to install a startup strainer in the blower or compressor intake pipe to protect the machines from ingestion of foreign objects. See Figure 12.6. As mentioned before in this text, strainer resistance should always be measurable – by installing ΔP gauges – where the permissible pressure drop should not exceed 0.75 psi or 50 mbar. If the startup strainer remains clean after 500 operating hours it may be replaced by a spacer ring or the mesh can be removed from the perforated plate.

It is assumed that the motor or prime mover furnished with the lobe blower or screw compressor set has been test-run and checked for rotation after having been uncoupled. The coupled unit should now be turned over by hand and it should be ascertained that there are no obstructions and resistance to turning.

Before starting and operating helical screw compressors, as typical displacement compressors, exhibit the following operating characteristics:

1. Increase of pressure ratio at constant speed:
   - Volume throughput decreases only slightly. They exhibit steep pressure ratio – inlet volume curves.
   - Power demand and discharge temperature increase.
   - Discharge temperature limits the maximum stage pressure ratio.
2. Increase of speed at constant pressure ratio:
   - Volume capacity and power absorption are approximately proportional to the rotational speed.
   - Torque increases only negligibly.

We ought to thoroughly understand Figure 13.7.

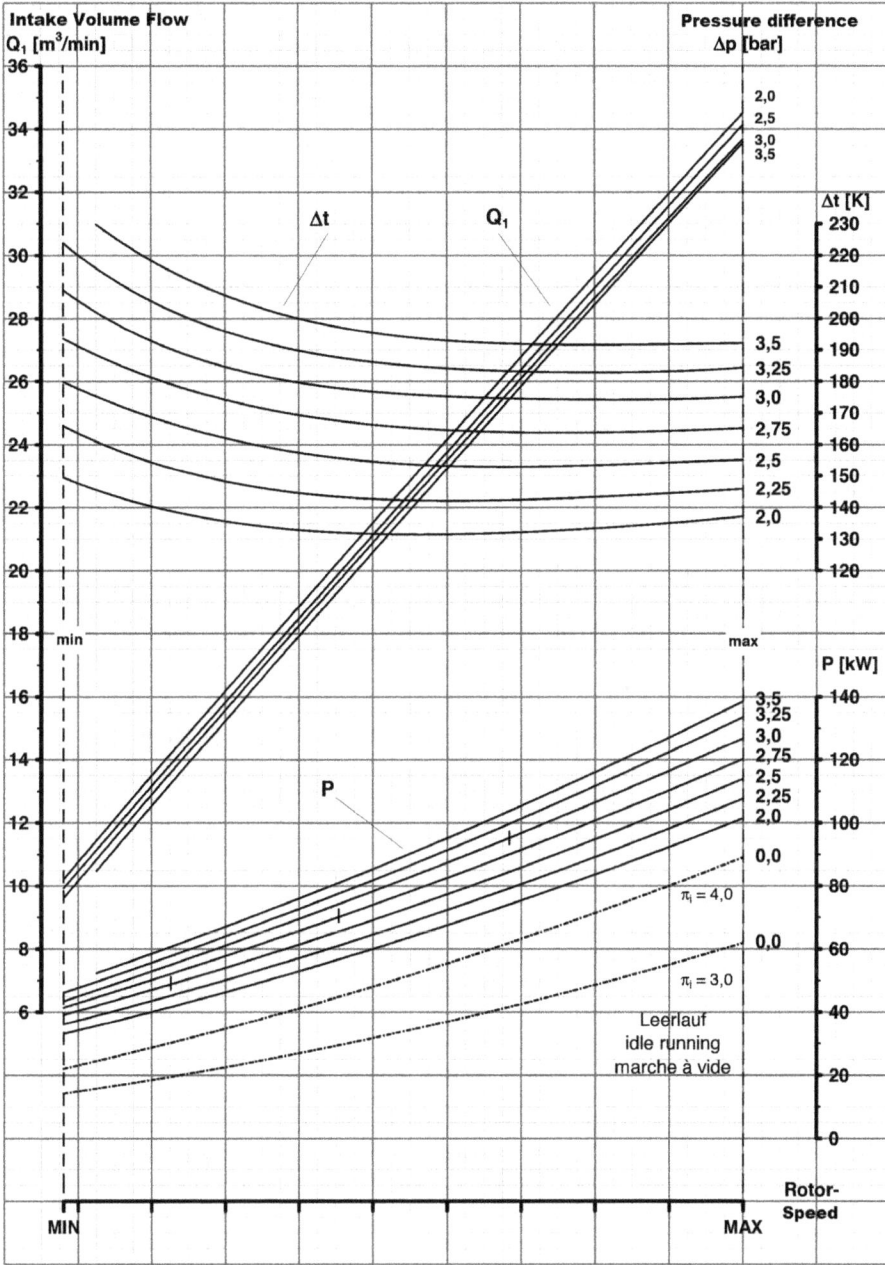

Figure 13.7: Typical screw compressor performance characteristic (courtesy of AERZEN).

As with other compression equipment, it would be well to know expected discharge temperature levels – not only for personnel safety reasons – but also as a performance indicator. In blowers, for example, conveying pipes can reach 480 °F or 250 °C and suitable lagging would have to be installed.

13.4 Initial Startup

There should always be the essential startup procedure – Table 13.4 – and an initial startup checklist – Table 13.5. During the initial startup phase and thereafter, potential risks are addressed by intensive condition monitoring:

– Performance indicators such as flow and efficiency trends which indicate the health of the rotors. A trend display of delivered flow is extremely useful because blowers and screw compressors are displacement compressors and will deliver a constant volume. Reduction of flow would be an indication of an increase of slip or internal leakage.
– Watching journal and thrust bearing where bearing temperature and casing vibration monitoring would be the indication of rotor support system condition.
– Monitoring of the lube oil system – pressures, temperatures and drainage flows as described in Chapter 5.
– Surveillance of the seal oil system if oil-lubricated contact seals are being used.
– Surveillance of the dry gas seal – DGS – support system. See Chapter 6.

Table 13.4: Typical startup procedure for helical screw compressor units.

---

1. Check that the AC and DC to the local panel is turned on by observing the local panel lights (if applicable).
2. Commission the lube/seal oil system as per procedure.
3. Start cooling water circulating through the compressor cooling jacket by fully opening the valve in the inlet line. Open the discharge valve in the cooling water drain line, maintaining the inlet water pressure at least above _____ psig or _____ bar,
4. Open the valves in the buffer gas line and check that the isolation valve on each differential pressure control valve are fully open and the bypasses fully closed.
5. Put the "Unload Test" switch in the test position and stroke the unloader valve. Leave the valve fully open and move the switch to the normal position.
6. Open the suction and discharge ROVs from the control center panel. Check that the lights on the local panel indicate that the valves are fully open.
7. Check that the pressure control valve is open so that the compressor is initially running on recycle.
8. Reset the start circuit and check that the local panel is cleared of all trip and alarm indications.
9. Start the motor. Listen for any unusual noise or vibration. The bypass valve should go close: in about 10 s after the motor is running, thus loading the compressor.
10. Observe all temperatures, pressure and sight-flow indicators for any abnormal readings.

Table 13.4 (continued)

11. Adjust the-cooling water to the in-service lube oil cooler, maintaining an oil temperature out of the cooler at about 110–120 °F or 43–49 °C.
12. Record operating data.

SHUTDOWN PROCEDURES FOR ROTARY SCREW COMPRESSORS

A.   Normal Shutdown Procedure

1.   Stop the compressor by pushing the emergency stop button on the local panel or in the control center.
2.   Depending on the reason for stopping, either close the suction and discharge ROV's from the control center or leave them fully open.
3.   Continue to circulate oil to the seals and bearings
4.   Continue to circulate cooling water to the compressor jacket.
5.   Before re-starting, the motor breaker should be racked out and the compressor barred over to determine if there is any binding. This also should be done before the compressor is re-started after a trip.

NOTE: DO NOT stop seal oil flow to seals by stopping the main and auxiliary oil pumps unless the casing is depressurized.

B.   Emergency Shutdown Procedure

1.   Stop the motor by using the local stop–start button next to the motor. The bypass valve should immediately go fully open when the motor is stopped.
2.   Depending on the reason for stopping, either close the suction and discharge ROV's from the control center or leave them fully open.
3.   Continue to circulate oil to the seals and bearings.
4.   Continue to circulate cooling water to the compressor jacket.

NOTE: DO NOT stop seal oil flow to seals by stopping the main and auxiliary oil pumps unless the casing is depressurized.

Table 13.5: Initial startup checklist for a two-stage gear driven helical screw compressor unit.

A. Getting the lube/seal oil system ready.	EPC	Owner	Comments
1. Perform 4 h uncoupled run-in of main oil pump motor.			
2. Align motor–oil pump and install coupling.			
3. Blow down steam lines to auxiliary steam turbine.			
4. Clean the interior of the lube oil reservoir.			
5. Flush all cooling water lines.			
6. Commission all instruments on oil system.			
7. Check oil system relief valves. Set pressure at _____ psi or _____ bar			
8. Make completeness check of steam turbine installation.			

Table 13.5 (continued)

9.	Perform uncoupled run-in of auxiliary turbine for four hours.
10.	Align turbine/pump and install coupling.
11.	Fill lube oil reservoir with flushing oil.
12.	Install temporary lines across all bearings and seals for flushing. Install a screen in the flange which connects the base mounted supply header and the field fabricated pipe.
13.	Blank-off bladder-type accumulators.
14.	Flush the oil filters and heat exchangers.
15.	Fill the false bottom of the oil reservoir with oil.
16.	Supply a temporary steam line to the heating coil in the oil reservoir.
17.	Heat the flushing oil to about 150 °F or 65 °C and start circulating. Both pumps should be used.
18.	After clean cloth and screens are obtained, remove the temporary lines and drain the reservoir of flushing oil.
19.	Clean the interior of the reservoir and fill with the lubricating oil, Teresso™ or equal.
20.	Install a screen with a cloth downstream of the transfer valve and circulate lubricating oil until a clean cloth is obtained*. Both oil pumps must be used.
21.	Replace oil filter cartridges with new elements.
22.	Complete piping of nitrogen line to the accumulator.
23.	Blow the nitrogen line.
B.	**Preparation for run-in**
1.	Remove, clean and inspect the suction female bearing and seal.
2.	Remove timing gear cover and check the timing gear for proper backlash and tooth contact. Backlash 0.0012–0.002" or 30–51 μm.
3.	Inspect the interior of the suction and discharge silencers.
4.	Inspect the compressor rotors through the suction and discharge flange openings. If possible, check the rotor/casing cold clearances at the suction and discharge. Use borescope if required.
5.	Remove the inspection cover on the speed increase gear and check tooth contact and backlash.
6.	Check that the gearbox thrust bearing has the proper clearance.

Table 13.5 (continued)

7.	Set the motor/gearbox alignment and install the coupling. If applicable, use _____ as coupling lubricant. Install vendor supplied fixture for solo run on coupling half.
8.	Start the lube oil system and run-in the motor uncoupled for 4 h – Solo Run.
9.	Install silencers on compressor.
10.	Chemically clean suction and bypass line to compressor.
11.	Check final cold alignment between gearbox and compressor and check without and with piping. Connect piping. Install coupling spacer and guard.
12.	Check that the temporary suction strainer is installed.
13.	Blow the buffer gas supply lines and feed lines to the individual seals.

C. Run-in on air

1.	Install a temporary line to supply nitrogen to the buffer gas connection. Nitrogen pressure should be at least _____ psi or _____ bar.
2.	Bar compressor over by hand to check that rotors are free. If any metal rubbing or sounds are detected in the compressor, determine the cause and correct before starting.
3.	Turn on buffer gas to compressor _____ psi or _____ bar
4.	Start main oil pump and check flows and pressures to bearings and seals. Oil temperature should be _____ °F or _____ °C before starting train.
5.	Open the man holes in _____ and cover with a wire screen.
6.	Block in the discharge line using valve _____ and open the valve in the _____ in. or _____ mm ø vent to atmosphere.
7.	Make sure main shutoff valve and bypass valve are open.
8.	Run driver for a period just long enough to bring the unit up to approximately 1/4 speed. Shut off driver, and observe the unit while it coasts down to a stop to make sure there is no undue vibration or noise. Bar unit over again to check for any binding.
9.	Start driver again and brig unit up to speed. Do not run for more than 2 min with zero discharge pressure.

Table 13.5 (continued)

10.	Throttle the bypass valve in the discharge line of the process gas piping and bring and let the discharge temperature rise to _____°F or _____°C over a period of approximately 15 min.
11.	Run for 30 min at above condition and observe operation of unit and auxiliary equipment. Check for any unusual vibration or noise.
12.	Stop the motor and let unit coast down to a stop. Record coast-down time.
13.	Check for any binding of rotating parts.
14.	Check and clean the screens if necessary.
15.	Start unit as above, full RPM and _____°F or _____°C discharge temperature.
16.	Run for 30 min and perform same checks as during the low discharge temperature run.

*Recommended cleanliness standard:** No hard particles discernable with fingers touching cloth. Maximum count: 12 particles per in² or ~2/cm² after a minimum run of 12 h since last changing cloth. Maximum size 0.001 in. diameter or 25 μm Ø. See also Figure 5.10.

Trending adiabatic temperature and efficiency is another useful indicator of performance. Again, care must be taken to ensure oil injection, cooling effects, slide and bypass valve positions are constant on oil injected machines.

Off-design operation should be avoided for long-term reliability. For example, if the compressor is run around 86 °F or 30 °C, more than rated discharge temperature, loss in area due to expansion would result in an efficiency loss more than 1%. In the long term, it can cause more deterioration and reliability issues.

During startup and initial run-in we should not forget to obtain vibration baseline data. Proposed additional health monitoring methods during operation[6] are:

– Airborne sound
– Structure borne sound
– Vibration analysis with emphasis on rotor and/or gear passing frequencies
– Motor current analysis
– Dynamic discharge pressure analysis

Appendix 13.A: Commissioning of lube and seal oil systems – large oil-free helical screw compressor

1. Check the oil level in the reservoir by the sight glass on the tank side. Fill the reservoir with Teresso™ or equal as needed.

2. Check that the block valve in the discharge line from each pump and the isolation valves around control valves are fully open. Close all valves in the bypass line around control valves.
3. All isolation valves for instruments should be open.
4. Start the motor driven oil pump. The pump discharge pressure should be about _____ psig or _____ bar.
5. Vent the oil cooler and filter which is in service and fill the idle cooler and filter by opening the valve in the bleed line which connects the two filters and cooler oil circuits. Vent the idle cooler and filter.
6. Check the pressure drop across the in-service filter. If the pressure drop exceeds _____ psi or _____ bar, the transfer valve should be switched to put the other filter in service. The used cartridges (for high AP filter) must be changed. Also, a high Δ P alarm should be activated on the local panel.
7. Open the valve in the main header of the nitrogen line to the accumulator and adjust the pressure regulator at the nitrogen bottles so that the downstream pressure is 220 psig or 15 bar.
8. Open the block valve in the steam line to the standby oil pump turbine. Adjust the valve in the bypass line around the steam cut-in valve so that the turbine is slow rolling.
9. Check that oil is flowing to the bearings of the motor, gearbox and compressor by using the sight glasses in the feed lines to each bearing (_____ sight glasses).
10. Check that the lube oil header pressure is _____ psig or _____ bar.
11. Check that the isolation valves around the seal oil differential pressure control valves are fully open. The bypasses around these control valves should be closed.
12. Check that the isolation valves around the buffer gas differential pressure control valves are fully open and the bypasses closed. Check that the valves in the lines from the suction and discharge nozzles to the differential pressure control valves are open.
13. Check that the isolation valves around the seal drain oil differential pressure control valves are fully open and the bypasses are closed.
14. Check that the valves in the lines which permit the sweet seal oil to drain back to the oil reservoir are cracked open.
15. Check that the three isolation valves for each seal oil drain trap are open and that the bypass valve around the trap is closed.
16. Close the valves which interconnect the seal oil trap so that the sour oil from each seal supplies a different trap.
17. Test the startup capability of the standby oil pump by switching off the electric motor driven oil pump. Check pump discharge pressure and oil supply header pressure.

18. Restart the motor driven oil pump and reset the solenoid for starting the steam turbine driven oil pump. The turbine should continue to slow roll.
19. Start cooling water circulating to the oil cooler which is in service. Adjust the water flow so that the oil temperature out of the cooler is about _____ to _____ °F or _____ to _____ °C This oil temperature should be maintained during operation.

References

1 Gea.com.

2 As per API 671.

3 Ohama, Takao; Koga, Takao; Kurioka, Yoshinori; Tanaka, Hironao, *PROCESS GAS APPLICA-TIONS WHERE API 619 SCREW COMPRESSORS REPLACED RECIPROCATING AND CENTRIFUGAL COMPRESSORS.* 2006, Texas A&M University, Turbomachinery Laboratories. Available electronically from http : / /hdl .handle.net /1969 .1 /163186.

4 API 619, *Rotary-Type Positive Displacement Compressor for Petroleum, Petrochemical and Natural Gas Industries,* Fifth Edition, December 2010 (American Petroleum Institute Publishing, Washington, D.C., U.S.A.).

5 Bloch, H.P. & C. Soares, *Process Plant Machinery,* 2nd Edition, 1998, Butterworth-Heinemann, Woburn, Mass., USA, a member of the Elsevier Group.

6 Dieter Franke and Mathias Luft, *Effective Diagnose und Überwachung von Schraubenverdichtern,* Industriepumpen + Kompressoren – Heft 2/2007, Pages 87–93.

Chapter 14
Commissioning and Startup of Turbocompressors

14.1 Introduction

As is evident from Figure 1.1, turbocompressors are applied in four different configurations:

14.1.1 Axial and Centrifugal Compressors

Axial compressors can handle large flows with relative low compression ratios. Gas movement and pressure generation is accomplished by multiple pairs of airfoils with one pair consisting of a stationary blade and a rotating blade. The rotating blade accelerates the gas in order to add kinetic energy and then the stator blade decelerates the fluid to generate pressure. Axial compressors have many pairs or stages of compression in a single casing, 15 stages of compression are not uncommon. Flows inside the axial compressor casing are parallel to the axis of rotation. Axial compressors in the process industry are not as numerous as centrifugal compressors. Axial compressors operate usually at moderate speeds between 1,800 and 3,600 rpm. They are found in most fluid catalytic cracking processes for oil refineries,

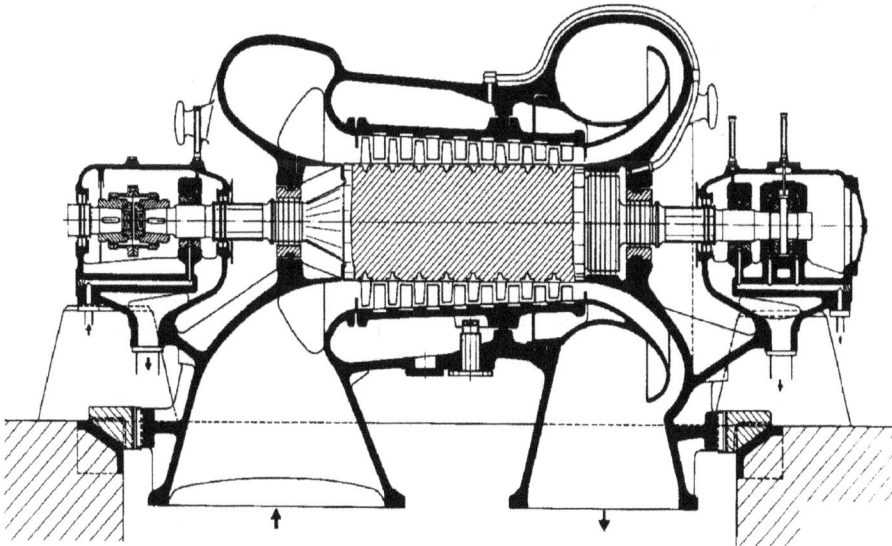

Figure 14.1: Typical axial process compressor (DEMAG).

https://doi.org/10.1515/9783110701074-014

as well as being the source of combustion air in gas turbines. Axial compressors can also be encountered in air separation process facilities.

Centrifugal compressors are smaller in size. Figure 14.2 shows a modern centrifugal process compressor ready to be installed. Centrifugal compressors work on the principle of pressure generation by a number of impellers arranged on a shaft. They are sometimes referred to as single shaft centrifugal compressors and run at speeds between 3,600 and 15,000 rpm. An eight-stage centrifugal compressor is generally the largest configuration applied in the hydrocarbon processing industries due to negative rotor dynamic effects when more than eight stages are used. When higher levels of compression are required, multiple centrifugal compressors are used in a series configuration. Generally, oil refineries and chemical plants are suitable for a single centrifugal compressor. LNG applications, for example, call for large, multiple compressor trains.[1]

Figure 14.2: Modern centrifugal process compressor (courtesy: Siemens).

Centrifugal compressors can be adapted to changing conditions. A case in point: For upstream applications in the HCPI, it is well known that the produced gas molecular weight will drop during the life of the reservoir. With this knowledge in mind the writers took two actions during the FEED phase for the example noted above to ensure that reinjection compressors would be fit for purpose for the life of the reservoir. The first compressor body was designed with the maximum allowed number impellers according to the OEM's "best practices." The current duty for the reinjection compressor was met by removing one impeller. The spare rotor was manufactured with all impellers installed. The gas turbine driver was sized to be suitable for the fully staged spare rotor when operating with the lower molecular weight gas as the reservoir aged and the gas leaned. Simply installing the spare rotor will ensure that this gas reinjection application will be suitable for the life of the reservoir.

14.1.2 Integrally Geared Compressors

are often referred to as multi-shaft centrifugal compressors as shown in Figure 14.3. They work on the centrifugal compressor principle, but consist essentially of a high-speed gearbox with a large "bull"-gear[2] driving multiple individual shafts with attached overhung impellers.

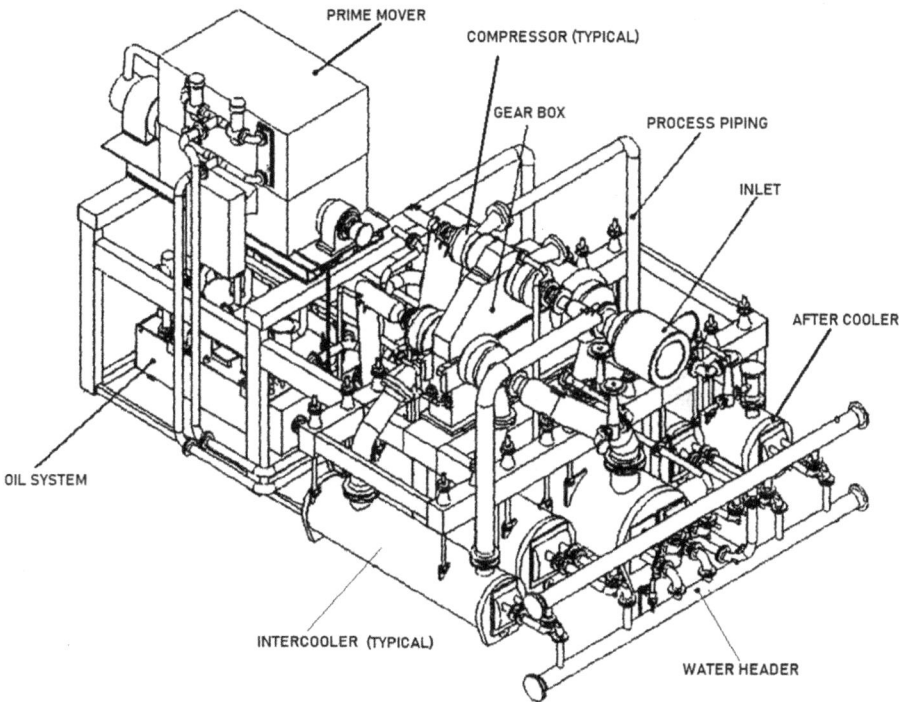

Figure 14.3: Typical two-stage integrally geared compressor (ATLAS COPCO).

The pinion gears are mounted to the rotating shafts that carry the impellers. The pinion speeds range to a maximum speed of 75,000 rpm.

Integrally geared compressors are available with many options for compressor stages. Each rotor can carry two impellers. Each impeller is called a compressor stage. Since every pinon rotor can have its own rotating speed the efficiency of this type of compressor can be optimized to match the process conditions. Integrally geared compressors are available with one to five pinion rotors mounted around the "bull gear" which allows one to ten compression stages per machine.

To date, these types of compressors have been built for a wide range of services, including gas-turbine-fuel gas (methane), hydrocarbon refrigeration gases, ammonia, carbon monoxide, carbon dioxide, syngas, gas-field gathering and others. In the past, these types of compressors might have been somewhat modestly sized machines. Today, however, we see flow rates ranging to 400,000 m^3/h; molecular weights from 6.5 to 58 MW; stage-pressure ratios as high as 2.6:1; inlet pressures from ambient to 45 bar; and discharge pressures to 200 bar.

One US manufacturer has designed integrally geared compressors for drivers up to 26,800 hp or 20 MW. This OEM and/or its predecessor companies have produced 12,000 such machines, with six high-speed stages available in a single integrally gear-driven frame (casing). However, as many as 12 stages have been incorporated in an integrally geared compressor designed by a European manufacturer, and drivers have, on occasion, become considerably larger than 20 MW.

Today's compressor efficiencies can approach 90%. Some manufacturers have designed and built their own gearboxes for decades. Many OEM factories have undergone expansions over the past 50 years. The integrallly geared compressors they manufacture are extensively shop-tested, and the overall designs are up-to-date. Seal designs have kept pace with modern dry-gas-sealing technology, as have their well-proven intercoolers.

Intercoolers on integrally geared plant air compressor packages, in particular, had displayed a history of fouling and plugging – however, not always due to their design.

In any event, there are many opportunities to use multi-stage, integrally geared machines where mere tradition may have steered us in the direction of positive-displacement machines. The design of these late-generations, dynamic compressors favors single-point train responsibility. As an additional feature, modern, integrally driven compressors often represent the best possible blend of standardization and customized design. They also fit a progressing modularization concept in industry which, in certain cases, offers many advantages over conventional project practices.[3] Multi-stage, integrally geared compressors certainly merit consideration and invite comparisons with every other type of compression machinery.

14.1.3 Expander–Compressors

They are often referred to as turboexpanders – expansion turbines as rotating machines similar to steam turbines. They are frequently just called expanders. They work on the principle of gas expansion. It is actually a prime mover in that it performs compression work by driving a centrifugal compressor impeller. Figure 14.4 depicts an expander–compressor.

Designers and ultimately owners of turbocompressors are generally guided by the industrial standard, API 617.[4] This standard covers the minimum requirements

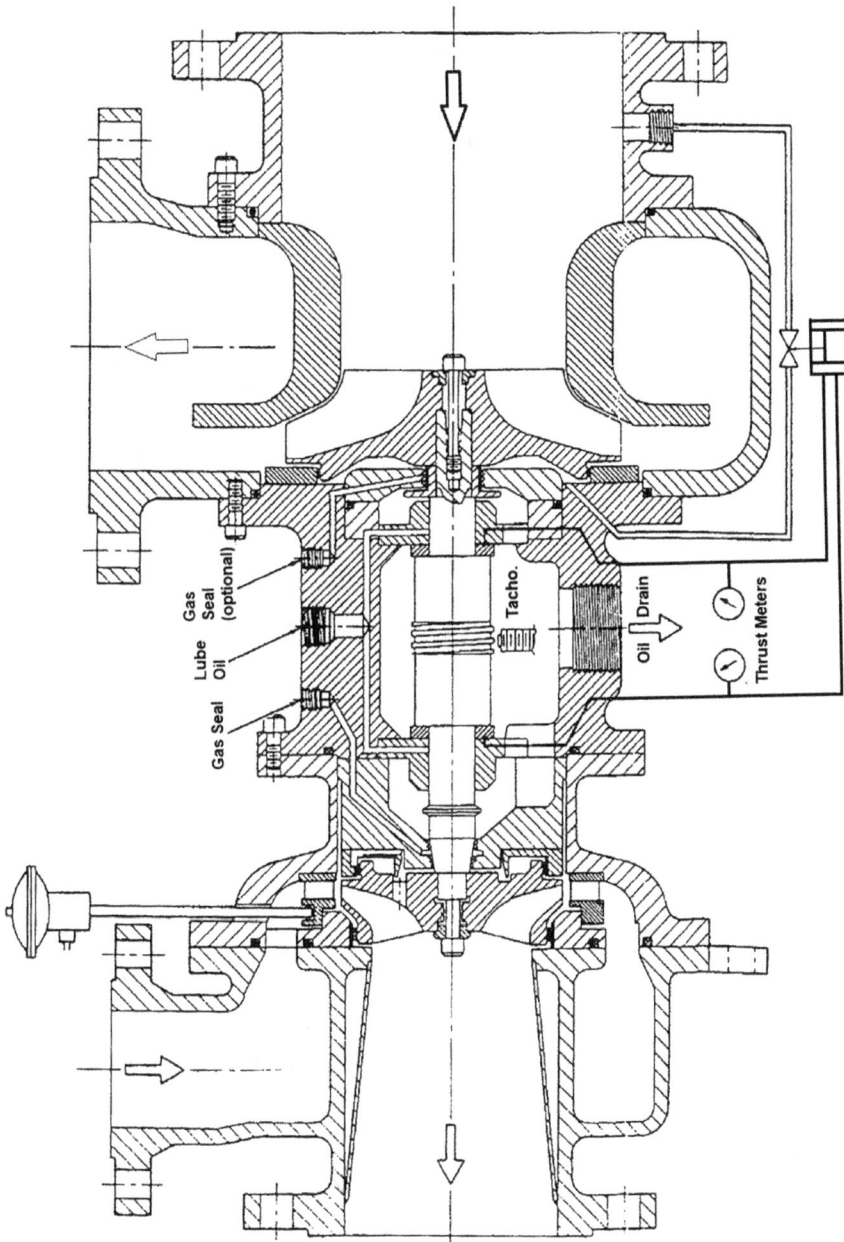

Figure 14.4: Typical turboexpander–compressor cross section (source: ROTOFLOW).

for axial compressors, single-shaft and integrally geared process centrifugal compressors, and expander–compressors for use in the petroleum, chemical, and gas industries services that handle air or gas. An additional standard is API Standard 672, Packaged Integrally Geared Centrifugal Air Compressors for Pertroleum, Chemical and Gas Industry Survives, 4th Edition, March 2004. The readers are encouraged to consult these standards as reliable information sources in case a deeper understanding of these machines is needed.

Commonly, the terms "expansion turbines" and "turboexpanders" specifically exclude steam turbines and combustion gas turbines. Turboexpanders can also be characterized as modern rotating devices that convert the pressure energy of a gas or vapor stream into mechanical work as the gas or vapor expands through the turbine.[5] Figure 14.5 shows a typical turboexpander application.

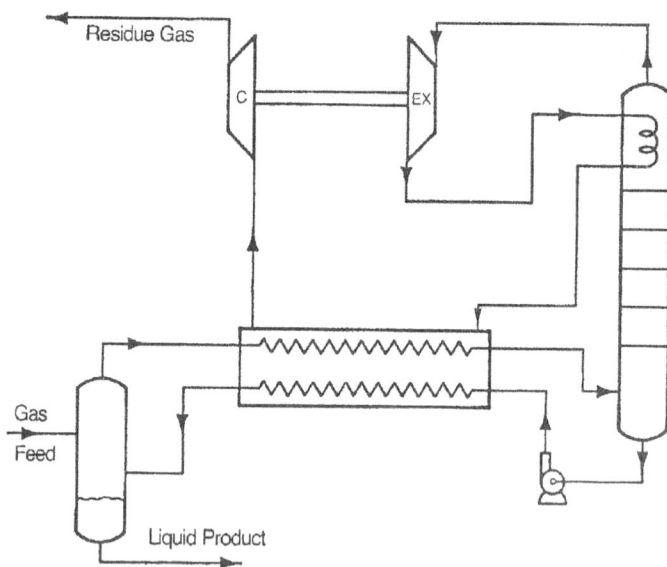

Figure 14.5: Turboexpander applied to the separation of propane and heavier hydrocarbons from a natural gas stream (from reference[5]).

Many more applications are covered in reference[5].

14.2 Commissioning

14.2.1 Risk Identification

Turbocompressors are widely used because they are reliable due to their robust and generally simple design. Once started up and operating, a turbocompressor, if it is fit for purpose, is highly reliable. Records abound of these machines operating in clean gas service for some 10 years without a forced outage incident related to mechanical components. A typical process turbocompressor train will most likely experience a forced outage of short duration mainly because of control and instrumentation problems. This is reflected in Figure 14.6 which is based on operating experience of a fleet of motor- and gear–driven compressor trains.

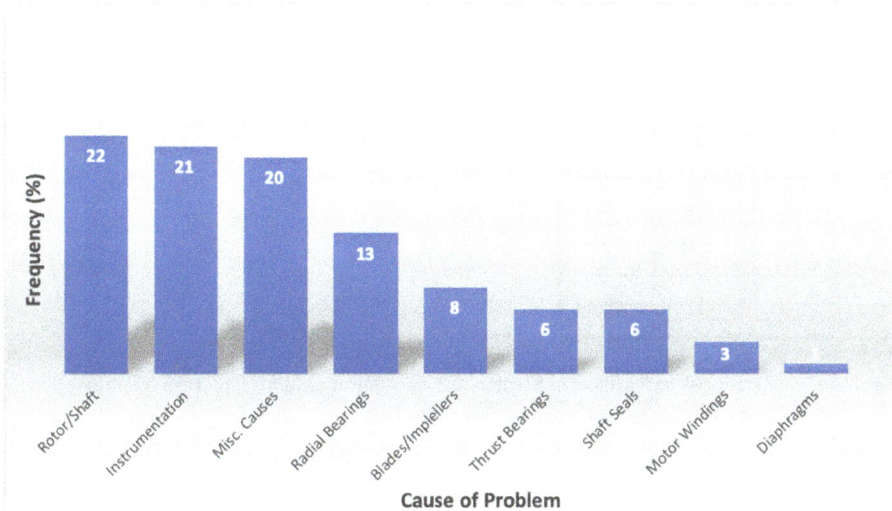

Figure 14.6: Risk profile of a motor-driven turbocompressor train.[6]

An example of an inherent-risk value (RIN) is displayed in Figure 14.7. It is based on a three-compressor string driven by a gear and an induction motor. The machines, schematically represented in Figure 14.8, feed natural gas to an ethylene plant. The risk number is high because of the consequence of an outage. For example, for a 9MMTPA oil refinery, the average estimated production loss per day can amount to approximately 1.7 million USD.[7] Incentives for a successful and timely startup are high. We must therefore, devote our time and energy to ensure commissioning and startup of this major turbocompressor train is successful.

Commissioning and startup related risks are discussed for review in Table 14.1.

We see that rotating parts of the machine are the predominant failure cause. Following the cause-and-effect chain, we find that cleanliness of the gas stream is a key factor in the trouble-free operation and reliability of any compressor. Contamination of the machine internals can be caused by:

- *Corrosive, sour or contaminated gases*. They require special provisions and operational considerations. Fouling due to contamination or process reactions can cause rapid degradation leading to rotor difficulties.

1	2
	7,800hp/5.8 MW
	Motor/Gear driven
	3-String CenrCmp.
Rel. Risk #	783
Rel.Risk Level:	HIGH
Complexity	27
1.1	6
2.2	4
4.2	8
5.1	4
6.5	5
7.0	0
Severity	29
Table 2.4.1	5
Table 2.4.2	8
Table 2.4.3	8
Table 2.4.4	8

Figure 14.7: Inherent-risk assessment of a motor- and gear-driven centrifugal process compressor unit.

- *Liquids.* Particular care should be taken when liquids are present in the gas stream. Unlike other compressor types, in particular reciprocating compressors, turbocompressors are more forgiving to liquid ingestion in the form of a mist. Nevertheless, one should avoid any intake of free liquids, because a turbocompressor is in danger of severe mechanical damage if suddenly deluged with liquid. For this reason, properly sized suction drums are required to trap any liquids, particularly where there is a possibility of condensation in the suction piping and passages.
- *Foreign objects.* Particles or other forms of contamination are also a concern. Ingestion of particles with in a size range of 30 microns or smaller is reasonable, 80 microns would be marginal, and 400 microns would be definitely too large. A significant exception to this rule would be when gas contains fine, but erosive, particles. In such cases, lower limits should be adopted. Turbocompressors can be washed on-stream to counteract the effects of fouling. Trial and error helps to find the optimum liquid wash and associated parameters, such as rate of flow. The flow should be enough to clean the machine without causing erosion.

In all liquid-injection applications, it is recommended to use tangential sprays where practical. Ideally, the liquid should be injected so it does not impinge on any surface to avoid erosion.

Obviously, there is a long chain of activities that lead to the point where commissioning and startup of a turbocompressor train can begin. It stands therefore to reason that the lead machinery person should become as early as possible intimately

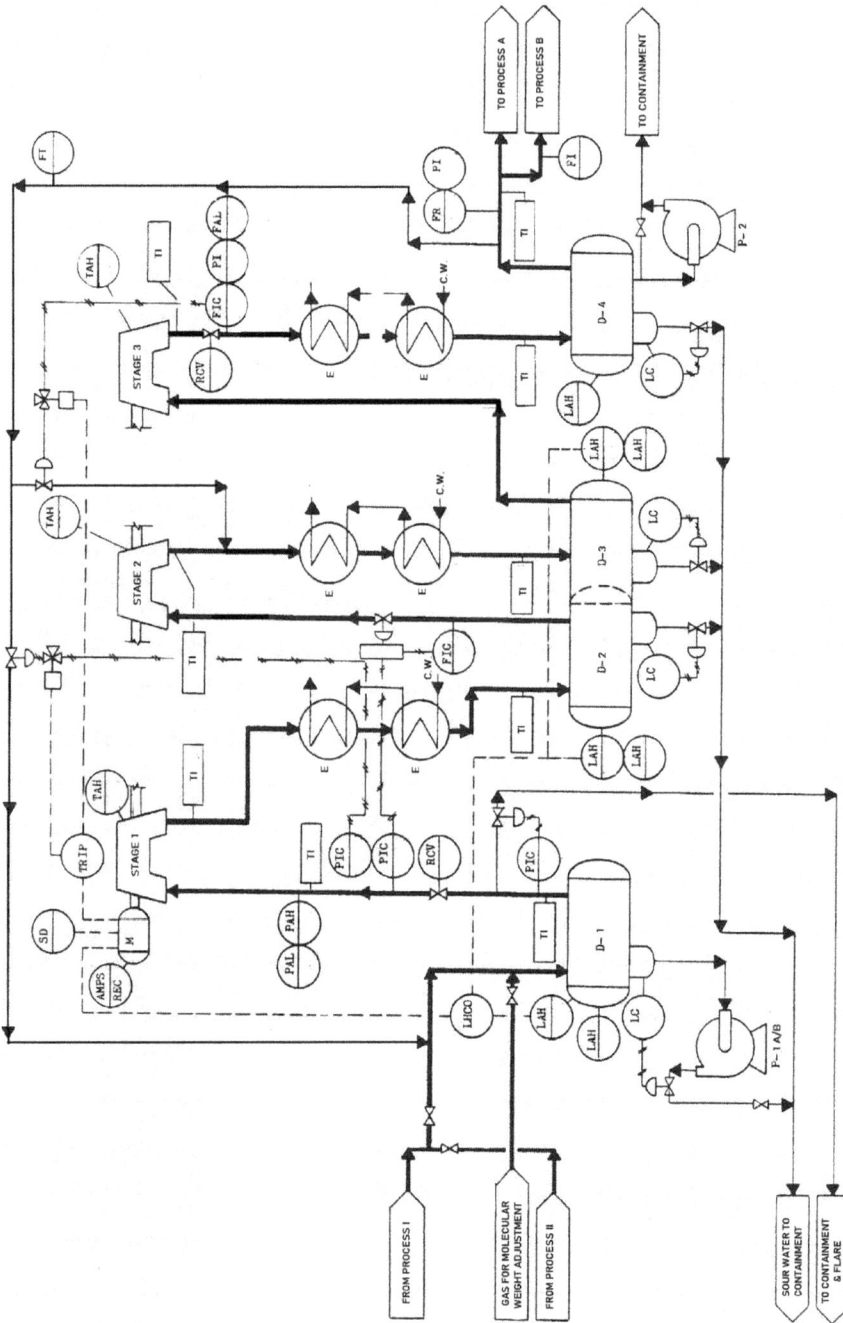

Figure 14.8: Simplified flow diagram for a feed gas compressor train.

Table 14.1: Project-related risk mitigation.

Phase	Risk	Most probable cause[1]	Suggested actions/mitigating measures
Rigging and lifting	Load tilting	QP	Review lifting plan
Handling, staging and storage protection	Inadequate protection resulting in rework and delay	OR, HK	The process starts with preparation for shipment by the OEM. API 617 has excellent instructions for preparation. Review plans for handling and storage protection.
Foundation and grouting	Low Risk	QP	Compressor packages are customarily furnished with flexible feet and can be therefore mounted with any need for a special foundation.
Piping	Ingestion of foreign objects after cleaning	OR, HK	Cleanliness and assembly to the machine should follow practices established as part of the project execution agreement between OEM and the owner organization.
Shaft alignment	Misalignment	QP	Shaft alignment starts with leveling the unit. A minimum permissible deviation from the horizontal line would be customarily 0.002 in. per foot or 0.2 mm per linear meter. Usually, if there is not a gear box in the train, axial or centrifugal compressors are considered the fixed machine and the driver is aligned to it.
Lubrication system	Wrong Oil Specification Mix-up Contamination	QP, DO, CO	Cleanliness of the lubrications system is of utmost importance. Table 14.3, describing FAT events, should be consulted to understand if the system is in need of additional cleaning judging from the recorded work done in the OEM shops.
			See also API RP 686, 2nd Edition, 2006, *Recommended Practice for Machinery Installation and Installation Design and / or* API RP 1FSC, 1st Edition, 2013, *Facilities Systems Completion Planning and Execution.*

[1]Refer to Figure 2.12.

familiar with the equipment in his care. A good approach would be to review the EPC's drawings before it is too late to implement any changes. Table 14.2 represents a check list that should be a guide for the conscientious professional to get familiar with the unit at hand before any commissioning takes place.

Turbocompressors offer a unique opportunity for risk mitigation by being tested in the OEM's facility prior to delivery at the installation site.[8] There are two types of factory acceptance tests (FAT). One, the mechanical run test and two, the aerodynamic performance test.[9] When specified by the purchaser, the aerodynamic performance test in the OEM facilities is usually executed following the rules set out in the ASME Power Test Code.[10] Here we would like to discuss the mechanical run test only. While it is not always possible to eliminate all faults and shortcomings during these tests, they can be an insurance against nasty surprises during commissioning and startup.

Table 14.2: Example for an EPC's drawing review – turbocompressors.

EPC's Drawing Review Centrifugal Compressors Designation: _____ Location: _____ Service: _____	
P&ID's	
1. Is there a KO drum on the suction and any cooled interstages?	_____ (Yes or no)
2. Do the recycle lines re-enter upstream of the KO drums?	_____ (Yes or no)
3. Are the KO drums equipped with a gauge glass, LHA and shutdown switches?	_____ (Yes or no)
4. Have all the alarm and shutdown switches specified been provided? Low-lube oil pressure alarm and auxiliary pump start-actuation_____ Low-lube oil pressure alarm and trip? _____ Low seal oil level or pressure alarm and auxiliary pump start up actuation _____ Low seal oil level or pressure alarm and trip _____ High seal oil level alarm high discharge temperature alarm _____ Others (detail)	_____ (Yes or no)
5. Are TI's and PT's specified for suction and discharge of each stage?	_____ (Yes or no)
6. (a) Is there a check valve in the discharge downstream of the anti-surge recycle?	_____ (Yes or no)
(b) If there are two machines in parallel, is there a check valve on the discharge of each?	_____ (Yes or no)

Table 14.2 (continued)

P&ID's	
7. Is there a flow meter on each feed and discharge stream from the machine? The mass flow on each stage is to be measured more than one meter is only required if the mass flow changes from section to section	_____ (Yes or no)
8. Have pressure taps been provided up and downstream of the temporary suction strainers?	_____ (Yes or no)
9. Will all instruments be changeable on the run and are shutdown circuits testable on the run?	_____ (Yes or no)
10. Are all lines to remote pressure gauges valved at the tie-in to the main line?	_____ (Yes or no)
11. Is there an isolatable closed circuit including a cooler at the machine so that the pre-startup run-in can be performed?	_____ (Yes or no)
12. If the machine is not in closed circuit and is not an atmospheric air machine, is there a suction flare release to dump the suction gas in the event of shutdown?	_____ (Yes or no)
13. If the process is "flow controlled," is the metering elements outside the recycle loop?	_____ (Yes or no)
14. Is the anti-surge metering element inside the recycle loop?	_____ (Yes or no)
15. On refrigeration machines the TIC's must close on driver trip. Do they?	_____ (Yes or no)
16. On motor-driven refrigeration machines, the casing pressure must be reduced to about 40 psig or 2.75 bar before starting to prevent driver overload. Liquid in the suction drums impedes the pressure reduction. Can the liquid from the drums be pumped into the accumulator?	_____ (Yes or no)
17. Is there a safety valve on the discharge of the machine if the downstream equipment cannot stand the machine's discharge pressure under the combined conditions of: Trip speed _____ High mol wt. _____ High suction pressure _____ Low temperature _____	_____ (Yes or no)

Layouts	
1. Are the main and interstage suction lines from the KO drums cleanable by the method proposed?	_____ (Yes or no)

Table 14.2 (continued)

Layouts	
2. By the time of this review the moment of inertia of compressor and driver should be available. Is the hydraulic energy inside the check valves less than 1.3 of the kinetic energy of the shafts? If not, the check valves will either have to be relocated or additional ones installed.	_____ (Yes or no)
3. Are there drain valves at the low points of the suction and discharge lines?	_____ (Yes or no)
4. Are all the check valves horizontal?	_____ (Yes or no)
5. Are all the check valves damped or equivalent?	_____ (Yes or no)
6. Will the recycle valves pass the compressor design flow?	_____ (Yes or no)
7. Are the control valve actuators adequate for contemplated operating conditions? (Users have had trouble on both straight control valves and butterflies.)	_____ (Yes or no)
8. If the machine is an atmospheric air compressor, does the anti-surge vent have a silencer and is the intake of a sound attenuating type?	_____ (Yes or no)
9. Is the foundation separate from that of reciprocating machines?	_____ (Yes or no)
10. Can the temporary strainers be removed without disconnecting any piping?	_____ (Yes or no)
11. Are the suction, discharge and compressor drains connected either to a blowdown system or vented to a safe place?	_____ (Yes or no)
12. Unless the compressor OEM has given dispensation, is there a straight section of at least three pipe diameters on the suction flanges?	_____ (Yes or no)
13. Are the pipe stresses and moments within the levels allowed by the vendor?	_____ (Yes or no)
14. Are all the piping supports and anchors as described in the piping stress calculation?	_____ (Yes or no)
15. Has sufficient allowance been made in the stress calculation for friction of supports?	_____ (Yes or no)
16. On refrigeration machines, are the liquid injection points at a sufficient distance from the drums to ensure vaporization of all the liquid?	_____ (Yes or no)
17. On refrigeration machines, do the liquid level control valves have blocks and bypasses?	_____ (Yes or no)

Table 14.2 (continued)

Layouts	
18. On refrigeration machines, it is necessary to adjust the TIC controllers during startup. On motor-driven machines, is there a single switch to commission the TIC's immediately after startup? Is it readily accessible from the platform?	_____ (Yes or no)
19. On turbine-driven machines are the TIC's readily accessible from the platform? (Adjustments to the set points are necessary during run-up.)	_____ (Yes or no)
20. Have MOVs been provided on all lines which can feed hydrocarbons to a fire at the machine and do not have block valves at least 25 ft. or 7.6 m horizontally from the machine? Has the electrical conduit and valve operator to such MOVs been fireproofed sufficiently to permit operation of the valves after a 10-minute fire?	_____ (Yes or no)

Operability and surveillability	
1. Are all instruments clearly visible?	_____ (Yes or no)
2. Has the operator safe and easy access to all bearings?	_____ (Yes or no)
3. Has the operator safe and easy access to the handwheels on the MOV's, the flow control devices and the recycle valves?	_____ (Yes or no)
4. Run through a startup sequence. Can all the operations required be done by one person?	_____ (Yes or no)
5. Is there safe access to the suction and discharge line and all casing drains?	_____ (Yes or no)
6. Can the oil drain sight glasses be readily seen?	_____ (Yes or no)
7. If there are overhead seal tanks, has the operator a clear view of the level gauge?	_____ (Yes or no)
8. If the oil level control has to be put on hand control using the bypass, will the operator be able to see the level gauge from his position at the valve?	_____ (Yes or no)

Maintainability	
1. Is there a suitable location where the casing top half can be put without interfering with maintenance? Note: If the machine has multiple casings or if it has a turbine driver, all top halves may be off at the same time.	_____ (Yes or no)
Is the crane big enough to carry the largest maintenance weight? Usually the top half of the largest casing.	_____ (Yes or no)
Can the top halves be moved to the storage area without passing over operating machinery?	_____ (Yes or no)

Table 14.2 (continued)

Maintainability	
If the machine is motor-driven, is there access to that end so that the motor rotor can be pulled, if necessary, using portable equipment?	________ (Yes or no)
Are the motor cooling ducts so positioned that they do not unnecessarily interfere with the crane movement?	________ (Yes or no)
If the compressor is at grade with overhead piping, can the piping spools be readily removed and swung out of the way leaving vertical lift for casing?	________ (Yes or no)
Have lifting provisions been made to facilitate this?	________ (Yes or no)
Can the rotors be removed to the maintenance shop without passing over operating machinery?	________ (Yes or no)
Are the motor cooling ducts so positioned that they do not unnecessarily interfere with the crane movement?	________ (Yes or no)
If the compressor is at grade with overhead piping, can the piping spools be readily removed and swung out of the way leaving vertical lift for casing?	________ (Yes or no)
Have lifting provisions been made to facilitate this?	________ (Yes or no)
Can the rotors be removed to the maintenance shop without passing over operating machinery?	________ (Yes or no)
Instrumentation issues	
Are the TI's installed in such a way that they will measure the correct temperature?	________ (Yes or no)
If the lines are two-phase, will they see the correct phase?	________ (Yes or no)

14.2.2 The Mechanical Run Test

of rotating equipment should be viewed as a minimal test to determine the rotodynamic acceptability and should be considered for equipment that is designed-for-purpose in contrast to equipment selected from a catalog. API Standard 617 includes specific procedures to follow during a mechanical test. Important test factors related to the rotodynamic behavior of the equipment are:

- Test speeds/duration
- Lube oil parameters (temperature, flow rate and viscosity)
- Rotor and support configuration

Mechanical run test of a compressor is mandatory and is essentially a vibration demonstration test. It also provides an opportunity to the owner's representatives to get familiarized with the equipment.

Mechanical testing provides information related to the critical speed location and some indication of shaft modes behavior. The modal information is limited to only those modes located below the maximum test speed achieved, the trip speed in most cases. Typically, this is only the first critical speed.

What isn't tested however can be significant. For example, subcritical critical speeds, frequently and incorrectly termed stiff shaft, operate below the first bending critical speed. These critical speeds can have high amplification factors and can be damaging if the separation margin is lost. Performing only a mechanical test will tell the user whether the mode is on or below the operating speed. The amount of separation remains untested and can only be inferred from the unverified analytical predictions.[11]

Operation of the compressor during the mechanical test should include a warm-up portion where the rotor speed is incremented in 10-min intervals carefully avoiding exclusions zones of critical speeds and blade natural frequencies. Following the warm-up, operation at trip speed is specified for 15 min followed by an uninterrupted 4-h run at maximum continuous speed. A coast-down from trip follows the 4-h run. The warm-up portion is included to ensure that the case and rotor are given time to thermally expand gradually to avoid creating unintended interferences leading to rub damage. The warm-up portion also permits examination of the rotor behavior at increasing speed intervals. Thus, faults can be detected at less energetic stages potentially avoiding rotor/stator damage and project delays.

Trip speed is included to ensure that vibration levels – and overall operation – are acceptable at this speed. The 4-h run portion is used to set the thermal conditions of the system. Bearing temperatures and vibration levels are important factors to watch during the test. Stable levels of each parameter need to be reached during the test. If any parameter shows signs of continual movement (increase or decrease), the test should be extended until stable levels are achieved. If not, the test should be rejected. Following the 4-h run, a coast-down from trip speed is performed. The coast-down is used to determine the overall and synchronous behavior of the rotor/support system. This data will also be used as the baseline for verification testing, if performed.

Operation at trip speed during the mechanical run test is important for several reasons. First, running to trip speed increases centrifugal forces on the rotor which may relax interference fits and permit the rotor's static shape to change. This may alter the balance state of the rotor. Second, reaching trip speed is necessary for trip testing in the field. It is convenient to reach these speeds during testing to identify any potential problems. Finally, trip speed operation can help identify any critical speeds occurring just above maximum continuous speed that would otherwise not be seen on the test stand.

Additional shutdown, startup and transient operation can be added at the beginning of the 4-hour test. When compared against the coast-down following the 4-h test, thermal transient behavior of the rotor can be examined and any changes to the balance state of the rotor can be identified. This may prove useful in diagnosing the Morton effect,[12] clearance closure of the radial bearings or fit relaxation due to operation at trip speed.

Lube oil parameters should simulate that used for the intended application. Bearing flow, lube oil inlet temperature and viscosity should be within the specified operating ranges set by the OEM for field operation. After stabilization during the 4-h run, lube oil inlet temperature can be varied from the minimum to maximum specified range to examine the effect on vibration levels and bearing operation.

Operation at the range limits should be held until steady-state conditions are achieved. The rotor configuration should be as close to the operating condition as possible. The major concern centers around the overhung weight associated with the coupling. Often mechanical testing is done at partial rated power levels. Smaller shop drivers need not have, nor in some cases could accommodate, the larger couplings of the job. It is essential to closely mimic the overhung weight of the job half-coupling. This may require that a simulator be added to the drive assembly to match the overhung moment. The opposite may be true for vendor's smaller casings where the test coupling's overhung moment is larger than the job coupling.

The startup team and particularly the senior machinery specialist must be familiar with the mechanical test results that can be found in the mechanical test run record as shown in Table 14.3.

Table 14.3: Example of a mechanical test run record – turbocompressors.

Mechanical run test
Centrifugal compressors
Designation: _____
Location: _____
Service: _____
The mechanical run test for centrifugal compressors is basically a balance check. In some cases data on the vibration characteristics of a machine will also be disclosed.
The basic procedure should be to run up to 110% of max. continuous for turbine-driven machines, and run for a minimum of 15 min. Then drop back to max. continuous speed and make the overall test 4 h. For motor-driven machines, max. continuous speed is design speed.

	Conditions	
Parameter	**Design**	**Test**
(a) Speed rpm	_____	_____
(b) L.O. inlet pressure	_____	_____

Table 14.3 (continued)

(c) L.O. inlet temperature[1]	_____	_____
(d) Max. vibration	_____	_____
(e) Calc. critical	_____	_____
(f) Max. noise level	_____	_____

1.   Do the first four test conditions match the design conditions to your satisfaction?

_____ (Yes or no)

2.   Is the actual critical speed within 5% of the calculated value?

_____ (Yes or no)

3.   Is the temperature rise across each bearing less than 60°F?

_____ (Yes or no)

4. Vibration
(a) Frequency survey when running at max. continuous speed. Note vibration at running speed and other frequencies.

Probe Location	Magnitude (mils / μ)	Frequency (cpm)
_____	_____	_____
_____	_____	_____
_____	_____	_____
_____	_____	_____
_____	_____	_____

(b) Is there undue vibration at critical frequency, also at _____
frequencies between 35% and 50% of running speed?   (Yes or no)
(c) Shaft and bearing vibration attenuation:

	A, Shaft	B, Housing	A/B, Attenuation
I.B. bearing	_____	_____	_____

5.   Is a thorough check being made for oil leaks? (Casing drains checked?)

_____ (Yes or no)

6.   Self-acting dry gas seals?

_____ (Yes or no)

O.B. bearing	_____	_____	_____
		Is the attenuation <4? (If no, mention in report)	(Yes or no)
	110% Design speed	Max. Continuous	Difference
I.B. bearing	_____	_____	_____
O.B. bearing	_____	_____	_____

Is the difference in vibration levels at these speeds less than 20%?   (Yes or no)
    If yes, have the following readings been taken – assuming a tandem seal arrangement?

Parameter	Design	Test
(a) I.B. Primary Seal vent flow	_____	_____
(b) O.B. Primary Seal vent flow	_____	_____

Table 14.3 (continued)

(c) I.B. Secondary Seal vent flow	_____	_____
(d) O.B. Secondary Seal vent flow	_____	_____

7. Oil Seals?

    _____
    (Yes or no)

7a. Is the oil collected from each seal drain less than 5 gallons per day or 19 L/day? (This is significant only on carbon seals or bushing seals with normal differential pressure.)

    _____
    (Yes or no)

If not, insist that it is lowered. However, operational of normal running seals at low pressure may take the outer bushing run hot and assurance that this is the cause should be accepted

7b. Is the seal oil outlet temperature less than 180°F or 82°C?

If not, insist that it be lowered. However, operation of normal running seals at low pressures may make the outer bushing run hot. An assurance that this is the cause should be accepted.

    _____
    (Yes or no)

8. Bearing inspection. Are all bearing surfaces showing normal running pattern? Demand replacement otherwise.

    _____
    (Yes or no)

9. Seals to be inspected only if it is the vendor's standard practice.

    _____
    (Yes or no)

10. Internal inspection to be carried out if a spare rotor is to be fitted and run (normally specified).

Is there absence of rubbing? _____

If rubbing is evident, demand a clearance check.

    _____
    (Yes or no)

11. Check the internal alignment and clearance data from final assembly.

(a) Is alignment good, clearances within tolerances_____

(b) Likewise, for a spare rotor if it has been fitted _____

    _____
    (Yes or no)

12. Were copies of vendor's test log sheet and final internal clearance diagrams obtained?

    _____
    (Yes or no)

13. When witnessing a test, always try to find out if any difficulties occurred in preparing for it. Such problems could be repeaters.

    _____
    (Yes or no)

14. After completion of mechanical running test, was casing along with seals pressurized to maximum sealing? (or seal design pressure)?

    _____
    (Yes or no)

15a. Gas pressure test following the mechanical run

Is the test being carried out with visible soap bubbles?

    _____
    (Yes or no)

Table 14.3 (continued)

15b. Gas pressure test following the mechanical run Is the shaft being rotated to check for freedom of seals?	_____ (Yes or no)
	_____ (Yes or no)
	_____ (Yes or no)

[1]Try to use equivalent to oil grade at site and maintain highest permissible lube oil inlet temperature.

Similarly, attention should be paid to the inspectors' reports. They may contain outstanding issues and reservations based on inspections by the purchaser's or owner's representative who had access to all vendor or OEM and sub-vendor plants where manufacturing, testing or inspection of the purchased equipment was in progress. He would have had insight into the suppliers' production organization and thus become aware of the risks arising from a potentially shallow OEM manufacturing depth. Additionally, an attentive and conscientious owner's representative will have sensed hidden problems in the OEM's shop and test operation that were not obvious, not communicated and therefore unknown to everyone involved.

This is also the point in time where the startup team has to assure that maintenance tools, such as strong arms and other special tools, specifically designed for the machine, are on hand. Refer to Figure 14.9 and Table 14.4. We alluded to this in an earlier chapter. The inspector will have made sure that during the assembly of the machine in the OEM's shop these tools and no substitutes were in effect used. He might also be aware of parts and component exchanges as well as design changes

Figure 14.9: Compressor maintenance tool ("strong arm" assembly) (source: Cooper-Bessemer).

Table 14.4: Example of a tool list for a centrifugal compressor.

Qty.	Description	Use	Remarks
1	Permanently installed crane above compressor train	To lift/handle heaviest parts.	By owner
1	Lifting beam	To lift/handle unit/skid – if required.	By OEM
1	Set of alignment tools (Laser)	To align train components.	By OEM
1	Coupling hub puller	To pull keyed or thermal expansion fitted coupling hubs.	By OEM – taps in hub should be preferred
1	Set of hydraulic pulling and mounting fixtures incl. hydraulic pump	To pull and mount hydraulic fitted tapered coupling hubs. To torque stud bolts.	By OEM – taps in hub should be preferred
1	Set of assembly sleeves	To cover and protect bearing and seal shaft surfaces during assembly.	By OEM
1	Pulling device for dry gas seal (DGS) package	To assemble/ disassemble gas face seal labyrinth.	By OEM
1	Thrust collar puller fixture	To pull damaged thrust collar – if not integral to the shaft.	By OEM – owner's engineer to ascertain tools have been used during shop assembly
1	Hydraulic impeller and installation tool	To pull impeller on overhung shaft designs.	By OEM – owner's engineer to ascertain tools have been used during shop assembly
1	Assembly fixture	To handle and lift overhung impellers – see also Figure 14.9.	By OEM – owner's engineer to ascertain tools have been used during shop assembly
1	Set of special tools (rails and saddle)	To remove aerodynamic bundle (rotor and diaphragms) from vertically split casings.	By OEM – owner's engineer to ascertain tools have been used during shop assembly
2	Lockwiring tools	Lockwiring is a technique to prevent threaded fasteners from loosening.	By owner

during assembly and testing in the vendor's shop. There has to be assurance and verification that these changes have found their way into drawings, records and spare parts inventory for the new unit to be delivered or already on site.

14.3 Installation

During and after commissioning, a turbocompressor is started up when all accessories and utilities are connected and ready. Before arriving at this point the following tasks have to be successfully accomplished:
– Shipping, handling and receiving at the construction site
– Storage protection
– Foundation and grouting
– Piping connections
– Shaft alignment
– Preparation and cleaning of lubrication systems

14.3.1 Storage Protection

The process starts with preparation for shipment by the OEM. Standards specify that the equipment shall be suitably prepared by the original manufacturer for the type of shipment specified, including blocking of the rotor when necessary. The preparation shall make the equipment suitable for six months of outdoor storage from the time of shipment, with no disassembly required before operation, except for inspection of bearing and seals. After unpacking the equipment for imminent installation, one manufacturer suggests the storage protection steps outlined in Appendix 14.A and Figure 14.A1. If longer storage periods are contemplated, the owner will consult with the vendor regarding the recommended procedures to be followed. For this case, our recommendation is as outlined in Appendix 14.B.

14.3.2 Foundation and Grouting

While problems with foundations and grouting in the context of turbocompressor trains are infrequent, steel structures, such as base frames and baseplates, play a major role in the reliability of turbomachinery packages. Their design and manufacturing have been closely related to some critical topics, such as allowable nozzle loads, vibration and dynamic behavior.

It is painful to see an expensive turbomachinery package fail only because of a faulty steel support the value of which amounts to less than 3% of the overall cost. If steel structures and the structural aspects of turbomachinery packages are neglected, many problems can result. Yet, it is possible to design and build efficient structures with high strength-to-weight ratios.

We must understand however, that all stresses, deformations and simulations are estimates. There can be many sources of errors and inaccuracies. Loadings may not be realistic. In many cases, actual forces and moments on a structure may be far more than those simulated, both in terms of number of loads and their complexities. Models, therefore, are usually simplified versions of actual structures. This introduces the risk of errors and inaccuracies.

So how to eliminate or mitigate the risk of ending up with a structure not fit for purpose? A good recommendation is to look at what others have done. If a structure for a turbomachinery package has worked well, why not simply use it again?

14.3.3 Piping

Turbomachinery always faces risks from the effects of inadequate piping. Our continuing concern about the consequences of poorly designed and equally badly braced small-bore piping is a good example. We have discussed this previously in Chapter 5. It cannot be sufficiently often mentioned how important it is that there must be a unified approach on the part of the commissioning team regarding cleanliness standards, piping support concepts, flange-up procedures and final piping quality acceptance.

14.3.4 Shaft Alignment

Final cold alignment of the compressor train should be documented and certified. There are many examples where startup delays were caused by inattention to the details of shaft alignment. Many experts attribute most bearing overload and destructive vibration in machinery to shaft misalignment.

A case in point: We were a group of engineers working in the technical planning and support department of a major gas pipeline company. Every morning we would review field reports for any indications of problems where our help might be required. During one week, we kept noticing how operations was seemingly having a difficult time with bringing a 25 MW gas turbine-driven pipeline compressor back on line. The reports showed that, after repair to some valves in the piping around the unit, it had been restarted and kept shutting down on high vibration. One of our colleagues suggested that it might be an alignment problem. His contention was flatly rejected by the rest of the group: "Impossible, that is the first thing they are checking, and we have procedures that call for regular alignment checks on these units." Some time went by and finally we were dispatched in order to try to help.

The first thing we did was to ask for alignment records for the unit. The data was not available even though someone in the operations and maintenance crew stated that the alignment had been checked prior to the first startup of the train and been found satisfactory. We felt somewhat uncomfortable when we insisted that alignment be checked again using a reverse indicator method with an appropriate plotting procedure.[13] The result was embarrassing for everyone involved because it showed that an angular misalignment between the gas turbine and the compressor existed, see Figure 14.10. It was grossly exceeding commonly accepted guidelines. This was no doubt the reason for the shutdowns, because, once the misalignment had been corrected, the compression train ran satisfactorily. During operation the shaft alignment of compressor units can be adversely affected by thermal bearing pedestal growth, by foundation and base plate settling, by excessive piping loads or other sometimes unpredictable changes.

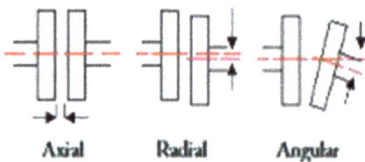

| Axial | Radial | Angular |

Figure 14.10: Misalignment modes.

While this incident showed how complacency is frequently found around the issue of machinery alignment it also brought to light that there were no guidelines for alignment tolerances on that particular site, nor were there procedures in place outlining shaft alignment steps.[14] Commissioning and startup audits should always include a critical look at the project's alignment practices.

14.3.5 Lubrication

Usually, the owner will have specified whether the seal-oil and lube-oil systems are separate or combined if and when oil-type shaft sealing is applied. Today, many turbocompressors are purchased with self-acting dry gas seals which do not require any seal oil and attendant support systems. Consequently, the owner only has to look after a lubrication system designed according to Standard 614.[15] The singular risk arising from lubrication systems is that they are not clean while we think they are. Other risks are:

- Charged with the incorrect oil which should be a hydrocarbon-based oil, viscosity grade 32, in accordance with ISO 8068, unless specified differently.
- Incorrect pressure setting
- Incorrect temperature setting

14.4 Startup

14.4.1 Review of Operating Characteristics

A compressor must meet not only its intended design pressure ratio and flow values, but also a multitude of other conditions. Even if its speed is fixed, the compressor, as is shown in a typical compressor performance map (Figure 14.11), may move gas anywhere within a wide range of flows at discharge pressures following a given pressure ratio versus flow characteristic. Flow and discharge pressure that the compressor will actually deliver are thus determined by the point or points where the two characteristics, pressure ratio–flow and system resistance–flow, intersect, as illustrated in Figure 14.12.

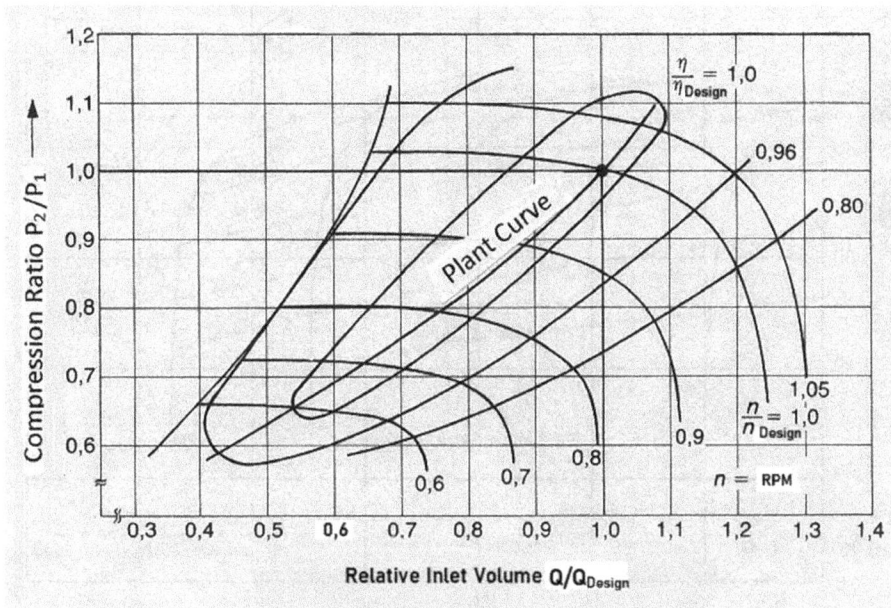

Figure 14.11: Typical performance map of a centrifugal compressor (source: Atlas Copco).[16]

In a few cases, such as ventilating systems, a turbomachine can be matched to a system without any further controls. Usually a periodic or continuous rematch, that is, an adjustment of compressor characteristics, is required for such reasons as:
1. Operator must meet certain discharge pressures and/or weight flows, rather than pressure ratio and volume flows. This calls for adjustments when the inlet temperature, the inlet pressure or the gas composition changes.
2. Multiple or changing system conditions must be met with one compressor.
3. Compressor must operate in parallel or series with other compressed gas sources.
4. Driver overload must be avoided at startup or at certain operating conditions.
5. Compressor instability – surge[17] – must be avoided.

Compressor performance control may be accomplished by:
–   Change of speed (rpm)
–   Throttling at discharge flange

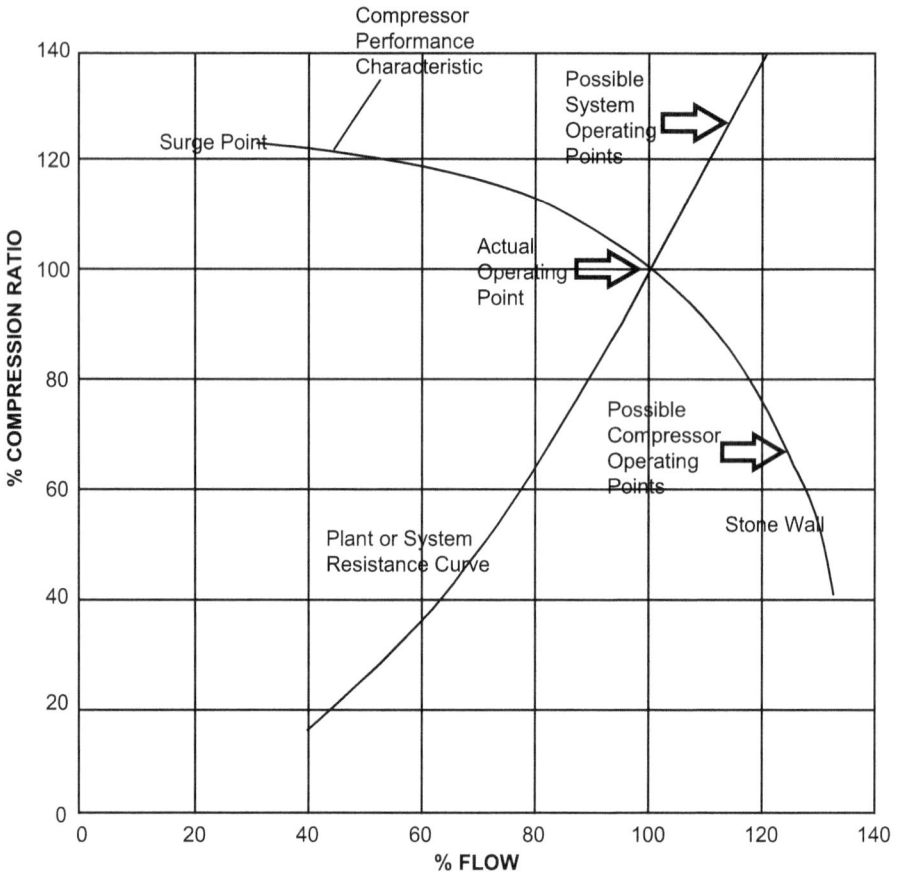

Figure 14.12: System versus compressor performance characteristic.

- Throttling at inlet flange
- Guide vane control
- Bypass (or blowoff) control
- *Discharge throttling.* This is a very simple device for reducing the head without changing flow range. It incurs a straight loss of compressor horsepower. Discharge throttling is used on low-horsepower fans and blowers.
- *Inlet throttling.* Inlet throttling is a widely used discharge pressure and flow reducing device for blowers and compressors. Horsepower savings are most pronounced on multistage compressors with relatively steep head – flow characteristics. Inlet throttling has the effect of steepening the head volume characteristic, rotating it downward around an imaginary, fulcrum point at high or and near-maximum pressure ratio, and thus provides good flow

- *Guide vane control.* Guide vane control on the first or on several stages produces a similar rotation of the head volume curve as inlet throttling does, but with significant improvements:
  a. It works both ways, for increases and decreases of pressure ratio and volume, by counter- and pre-rotation setting of guide vanes.
  b. It incurs no direct efficiency loss. Impeller pressure rise capability changes are obtained almost loss-free through conversion of static pressure and temperature into velocity pressure.
- *Bypass control* – surge control – is one of the controls used to avoid compressor "surge" or "pumping." On axial compressors this surge may be termed "stall." Surge cycle frequencies may vary from a few cycles per minute to 10 or more cycles per second; the attendant flow reversals inside the compressor and system often cause severe noise and vibration or shaking of pipes. A few seconds or even minutes of surge will not usually harm the compressor, but operation in surge is inefficient and may lead to excessive heating and ultimately component damage.
- The surge limit, that is, the minimum flow for stable operation at any given speed, can be shifted downward to some extent, maybe 10–20%, by intake throttling or guide vane control. The operator, then, must simply make sure that the compressor flow will always be above the surge limit. If the process calls for less flow on occasion, then the difference between demand flow and minimum stable compressor flow must be recirculated around the compressor in a throttle-controlled bypass, or – in the case of air blowers – blown off through a valve.

14.4.2 Starting Up

The first startup may last from several days to many weeks depending on the preparation and complexity of the system associated with the compressor. Normally the startup team will follow a checklist as shown in Table 14.5 before rolling the equipment. This table covers the commissioning and startup procedure for a motor-driven turbocompressor train as an example for a typical procedure written by the startup team in cooperation with the OEM FSR. This procedure is recorded in Table 14.6. It should be obvious that the number of tasks and therefore, risk exposure, increases with the complexity of the prime mover. We think for a moment about a large steam turbine driver that requires a slow warm up roll where we must make sure the compressor's self-acting dry gas seals are designed to withstand slow rotation. Similarly, motor-driven machinery trains can present a problem by the rapid sequence of events. It does not take long for a constant speed motor driver to come up to operating speed, and a careful plan has to be in place to assure proper sequencing of valve positions and other maneuvers. Appropriate manning and portable communications equipment at valve, switch and control stations has to be provided. Otherwise, operating personnel in process facilities using exclusively steam turbine drivers for their compressors are astounded when they are listening to the experience of their counterparts around motor-driven units. How can you make adjustments with fast-starting motor drivers?

While this reflects a typical situation in the downstream business of the HCPI, there is a high level of automation in mid-stream operations of that industry. Here an initial startup of a gas turbine–driven pipeline compressor train or pump station equipment is often a fully automatic event by way of telemetry with a few responsible persons on stand-by in case something goes awry.

In the meantime, many other activities have been taking place concurrently. The startup team has been working on commissioning plans similar to Figure 14.13.

Table 14.5: Example of a typical commissioning and startup checklist for a motor-driven turbocompressor train in a petrochemical process plant.

Note: Cross out items not applicable.

1. Are specific operation and maintenance manuals, parts books, special tools & spares available?	_____ (Yes or no)
2. Have the vendor drawing and data requirements (VDDR) been fulfilled?	_____ (Yes or no)
3. Are the items/documents listed in the project VDDR form readily available?	_____ (Yes or no)
4. Receiving, storage protection, foundation, grouting, piping and alignment checklists completed and attached.	_____ (Yes or no)
5. Control loops functionally tested and correct, all setpoints set and verified.	_____ (Yes or no)
6. New or calibrated gauges supplied.	_____ (Yes or no)
7. P&IDs verified.	_____ (Yes or no)
8. Record coupling sizes, spacer lengths, and serial numbers. Check that hubs are match-marked.	_____ (Yes or no)

Motor-Driven Train Mechanical Check-Out

	EPC	OWNER	Comments
1. Reconnect heaters on large motor driver.			
2. Dismantle shop seal and install operating seal. Record clearances.			
3. Inspect all bearings. Record all clearances.			
4. Inspect all auxiliary equipment and instrumentation for completeness and conformance with final issue of bill of materials and electrical device list.			
5. Was presence of temporary suction strainer and correct mesh size for test run ascertained?			

Table 14.5 (continued)

	EPC	OWNER	Comments
6. Was mechanical strength, workmanship (loose wire or bad welds) and mesh size of all strainer elements checked before installation?			
7. Are pressure gauges installed to measure pressure drop?			
8. Review starting, operating and emergency shutdown procedures in detail with operators.			
9. Review compressor control system and surge protection.			
10. Check and record all gap settings on vibration and thrust probes. Plot calibration curve.			
11. Commission lube, seal oil and buffer gas systems according to documented procedure.			
12. Lube oil console and self-acting dry gas support system was commissioned according to documented procedure.			
13. If equipped with liquid film seals, review sour seal oil traps, sight glasses, vents. If there are inconsistencies, disassemble and check operability of trap floats.			
14 Is compressor system settling-out pressure known? Settling-out pressure: _____ psi/kPa			
15. Perform static seal test to check seals.			
16. Install any temporary piping and instruments required for field mechanical test run. Decide on what initial run-in gas to be used.			

17. Refer to electric motor commissioning checklist. Electrically check out compressor motor driver. Run in for 4 h. Check:
  – Acceleration time
  – Rotation
  – Vibration
  – Noise
  – Temperature
  – No-load current
  – Coast-down time
  – Record data			

18. Install coupling and coupling guard between motor and compressor.			
19. Match-mark coupling hubs, if not already done so by manufacturer.			
20. Recheck and simulate all trips and alarms. Commission vibration monitoring system.			

Table 14.5 (continued)

	EPC	OWNER	Comments
21. Conduct dry run sessions with operators to familiarize them with controls.			
22. Final check of piping for valve positions and blinds, operation of RCV's.			
23. Leak test compressor casing.			
24. Check for completeness of piping.			
25. Startup compressor. This test run requires close supervision by OEM FSR, EPC representative, and Owner's Engineer. Complete manual log sheets of all field flows, temperatures, pressures, vibration, speed, etc. (Even though data is available in DCS).			
26. Shut down and check hot alignment.			
27. Dowel after final alignment check.			
28. Inspect temporary strainers and piping for cleanliness.			
29. Connect all process piping.			
30. Fill system with process gas.			
31. Run compressor on total recycle and obtain data to evaluate performance. Check operability of inlet guide vanes.			
32. Obtain reservoir oil sample. Inspect for discoloration. Send to lab for viscosity check at 100 °F or 38 °C and moisture content. Have spectrometric analysis performed.			

Table 14.6: Comprehensive run-in procedure for motor-driven turbocompressor trains.

Verify that all applicable steps of the Pre-Startup cheek list have been completed. Bring any omissions or deviations to the attention of the startup advisor before proceeding as follows:
A. *Piping preparation (case example – helium gas)*
   1. Isolate the circuit after making up a pertinent flow sheet copy and discussing with process and EPC personnel.
     a) Close appropriate block valves in process lines.
     b) Close appropriate bypass lines.
     c) Close block valves in casing drain lines and discharge lines after draining all liquids.
   2. Install temporary block valves for throttling, as applicable.
   3. Identify block valves used for helium charging and air venting.
   4. Identify pressure gauges for monitoring loop pressure.
   5. Verify that run-in loop is fully isolated.

Table 14.6 (continued)

B. *Checkout and startup of lube oil system – seal oil system if applicable.*
   1. Insure level in lube oil reservoir is full to the maximum level, 2 in. or 50 mm from the top of the reservoir.
   2. Check nitrogen pressure in lube oil accumulator. Precharge to____ psig or bar on gas side of bladder before lube oil pressure is established.
   3. Heat lube oil in reservoir to 95 °F or 35 °C by using steam coil in bottom section of reservoir before starting either of the lube oil pumps. Low-temperature high-viscosity oil may overload the motors. The turbine-driven lube pump can be run at a slower speed if the oil is colder. However, care must be taken to avoid collapsing the filter elements in the oil systems from high differential pressure if the oil is not at normal temperature. After the oil temperature is up to 110 °F or 43 °C, shut off the steam to the reservoir heating element.
   4. **Caution:** Do not start lube oil system without prior having admitted buffer gas or Nitrogen into the gas seal.
   5. Follow the normal checkout procedure for starting up any positive displacement pump. Vapor can be bled from each pump casing through a vent line.
   6. Place motor drive pump in service with proper valving through the oil cooler filter. Set oil pressure control valve at _____ psig or bar. Check for leaks.
   7. Place the second motor-driven spare pump in the automatic standby condition.
   8. Purge air from all coolers and filters by bleeding air through the high point vents.
   9. After flow, temperatures and pressures in the lube and seal oil system have stabilized, place the lube oil turbine-driven pump in automatic standby condition.
       To place the turbine driver, _____, for the lube oil pump in automatic standby, the following must be done:
       a) First open the drain in the steam line to the turbine and drain until free of steam condensate.
       b) Commission the steam trap in this system on the supply header.
       c) Before opening steam admission block valve, be sure turbine exhaust block valve is car-sealed open (CSO) to the _____ psig or _____bar system, condensate can drain from exhaust header, and the automatic start steam inlet valve is closed. If not, manually reset the valve.
       d) Allow the turbine to warm up by cracking open the small bypass around the automatic start valve in the steam inlet header but do not open the valve enough to slow roll the turbine. Keep the casing drain bypass around the steam trap cracked open while the turbine is in automatic standby.
       NOTE: When placing a lube and seal oil cooler or filter in service, always purge unit of all air by displacing the air with oil at the high point vent. Some designs have a small-bore line or tubing continuous bleed vent.
   10. Check the governor control operation of the lube oil pump turbine to be sure the governor valve shaft moves freely. This can be done by visually observing stem movement when the turbine comes up to speed.
   11. Check the various automatic startup features and operation of all pumps by actually starting the pumps by the following device:
       a) Auxiliary turbine-driven lube oil pump.
   –   With motor-driven pump running, briefly open discharge bypass back to the reservoir. Observe PI in pump discharge line. At _____psig or bar, pump should start.

Table 14.6 (continued)

- With motor-driven pump running, briefly block lube oil pressure permissive start switch located near the end of the lube oil header, and slowly open the bleed valve. Pump should start when the pressure has dropped to _____psig or _____bar.
  b) Spare motor-driven oil pump
- While it is running, place its motor start switch in the STOP position. With its motor start switch in the AUTO position, it should start immediately.
- With it running, briefly open discharge bypass back to the reservoir. Observe PI in pump discharge line. At _____psig or _____bar, pump (start switch in "AUTO" position), should start immediately.
12. While doing the above checks, make sure that all low-lube oil pressure alarms, pump running alarms and shutdown devices function properly.
- Main Pump running alarm
- Aux. pump running alarm
- Lube oil header low pressure alarm
- Lube oil header shutdown and shutdown alarm
13. Commission the lube oil temperature controller and set temperature to _____°F or _____°C.
    NOTE: This temperature should, be rechecked as soon as the compressor is running to assure instrument is controlling at the desired temperature.
14. Check the lube oil coast down tank for being filled to capacity and note overflow at the sight flow indicator.
15. Check the lube oil filter high differential pressure switch to ensure that it functions at _____psi or bar delta-P. This can be accomplished by briefly blocking in the instrument and bleeding a small amount of oil from one of the two sensing legs.
16. Check each sight 6glass in oil return lines from compressor, gear and motor to assure oil is flowing before starting the machine. There are sight glasses for lube oil flow from each end of compressor and driver.
C. *Getting the dry gas seal support system ready. Refer to P&ID _____.*
   Note: It is advisable to have all instrumentation and devices such as differential pressure transmitters, tested and commissioned except. Neutralize shutdown functions to avoid nuisance trips at startup. The stainless-steel piping has been cleaned as described in Chapter 5 and purged. All filters are clean and installed.
   1. Block in dual filter set on I.B. and O.B. gas seal feed.
   2. Admit seal gas to the pre-filter by opening the seal gas main valve from helium storage bottles.
   3. Blow down filter by momentarily opening the drain valve.
   4. Activate the seal gas booster by allowing filtered instrument air to the booster drive. Pressurized seal supply gas will be delivered to the two dual filter sets.
   5. Blow down both dual filter-sets by sequentially and momentarily opening and closing the blow-down valves.
   6. Slowly open one valve on I.B. and O.B. primary seal gas feed. Observe flow in rotameter. Max. flow rate to be _____ l/min.
   7. Observe rotameter in primary seal exit or flare line. There should be no flow.

Table 14.6 (continued)

D. *Helium purging and filling**

Note: Helium purging and filling must be preceded by checking and commissioning of the lube oil and the self-acting dry gas seal system. Proper functioning of the compressor dry gas seals will keep helium losses to a minimum.

1. Tightness test the isolated run-in loop with plant air at 40 psig or 2.75 bar pressure and eliminate leaks.

   * The quantity of helium required for run-in was calculated to be _____ lbs. or _____kg. Allowing a 50% excess, the maximum anticipated usage will be _____ lbs. or _____kg of helium.
2. After depressuring to atmospheric pressure, hook up helium bottles to any accessible valved pressure tap (near compressor nozzles) or casing drain.
3. Admit helium to the loop and pressurize to 25 psig or 1.7 bar.
4. Depressure to 1.0 psig or 7.0 kPa.
5. Readmit helium to the loop and pressure a second time to 25 psig or 1.7 bar.
6. Depressure to 1.0 psig or 7.0 kPa (second purge step).
7. Introduce helium for a third time, pressurizing for a startup pressure of _____psi or _____bar.

E. *Motor run-in (also refer to motor pre-start checklist – Chapter 7)*

1. Disconnect coupling between driver and gear.
2. Install coupling idler adaptor.
3. Megger motor windings.
4. Verify that lube oil is flowing to the motor bearings.
5. Axially displace rotor to each end of travel and rotate by hand to assure clearance and freedom.
6. With assistance of instrument specialist, ascertain that the following motor shutdown switches are in fact fully functional.

   a) (low-lube oil header pressure)
   c) (isolation valve closed)*
   d) (isolation valve closed)*
   e) (isolation valve closed)
   f) (remote manual emergency)
   g) (first stage suction low temperature)
   h) (first stage suction high pressure)
7. Arrange for mechanical specialist coverage to take motor vibration readings and to observe train vibration levels.
8. Arrange for electrical engineering S/U advisor to stand by.

   *Checkout of these shutdown devices may be deferred until runs re are made with the compressor coupled up.
9. Bump motor to assure proper rotation.
10. Prepare for immediate and instantaneous shutdown by actuating a selected shutdown device.
11. During coast-down of motor, take shaft run-out pictures on outboard end inboard bearings. (This may require bumping twice, but do not bump a third time or restart until at least 45 min have elapsed).
12. Start motor for minimum 4-h test run.
13. Check and record vibration, uncommon noises, bearing temperature, and oil flow.
14. Check winding temperature as motor continues to run.

Table 14.6 (continued)

15. After 4 h or when all parties agree that motor is operating satisfactorily, shut the motor down.

F *Motor–gear run-in*
1. Couple motor and gear after verifying that coupling is at center of its limited end float when motor is at its geometric center.
2. Prepare for immediate shutdown by actuating a selected shutdown device.
3. Start motor–gear for 4-h test run, minimum.
4. Check and record vibration, uncommon noises, bearing temperature, oil flow, and excessive foaming of gear oil and adequate drainage.
5. After 4 h, or when all parties agree that gear is operating satisfactorily, shut the motor down.

G Compressor operation on helium
1. Reconnect the coupling elements between drive motor and gear after briefly rechecking the alignment and ensuring that equipment rotates freely.
2. The lube and seal oil system was commissioned prior to filling the loop with Helium and should still be operating per outline "B" above.
3. Establish flow of cooling water to exchangers. Verify that vibration data acquisition is in place and instrument and electrical personnel are standing by and are ready for the startup.
5. Verify that operator is ready to log all instrument readings.
6. Clear enunciator panel – if applicable.
7. Momentarily energize the motor and listen for compressor binding or rubbing.
8. Start motor and run compressor.
9. Once the machine is started, adjust temporary throttle valve from low stage Discharge to high stage suction to maintain _____ psia or _____bar Suction pressure. This must be done in conjunction with adjusting the _____block valve in the bypass around the compressor. Tabulated below is the expected data when operating a turbocompressor on helium:

Compressor first section (___ impellers)						
Q CFM/m^3/h	P$_1$ (PSIA/bar)	T$_1$ (°F/°C)	P$_2$ (PSIA/bar)	T$_2$ (°F/°C)	Power BHP/kW	RPM

Provide this information, as applicable. Compressor vendor or OEM, process engineering and EPC assistance may be required to obtain complete data.

Compressor second section (___ impellers)						
Q CFM / m^3/h	P$_1$ (PSIA/bar)	T$_1$ (°F/°C)	P$_2$ (PSIA/bar)	T$_2$ (°F/°C)	Power BHP/kW	RPM

Table 14.6 (continued)

Provide this information, as applicable. Compressor vendor or OEM, process engineering and EPC assistance may be required to obtain complete data.

Note: Lower concentrations of helium will cause higher temperatures.

10. Inspect for oil, water and gas leakage. If sizeable leaks occur, shut down the compressor.
11. Listen for unusual noises or rubbing. Be alert for unusual or increasing vibration levels. If noticed, shut down the compressor immediately.
12. Check the oil level in the reservoir. If low, determine the reason for the low level and add the required amount of clean lube oil.
13. Observe the operation of the complete lube oil system. Log all temperatures, levels, and pressures and compare readings with those listed in the "Normal Range" column of the compressor log sheet. (These values should be the same as those Listed in the "Range" column of the compressor instrumentation table in a typical operating manual).
14. Check the shaft seal area to be sure there is no excessive leakage.
15. Look for indications of lube oil foaming in gear sump and drainage problems.
16. The compressor must not run unattended following initial startup.
17. Assemble all pertinent log sheets and submit for startup advisor's review.
15. Run compressor for 8 h, then shut down motor by actuating a selected shutdown device (process personnel to define which one).
19. Take vibration data during compressor coast-down and submit for review.
20. Do not shut down lube oil system and gas seal supply until decision is reached to discontinue helium testing.
21. Inspection condition of compressor journal and thrust bearings – don't touch just look.
22. Remove temporary strainer inserts from compressor inlet piping.
    Note: The internal inspection of the gearbox could be scheduled for this point in time. See Chapter 11, page 318.

When starting a centrifugal compressor train, several different methods are available. The method used is a function of the system as well as of the type of prime mover used. Consequently, in order to address specifics, we should go through the startup of a steam turbine-driven centrifugal compression system such as schematically shown on Figure 14.14. In order to start this system, cooling water would have to be circulated through the heat exchanger or cooler, the bypass valve opened and the unit brought up to operating speed.

The procedure for bringing a steam turbine–driven train up to speed is usually determined by the manufacturer's recommended steam turbine starting procedure which should always be followed. Typically, this involves introducing steam into the turbine until breakaway[19] occurs and a speed of approximately 500rpm is established. After warming up, the unit's speed is gradually increased. Care must be taken when approaching the critical speeds – of turbine and compressor – which should be passed through rapidly. The ability to gradually increase speed is a unique feature of a steam turbine, as it gives the operator time to make checks and adjustments as the speed is increased.

Figure 14.15 shows a typical pressure ratio characteristic which might be experienced as the steam turbine is brought up to speed.

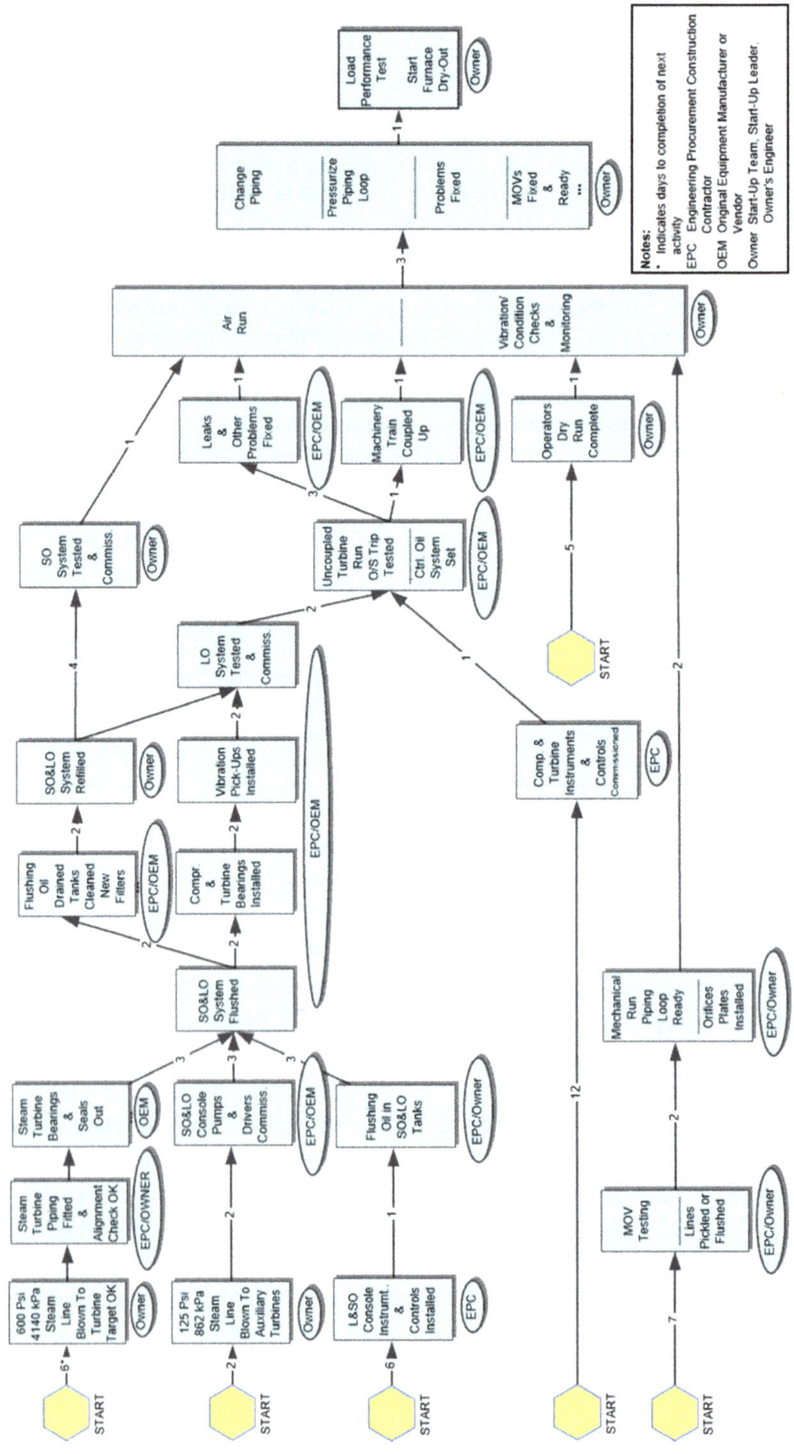

Figure 14.13: Example of a critical path diagram for commissioning and startup of a turbocompressor train.

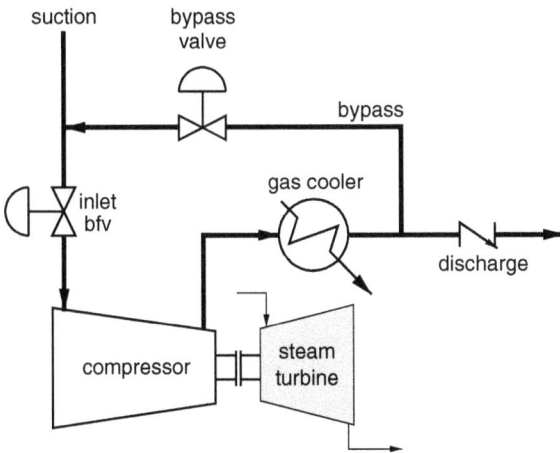

Figure 14.14: Schematic of a steam turbine–driven turbocompressor unit.[18]

This starting method deserves the following comments:
- By providing a large bypass or recirculation system around the compressor, the pressure ratio across the unit can be kept low.
- Low pressure ratio and corresponding high flow mean very little chance for surge[20] to occur.
- Actual stonewall flow,[21] shown as dashed lines, is very hard to predict as stonewall is not only a function of the compressor's design but also a function of the gas being compressed. Therefore, the amount of flow at low pressure ratios with a given unit may vary considerably from values shown.
- Once approximately 100% speed is reached, Point "I", the bypass valve can be slowly closed, causing the compressor performance to move up the 100% speed line to the design point.
- In order for the compressor to follow the path outlined above requires that the steam turbine must be capable of providing the required power. The power that must be delivered for this method of starting is shown on Figure 14.16. Here the compressor speed torque characteristic approximates a log function. It starts at zero torque and zero speed neglecting breakaway torque. The compressor speed torque curve ends with 120% torque at 100% speed, Point "I". While 120% torque is not the worst case, it does represent a reasonably safe number for most applications. Point "I" also corresponds to the power requirement for the Point "II" on Figure 14.15 where the volume versus pressure ratio curve first intersects the 100% speed curve.
- The power required for starting is provided by the steam turbine and its speed torque characteristic is shown by the dashed line on Figure 14.16.
- The power required by the compressor lies below the steam turbine curve up to approximately 96% speed. Thus, instead of reaching the 100% speed characteristic at Point "I", the unit reaches only 96% speed. This is no problem for our

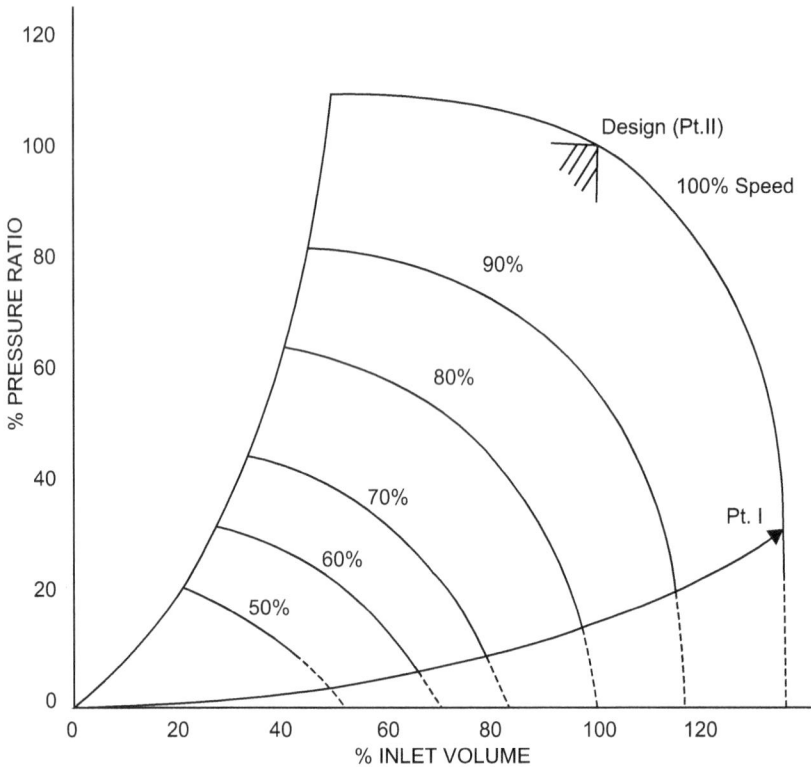

Figure 14.15: Typical pressure ratio characteristic for a turbocompresssor.

simple system because as we close the bypass valve, the discharge pressure will increase and the power required will drop off toward the design point, thus allowing the speed to reach 100% before the design point is reached. At some point prior to reaching design point the check valve will open allowing flow into the process. Further closing of the bypass will reduce the flow through the compressor until the design point is reached. When the bypass valve is completely closed, full compressor flow is going into process.

– For the starting sequence discussed, it was assumed that the compressor inlet pressure, temperature and molecular weight were at design conditions. If this is not the case, these changes should be factored into both the volume versus pressure ratio and speed torque curves used in preparing the startup plan.

Another type of system frequently encountered is the motor-driven centrifugal compressor unit. Unlike the steam turbine unit, the motor-driven unit usually has lower starting torque capabilities due to the requirement for limiting current inrush and heat build-up. Normally, applicable to a constant speed motor, the time it reaches

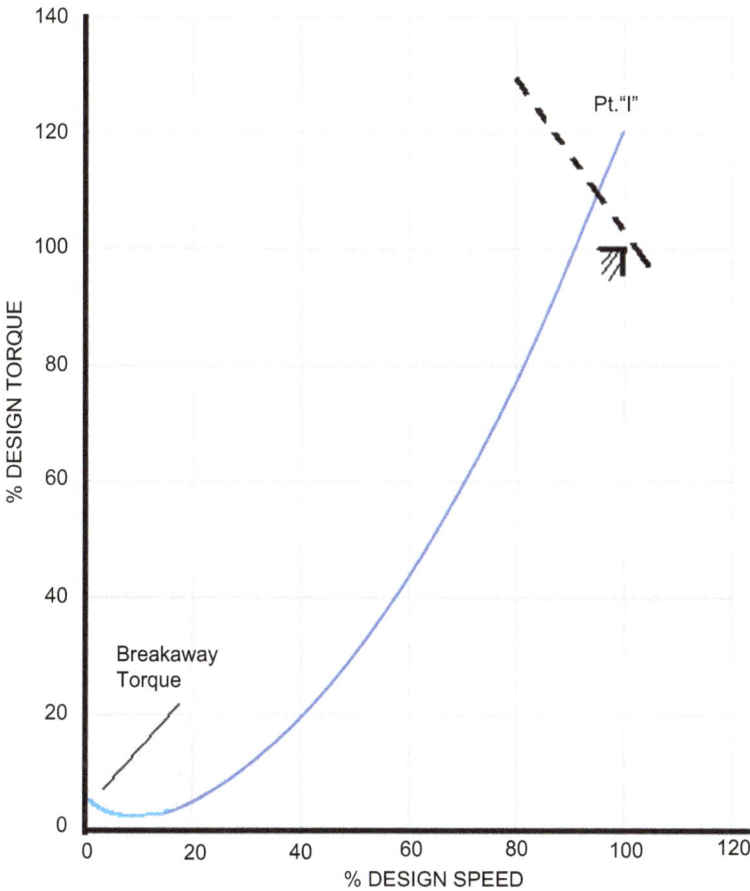

Figure 14.16: Speed–torque characteristic – steam turbine–driven turbocompressor.

operating speed is limited to less than 20 s. Figure 14.17 represents a typical simple motor–driven centrifugal turbocompressor train system.

Similar to the steam turbine-driven unit, we start coolant circulating in the gas cooler and open the bypass valve prior to starting. The similarity ends here. In order to minimize current inrush, the inlet butterfly would be almost closed or cracked open. If the start button is now pushed, the machine should come up to 100% speed within approximately 15 s.

Typically, we like to think that the starting characteristic is the one shown in Figure 14.8, where the flow remains to the right of the surge line during immediately following the starting process. Normally surging during acceleration is avoided by "cracking" the recycle valve open. However, it may not be open sufficiently to keep the machine out of surge on reaching 100% speed. Thus, the valve can be opened more to bring the compressor out of surge and into the stable operating region.

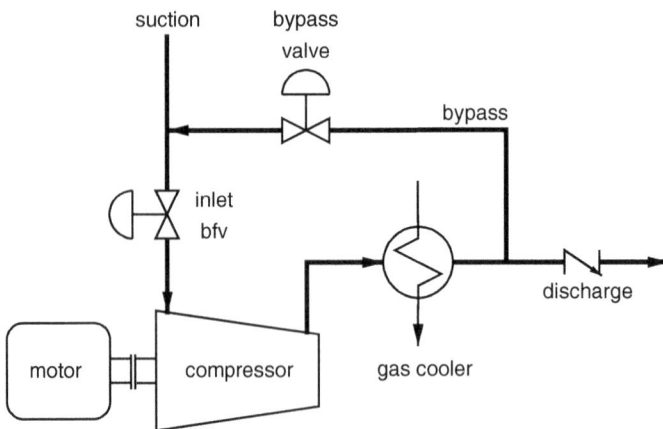

Figure 14.17: Schematic of a motor-driven turbocompressor unit.[22]

With process piping connected and charged with process gas, the proper functioning of the surge control valves, and associated surge control system should be confirmed during this time period.

Let us assume the curve depicted from zero to Point "II" has been followed and we are now operating at that point. From a review of Figure 14.16 and the valve settings we recall that the bypass valve is open. Thus, the compressor discharge should be very near suction pressure. If the compressor is developing the noted pressure ratio, the compressor suction pressure then has to be lower by this amount. If the inlet pressure temperature and molecular weight are at design conditions before the inlet butterfly valve, then almost the entire design pressure ratio will go into lowering the inlet gas density. Thus, both the reduced volume and lower gas density combine to effectively reduce the power, or torque, required at Point "II". Figure 14.19 shows a speed torque display for the starting condition discussed. Point "II" on Figure 14.19 corresponds to Point "II" on Figure 14.18.

With the compressor now operating at 100% speed, the butterfly valve can be slowly opened to increase the compressor flow. Concurrently, the bypass valve can be slowly closed to increase the discharge pressure. The discharge pressure will continue to increase until the check valve opens and gas starts flowing to process. The bypass valve can then be closed, and compressor load regulated by the inlet valve.

Going back to Figure 14.19, one must note the large difference in torque between that required by the compressor and that developed by the motor. This difference in torque provides for the acceleration of the system. Thus, the smaller the difference, the longer will be the starting time.

We hope it has become apparent that, as we manipulate the by-pass and the butterfly valve we are changing system resistance or demand-load represented by the so-called plant curve – the slanted parabolic curve we used in Figures 14.11 and 14.12.

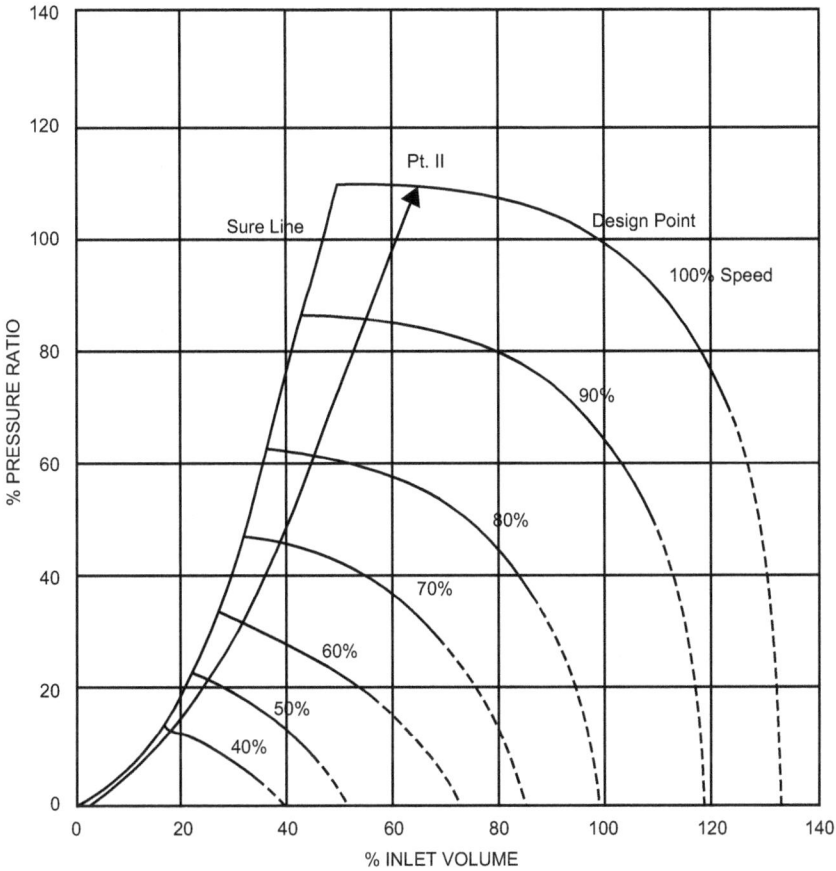

Figure 14.18: Inlet volume vs. pressure ratio characteristic for a motor-driven turbocompressor.

Having discussed two approaches to starting a turbocompressor at some length, one should be ready for more complex systems. While we will not discuss them here, the following points should be taken into account:

- Probably the most important thing to define is the condition which prevails at the end of the speed increasing cycle, Points "I" and "II". Once this is defined along with the various pressures and temperatures, the approach to starting can be determined.
- Below 30% speed very little pressure ratio is developed across the compressor. Also, flow is relatively low.
- When warming up the steam turbine at less than 30% speed, flow control or cooling of the compressor flow is seldom required. Other concerns are present: Will self-acting dry gas seals, if installed, last through the slow roll?

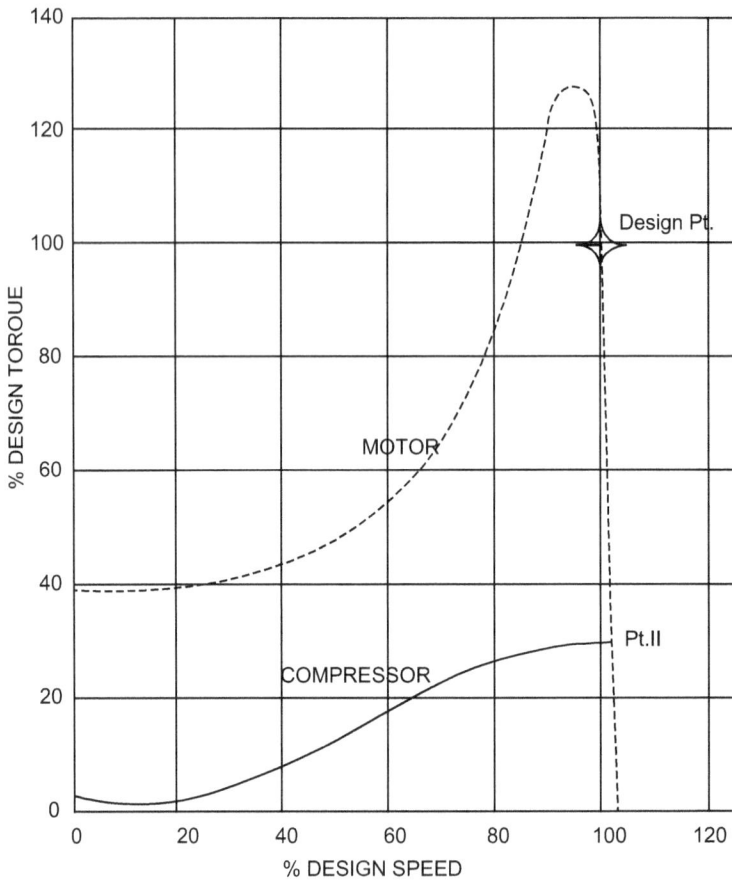

Figure 14.19: Speed torque characteristic for motor-driven turbocomperssor unit.

- On refrigeration units, the discharge flow normally goes to a condenser where the gas is condensed into liquid at a fixed pressure and temperature. In order to get the machine on stream, three things have to happen:
    1. Inlet pressure must be kept low to provide low flash temperatures.
    2. At the same time, suction temperatures must be kept low so that pressure ratios can be realized.
    3. Discharge pressure must be reached so condensing starts.

Thus, for a typical refrigeration system, once 30% speed is reached, the speed should be increased rapidly and at the same time the suction drums quenched with liquid in order to provide cooling. At the same time, gas may be blown off to keep suction pressures near design. As speed increases and inlet temperature falls, the blow off can be closed and the gas diverted to the condenser. By-passes must be controlled to maintain flows between surge and overload so that pressure ratios can he realized as design speed and temperature are approached.

On a compressor with side-loads or one which is inter-cooled, it is often a good idea to establish flows and pressures across compressor sections starting at the suction and working toward the discharge.

14.4.3 What We Have Learnt About Starting Up a Tubocompressor Train

While we have not talked much about variable frequency drives (VFDs) as means of speed control for constant speed motors, starting methods of compressor trains equipped with VFDs would be the same as practiced with steam turbines: Motor-driven trains without VFDs are started up from the left side of the curve – almost in surge – to prevent overload, whereas steam turbine or VFD-driven trains are started up from the right side of the performance curve while almost in choke.

14.5 Initial Run

During the first hours and days of operation, all monitoring equipment and sensors should be observed to establish the signature of parameters, such as temperature and vibration. This is used as a baseline for future condition monitoring and plays a major role in the overall operation and performance of the unit.

Sometimes, the engineering team and perhaps OEMs do not adequately communicate the bounds of the operating envelope to the operations team. The operations team cannot be expected to operate a machine if the operating envelope is poorly defined.

Sometimes, the provided performance curve is inaccurate and not certified during shop tests or site performance tests. Such inaccuracies in turbocompressor performance curves may be the root cause of problem when the operating point is too close to the surge line or too close to the end of curve, the region of high flow accompanied by low pressure rise.

Operating the compressor outside the defined operating envelope is another risk. This can happen due to production and operational pressures, for example. It is important to avoid such a mode of operation.

If this is not possible operations must carefully consider the risks involved. In some cases, it may only reduce efficiency with no significant problem to compressor reliability. But in many other cases, operation outside the allowable envelope is a factor that could seriously impact the reliability and life of the equipment.

In conclusion, the operating point should preferably be as close as possible to the best efficiency point – BEP – to benefit from the highest efficiency possible and minimize impact on reliability. All compressors will tend to experience a rise in operating temperature when operating off-design or far away from the BEP. In most cases, the operating temperature has an upper limit. Once this limit is exceeded, deterioration may take place.

14.6 C&S of Integrally Geared Compressors

Integrally geared compressors will follow basically the routines described above – often these packages are thoroughly instrumented with advanced controls and require therefore, a high degree of OEM field service assistance for the initial startup.

14.7 C&S of Turboexpanders

Turboexpanders, similar to other turbomachinery, need auxiliaries such as:
– Control gear, e.g., for modulating gas admission devices
– Seals and seal support systems
– Magnetic bearings and electric support system – if applicable
– Conventional journal bearings and lubrication oil supply system.

Commissioning activities would be very similar to those of other turbomachinery with utmost attention paid to the cleanliness of piping leading into the turbine. Similarly, oil systems would require flushing as covered in the preceding chapters. Initial startup modalities would depend from the application. Frequently expanders are tightly integrated into the process so that the process has to be started first and the expander gradually switched in.

Figure 14.20 represents a hot gas expander application for power recovery. The system is part of a reactor that produces hot effluent at conditions having a high thermal energy level. The energy recovered by the expander is used to drive the motor working as a generator and the large axial air compressor that provides pressurized air for the process and other users.

The startup of this machinery system is intimately connected to the process in that it has to be started first by energizing the electric motor and powering up the steam turbine to drive the axial compressor supplying air to the process. Once hot process gases are being generated and become available, the expander is gradually phased in starting to supply the power for the air compressor and the motor – now turned generator – while the steam turbine assists. It should be obvious that the sequence just described may vary as sometimes the design does not provide a motor/generator. In this case the steam turbine alone would have to send the train into motion.

There should be no doubt that, dealing with such a system, it would be of utmost importance to have suitable operating procedures in place. These procedures

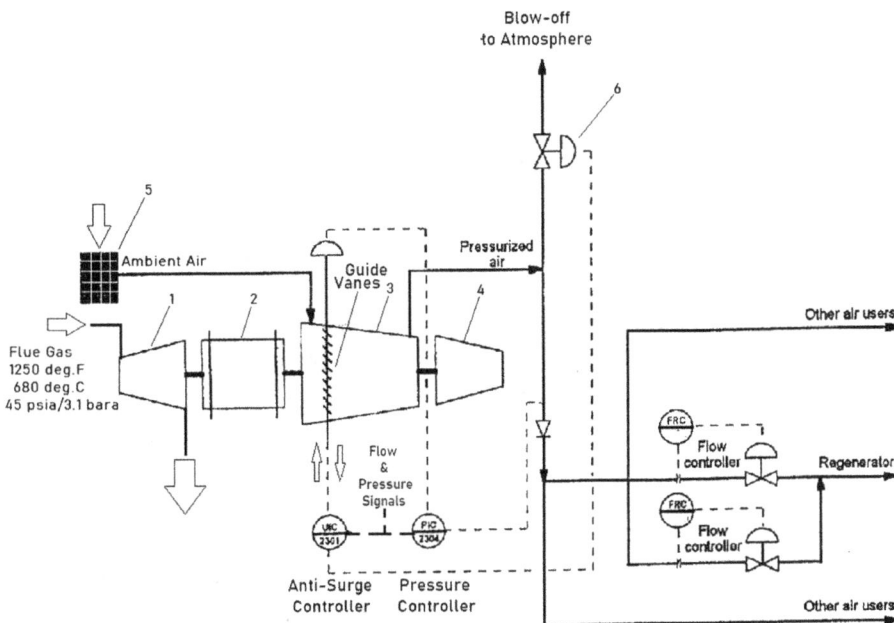

Figure 14.20: Simplified flow plan of a hot gas expander application: 1, hot gas expander or power recovery turbine (PRT); 2, electric motor/generator; 3, axial air compressor; 4, steam turbine; 5, air intake filter house; 6, surge control and blow-off valve.

must be specific to the equipment and process at hand. It would cover the different startup phases such as:
- Commissioning and startup of the lubrication system shared by all four machines
- Preparing the steam turbine for startup
- Warming up the steam turbine
- Increase train speed to an rpm range at the lower limit of the critical speed region
- Arrive at steady state indicated by temperature stabilization
- Increase speed and traverse rapidly through the critical speed range
- Waiting for steady state at approximately 90% of nominal speed
- Waiting for reactor process to start producing hot gas
- Ramping up to full load and rated speed with power input by the expander.

Appendix 14.A: Staging and storage protection of turbocompressors

1. The machine's main suction, discharge pipes or nozzles and casing drain lines must all be fully closed using the valves which are provided on the loop line. Blind flanges must be also provided at the balance connector line(s) – in case

that connection is made to the main gas piping – and at the drain and vent flanges of seal oil drain trap(s) – if applicable.
2. Dry nitrogen or another inert gas must be supplied from the balance connector (s). It may also be supplied from another connector(s).
3. Internal pressure must be observed by providing a U-tube manometer on the gas balance line or on another suitable location.
4. Gas vent lines (air breather) of the compressor(s) must be plugged with sealing tape.
5. Blind flanges must be provided at the bearing vent flanges.

Cautionary Note: Various countries have different safety regulations guiding the use of nitrogen for industrial purposes. In the United States, OSHA (Occupational Safety and Health Administration) tasks industrial manufacturers to maintain a safe working environment for employees. For example, OSHA 29 CFR 1910.146 sets guidelines for confined spaces that could contain higher than normal concentrations of nitrogen gas.

Regardless of industrial location, it is vital that all production process managers conduct nitrogen gas risk assessment exercises to determine their level of exposure, and institute adequate preventive measures. Also, all personnel must be trained on the proper use of personal protective devices as well as proper actions to carry out in case of accidental hazardous exposure.

Appendix 14.B1: Longer term preservation steps for turbocompressors

1. If available connect to oil mist – if not available perform steps 2 to 5.
2. Fill the casing and all bearing housings with Product A[23]. Circulate Product A through the entire system for at least 1 h.
3. Blank or blind oil return header.
4. Seal the shaft openings with silicone rubber caulking and then tape.
5. Fill the bearings housings with product A and run the turbine-driven oil pump at a reduced speed for at least one hour.
6. Fill the oil console with product A.
7. Fill the compressor with nitrogen when it is at an ambient temperature. Turn off heat tracing.
8. Coat all exposed machine parts including couplings, with product C.
9. Tape all shelf protrusions repeat this step after periodically turning the shaft

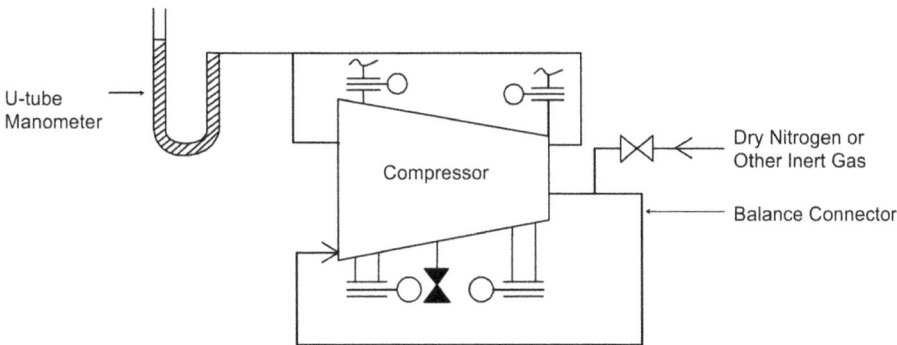

Figure 14.A1: Protecting turbocompressors during staging and installation. (Mitsubishi).

References

1 Amin Almasi, LARGE COMPRESSOR TRAINS FOR LNG APPLICATIONS, Simple and reliable design makes centrifugal compressor features ideal for large refrigeration systems, especially in LNG plants., COMPRESSORTECH^{TWO}, NOVEMBER 2011, Pages 36 to 44.

2 The bull gear is connected directly to the driver by a coupling. Driven by the bull gear are the pinion gears. The pinion gears are part of the rotors that carry the overhung impellers.

3 CATALIN EFTIMIE, *Standardized modularization: Drivers, challenges and perspectives in the oil and gas industry*, Hydrocarbon Processing, APRIL 2017, Pages 11 to 14.

4 API 617, Axial and Centrifugal Compressors and Expander-Compressors, 8th Edition, September 2014. (American Petroleum Institute, Washington, DC, USA)

5 Heinz P. Bloch, Claire Soares, *TURBOEXPANDERS AND PROCESS APPLICATIONS*, Gulf Professional Publishing, an imprint of Butterworth-Heinemann, Boston, New Delhi, 2001, ISBN 0-88415-509-9.

6 Geitner, F.K. & Bloch, H.P., Series *Practical Machinery Management for Process Plants*, Volume 2, *Machinery Failure Analysis and Troubleshooting*, Fourth Edition, Butterworth-Heinemann, an imprint of Elsevier, Oxford, UK, www.books.elsevier.com, 2012, and preceding editions, 743 Pages, pp. 346–349.

7 Arun Kumar, Mohit Sabharwal, HPCL-Mittal Energy, *"Save your centrifugal machinery during commissioning"*, 43rd Turbomachinery & 30th Pump Users' Symposia, Houton, TX, September 23–25, 2014.

8 W.E. Forsthoffer, *Getting the Most out of Turbomachinery Shop Tests*, Turbomachinery International • January/February 1996, Pages 40 to 42.

9 Hamad K. Al-Ruzihi and Eyad M. Al-Khateeb, USER VALIDATION TIPS, *CENTRIFUGAL COMPRESSOR FACTORY PERFORMANCE TESTING*, gascompressormagazine.com | NOVEMBER 2020, PAGE 26 to 28.

10 ASME PTC-10(2014), Performance Test Code on Compressors and Exhausters, New Yark, NY: American Society of Mechanical Engineers.

11 John A. Kocur and C. Hunter Cloud, *"Shop rotordynamic testing: options, objectives, benefits and practices"*, Tutorial, Texas A&M University, Turbomachinery and Pump Symposia (2013).

12 Ibid.

13 Today, most alignment work is accomplished by laser-optics equipment. See also Chapter 5.

14 Bloch, H.P. & Geitner, F.K., Series *Practical Machinery Management for Process Plants*, Volume 3, *Machinery Component Maintenance and Repair*, Fourth Edition, ELSEVIER, 2019, and preceding editions, 650 Pages, pp. 205–263.

15 API Standard 614, Lubrication, Shaft-Sealing, Control-Oil Systems and Auxiliaries, 5th Edition, April 2008.

16 Willi Bohl, Strömungsmaschinen, Vogel Verlag, Würzburg/Germany,1994.

17 Surge in an axial or centrifugal compressor is an unstable flow phenomenon that can cause significant damage. It occurs in low flow regions. In this flow region the friction forces imposed by the rotating assembly on the fluid exceed the kinetic energy forces imposed on the fluid by the rotating assembly. Consequently, an unstable velocity profile ensues. This allows the fluid to flow backward toward the suction of the compressor. If the compressor is allowed to operate in the surge region significant damage can occur. Operating in surge should be avoided if one wants to have reliable compressor operation.

18 This Figure is for instructional purposes only. Usually the so-called kick-back or recycle line should enter the compressor suction stream inside any block valve.

19 Breakaway is dependent from the friction of bearing to shaft stiction. It is difficult to calculate.

20 See Reference 17.

21 Stone wall can also be defined as the point where no further increase in compressor flow capacity is possible.

22 See Reference 18.

23 See Appendix 5.A.

Chapter 15
C&S of Fans and Centrifugal Blowers

15.1 Introduction

Fans and blowers are very important machines in any industrial commercial, residential and plant site. They are part of the family of turbomachinery used to transfer energy to a flowing fluid. Figure 15.1a and b illustrates two basic types of industrial fans: *Radial* and *axial* design principles. While radial fans and blowers are also often referred to as centrifugal fans and blowers, both, radial and axial types are installed and used in many different configurations, rotation and discharge positions. The terminology used in fan engineering and application is replete with jargon and synonyms.[1]

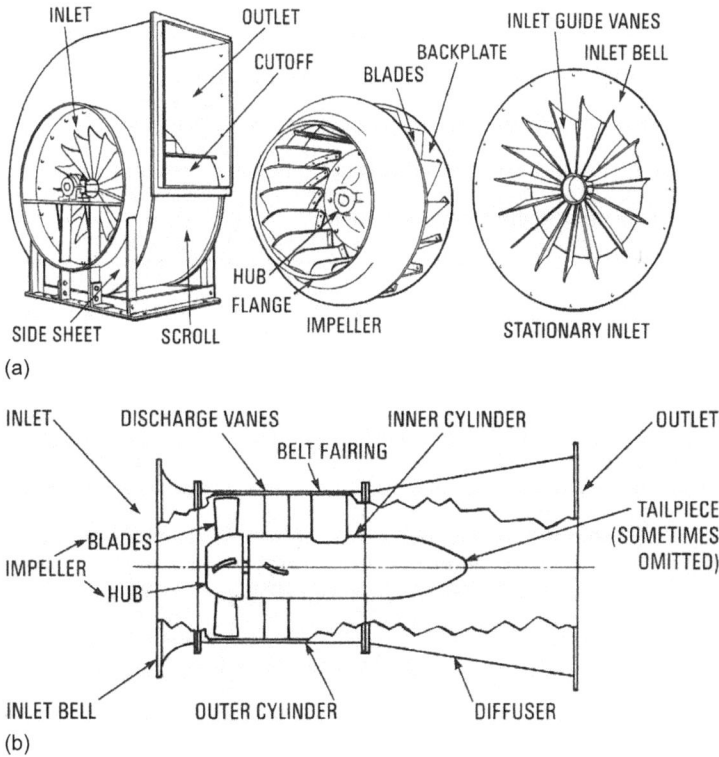

Figure 15.1: (a) Exploded view of a centrifugal (radial) fan and (b) cutaway view of an axial fan (source: Buffalo Forge Co.).

https://doi.org/10.1515/9783110701074-015

Fans and centrifugal blowers – as opposed to positive displacement blowers dealt with in Chapter 13 – provide large air or gas flows to various processes. Fan static pressure rise is limited to a differential not exceeding 130 in. or 330 cm of water column from a single impeller.[2] Flow rates can range from 200 to 2,000,000 cubic feet or 5.7 to 57,000 cubic meters per minute. Power requirement stretches from fractional horsepower to around 4,000 hp or a few watts to some 3 MW. While fans generate only a small pressure rise, centrifugal blowers are capable of pressure increases of up to some 3 psi or 21 kPa. Positive displacement blowers used in the transfer of polymers, for example, are capable of pressure increases of 15 psi or 1 bar and power requirements of approximately 300 hp or 224 kW. Generally, if no standard designation is apparent for centrifugal machines, Table 15.1 applies.

Table 15.1: Turbomachinery defined by pressure ratio.

Designation	Pressure ratio
Compressor	>1.3
Blower	1.1–1.3
Fan	<1.1

Important fan versions and applications in the process industries are:
- General air movement fans
- High pressure radial fans in series for pressure increase
- Radial fans in parallel for high air flow and redundancy
- Pipe ventilation of large electric motors
- Forced draft (FD) fan in chemical process furnace installations
- Induced draft (ID) fans on chemical process furnaces
- Balanced draft fans – FD and ID fan – in furnaces
- Axial flow fans with vertical shaft centerlines in air-cooled heat exchangers
- Axial flow fans with vertical shaft centerlines in cooling towers
- Transfer of polymers between the polymerization unit and the finishing facility in the petrochemical industry.

15.2 Commissioning and Startup Risks Assessed

Fans are relatively simple machines that are reliable when they are designed and built fit for purpose. They are subject to fouling and sensitive to foreign objects that will cause their rotating parts to lose their balance. During installation, they should be protected from the environment as they are as vulnerable to the effects of corrosion as any other process machine. They can be somewhat maintenance intensive as they are frequently belt driven. Belts require regular, operating time dependent maintenance intervention. Table 15.2 is a record of concerns to be dealt with by the startup team.

Table 15.2: Industrial fans and blowers – mitigating the risks.

Phase	Risk	Most probable cause[1]	Suggested actions/mitigating measures
Rigging and lifting	**Low** – Risk a function of degree of disassembly for shipment	QP	– Investigation required
Handling, staging and storage protection	**Medium** – Protection inadequate for long-term storage	OR, HK	– Provide proper storage space – proper boxing of parts and components
Foundation and grouting	**Medium** – Risk of vibration from inadequate foundation	QP	– Blower packages are sometimes furnished with flexible mountings. These types of packages do not need a special foundation. However, if the blower was specified as a separate unit, then the traditional installation of securing its feet and mounting it to the concrete foundation with proper dowels and bolting would be required. – Follow OEM instructions.
Piping and ducting	**Medium** – Foreign objects in ducting, e.g., construction debris – Improper specification of expansion joint	OR, HK	– Degrees of cleanliness need to be established as part of the project execution agreement between OEM and the owner organization. – Ensure expansion joint is suitable for the operating conditions especially high-temperature operations. – Pay attention to dampers/louvers system are installed properly and function checked.

Table 15.2 (continued)

Phase	Risk	Most probable cause[1]	Suggested actions/mitigating measures
Shaft alignment	**High** – Risk of vibration and excitation of ducts and support bases.	QP	– Shaft alignment starts with leveling the unit. Blower, fans and their drivers should be accurately leveled. A minimum permissible deviation from the horizontal line would be customarily 0.002 in. per foot or 0.2 mm per linear meter.
Lubrication system	**Low** – Contamination	QP, DO, CO	– Cleanliness of the lubrications system is of utmost importance.
Startup and initial operation	**High** – Running in unstable region – Experiencing vibration due to unbalance		– As contingent action, arrange for on-site fan balancing specialist.

[1] Refer to Figure 2.12

15.3 Commissioning

Pre-commissioning and commissioning of fans and blowers would generally follow the instructions contained in the pre-commissioning checklist presented in Table 15.3. Check-out and run-in activities of a FD fan unit would take place according to Table 15.4. Similar instructions for an air-fin fan – Figure 15.2 – are contained in Table 15.5.

Both types of fans are indispensable in processing facilities. In one case, FD fans are important components of efficient furnace operation. In the other case, air-cooled heat exchangers with their built fans are vital for plants that do not have access to any other cooling medium than ambient air.

Table 15.3: Fan pre-startup checklist.

Petronta Chemicals		
	Leondorf Olefins Plant 1	
Area:	*Machinery checklist* *Centrifugal blower*	
Procedure: SOP 951	*Equipment:*	

Purpose:	–	This checklist is designed to be a stand-alone document. It is a live document that should be used throughout the turnover/commissioning period. It should only be used by machinery startup personnel.
	–	This checklist assumes that the project team has witnessed and documented the relevant construction MQA witness points. (leveling, grouting, flange alignment, shaft alignment, pre-TO preservation, etc.)
	–	This check list is intended to be a "base list" of machinery checks. Not all checks will be appropriate to all pieces of equipment. Where a check is deemed to be inappropriate this must be ticked off as N/A.

Turnover

The following items have been checked	Yes/no or NA
1. Check that each piece of machinery is correctly identified as per the PID. Ensure the nameplate is legible with correct details including serial number.	_____
2. Check that machinery turns freely (depends on equipment type). Only approved barring tools may be used. UNDER NO CIRCUMSTANCES MAY PIPE WRENCHES BE USED TO BAR EQUIPMENT OVER. Inspect any exposed shaft for damage or corrosion.	_____
3. Check that insulation and personal protection is complete as per the PID.	_____
4. Ensure that all cover bolts are tight.	_____

Lubrication

1. Review the preservation documentation records. If the record is not complete or there is any question on the state of the equipment during preservation the machinery engineer must be informed immediately.	_____
2. Check the lubrication requirements of the equipment. Ensure that all oil and grease levels are correct and that transportation/preservation oils and greases have been correctly flushed clean.	_____
3. Blower bearing if oil lubricated.Type: Quantity: _____	_____
4. Motor bearing oil Oil type: Quantity: _____	_____
5. Check that constant level oil bottles are fitted, filled and adjusted to give the correct level in the bearing housing. Also ensure that oil breathers are clear.	_____

Table 15.3 (continued)

Piping and Civil	
1. Check the equipment vent and drains are correctly installed and piping is as per the PID's.	____
2. Ensure that the machinery hold down bolts are tight, jackscrews are removed. Confirm bolt torque where required by vendor.	____
3. Check base plate grouting certificate and inspect the visual state of the grouting. Ensure that the base plate DRAIN RIM is correctly sloping also check that no voids are present.	____
4. Check that pipe supports adjacent to the machine are correctly installed and un-chocked (pre alignment check)	____
NOTE: Pay close attention to spring pipe supports, check the settings.	
5. Check all small-bore pipe work is correctly installed. Check for gussets, supports and seal welded connections as appropriate.	____
6. Check that suction and discharge piping is correctly installed as per the PID.	____

Alignment and coupling/sheave	
1. Check machinery alignment records. Machinery personnel must witness the final alignment readings. Machine to driver alignment readings must be taken before, during and after suction and discharge flange bolt tightening. Obtain copy of record for blower piping alignment (gap).	____
2. Check and record: That alignment jacking-screw is fitted as required.	____
3. The number of shims used. The maximum number permissible is 3.	____
4. Coupling spacers are removable without moving pumps or drivers. DBSE to be checked on final alignment.	____
5. The coupling spacers are mounted with the special bolts supplied by the vendor. Different bolt lengths are not allowed.	____
6. Ensure the coupling guard is fitted properly and the coupling does not rub against it and all bolts are tight.	____

Instruments and electrical	
1. Ensure direction of rotation arrow is fitted to the equipment and is correct. Driver rotation must match this direction and must be verified during the light load run.	____
2. Ensure all condition monitoring location is determined on equipment.	____
3. Check that all machinery protection guards are in place and are securely fixed.	____
4. Record of driver 4 h light load run	____
5. Ensure that all other disciplines have signed off their signed off their pre-commissioning checklists.	____

Table 15.3 (continued)

Pre-startup	
1. Check oil level is correct	_____
2. Ensure seal support system is properly lined up.	_____
3. Confirm suction/discharge piping has not been removed for flushing. If so, verify piping nozzle to equipment checklist has been filled out.	_____
4. Air blow/flush all seal system piping internal to ensure all piping internal are clean.	_____
5. For pressurized seal, confirm seal pot/accumulator pressure per seal data sheet	_____
6. Ensure that the system is set up to take baseline vibration data.	_____
7. SOP approved sludge cup installed	_____
8. Break oil mist tubing and confirm oil mist to each point	_____

General remarks

1.1 _____

_____

_____

_____

_____

_____

_____

_____

_____

_____

_____

_____

_____

_____

_____

_____

NOTE:
- Not all checks will be appropriate to all pieces of equipment. Where a check is deemed to be inappropriate this must be ticked off N/A
- The checklist is a live document that should be used throughout the construction/turnover/commissioning period. It should only be used by machinery startup personnel.
- Any items there are not cleared (labeled as NO) must be included on the turnover punch list and must be resolved prior to equipment startup. When the item is cleared from the punch list, the date must be entered in the remarks column with the comment "Punch List Cleared."

Table 15.3 (continued)

Petronta Chemicals	SOP MACHINERY PROCEDURE

VALIDATION SHEET

EMPLOYEE NAME: _____

DATE: _____

CHECK THE APPLICABLE BOX BELOW:

[] This procedure was performed / reviewed as written (circle one), and is accepted as valid and technically correct.

[] This procedure, as written, requires the following revisions to be technically correct:

PAGE #	COMMENT
_____	_____
_____	_____
_____	_____
_____	_____
_____	_____
_____	_____

EMPLOYEE SIGNATURE: _____

SUPERVISOR SIGNATURE: _____

Please return this sheet to the appropriate SOP mechanical FLS.

** End of Procedure **

Table 15.4: FD fan check-out and run-in instructions.

COMPANY LOGO	PROCESS UNIT
SOP NO.	
Purpose:	The purpose if this SOP is to ensure operating reliability of forced draft fans in our facility. Further, we want to make sure the machine is safely commissioned and started up.
Safety, Health Environmental Consideration	
References	
Responsibility and Assistance	
Prerequisites	
Materials/Tools to be used	

Table 15.4 (continued)

Remarks	A. Pre-commissioning check	
	1. Check nameplate details against data sheet.	_____
	2. Foundation and grout have been checked satisfactorily.	_____
SHE risk!	3. Fan rotor is installed for proper rotation.	_____
	3. Checked for excessive piping strain. [] Adjustments made	_____
	4. Final cold alignment to standard tolerance.	_____
	5. Axial end clearance has been checked and recorded.	_____
	6. Check fan/blower and driver for freedom of movement.	
	7. Driver/gear solo test was performed and accepted.	_____
	8. [] Gear unit checked per gear checklist. [] Belt drive checked and belts properly tensioned.	_____

	9. Bearing lubrication: [] Oil mist
	     [] Standard Oilers
	     [] Grease
	System checked: clean and ready to start
		_____

SHE risk!	10. Belt or coupling guard in place and secured.	_____

	11. [] Guide vanes
	[] Damper
	  and control system is manually operable.
	[] auto controls will not be commissioned until after
	  startup of furnaces and boilers or other downstream
	  equipment.
		_____

	12. Mechanical integrity of inlet duct is verified.	_____
	B. Pre-startup checks	
	1. Operator is available for run-in.	_____
	2. Inlet ducting has been thoroughly checked for foreign objects/ construction debris.	_____
	C. Forced-draft fan startup and run-in	
	Note: Run fan drivers for mechanical run test after preparation as specified for pump and fan drivers. Arrange for coverage by OEM representative (FSR) if performance testing is contemplated.	_____
	1. Open discharge dampers completely and close inlet guide vanes to 10%.	_____

Table 15.4 (continued)

	C. Forced-draft fan startup and run-in	
SHE risk!	2.	[] Motor drive [] Turbine drive [] Direct [] Belt [] GBX [] Other _____ Start fan and allow to come up to speed, then shutdown unit and observe coast-down time. On turbine driven units, slow roll turbine at 500–1,000 RPM for about 10 min or until any condensate has been eliminated from the turbine. Bring unit up to minimum governor speed.	_____
	3.	If everything appears satisfactory during Step 2, start fan and allow to run for about 1/2 h. Adjust turbine drive units to design speed. During this period, continually check for noise, vibration heat and other unusual signs — use your five senses.	_____
	4.	After satisfactory 1/2 h run, record performance data on log sheet. With damper 100% open: Vibration: [] ips (peak) [] mm/s _____ Motor amps: _____ FLA: _____ Brg. Temperatures: [] °F [] °C _____	_____
	5.	If after 4 h of continuous running the unit appears mechanically satisfactory, arrange for a 24 h continuous run under operator surveillance.	_____
	6.	After 24 h run, shut down unit and perform hot alignment check; correct alignment as determined by hot check.	_____
	7.	Dismantle and inspect bearings.	_____
	8.	Dowel fan and driver.	_____

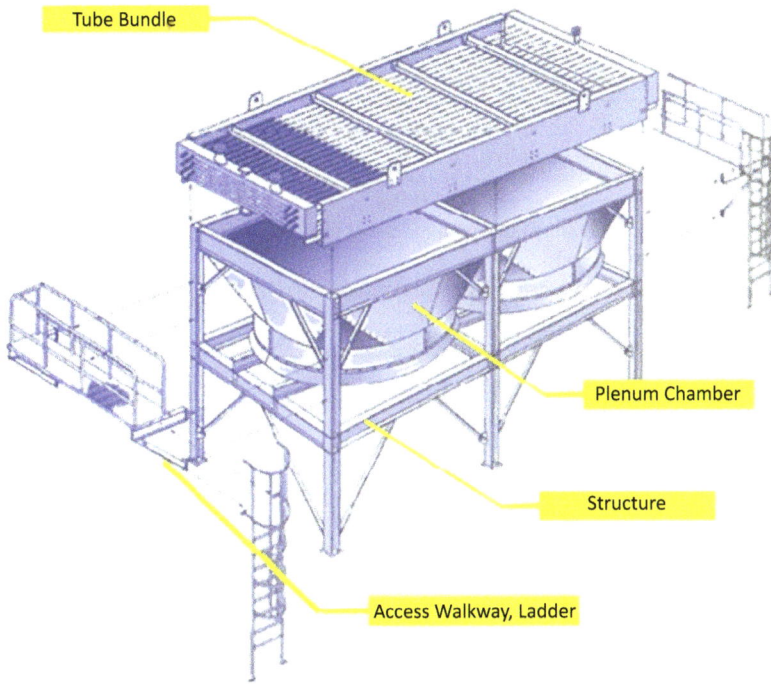

Figure 15.2: A typical air-cooled heat exchanger with cooling fan (fin fan) inside.

Table 15.5: Air-fin fan check-out and run-in instructions.

COMPANY LOGO PROCESS UNIT
SOP no.
Purpose:
Safety, health environmental consideration
References
Responsibility and assistance
Prerequisites
Materials/tools to be used

Table 15.5 (continued)

Remarks	A. Installation inspections	
	1. Check nameplate data against specifications.	_____
	2. Gear box internally inspected; clean and filled with specified oil.	_____

	3. Fan blades have adequate end clearance.
	   a. Check each blade at one point on the shroud by rotating the fan
	      (number blades).
	   b. With fan in one position check each blade tip clearance.
		_____

	4. Check match marks between hub and blades.	_____
SHE risk!	5. Fan rotor is installed for proper rotation.	_____
	6. Check and record blade angle.	_____
	7. Blade tightness in hub checked.	_____
	8. Driver/gearbox coupling installation checked.	_____
	9. Driver/gearbox alignment checked.	_____
	B. Pre-startup checks	
	1. Driver has been run-in satisfactorily.	_____
	2. Gearbox lube level is OK.	_____
	3. Fan rotates freely.	_____
	4. Process operator is available for run-in.	_____
	C. Fan startup and run-in	
	1. Start fan and allow to come up to speed, then shut down and time coast down (double check rotation).	_____
	2. If fan appears satisfactory during step 1, restart and allow to run for 4 h. Check vibration, temperatures and motor amperes at start of run and record. Stand by watching for any signs of problems. At the end of run, record vibration, gear and driver temperatures and motor amperes.	_____
	3. After run-in, flush and refill gear box with clean oil.	_____

15.4 Startup

Starting up a process fan or blower – or getting it on line – is usually a simple process. It involves energizing the driver and allowing the unit come up to the desired speed. Figure 15.3 depicts a fan characteristic – system resistance interaction.[3] It cautions the operator to stay away from unstable flow regions similar to the effect we described in Chapter 14 dealing with turbocompressors.

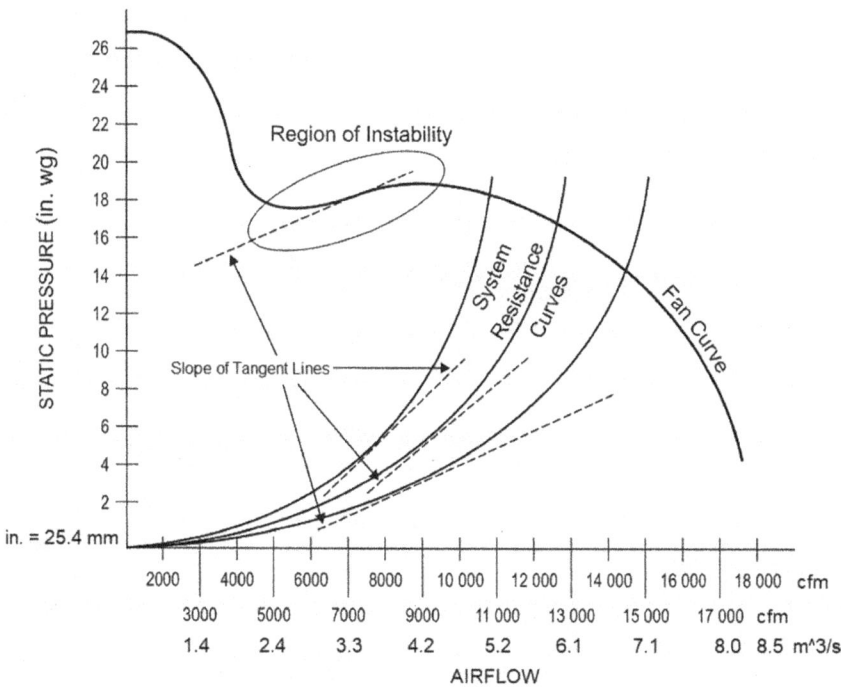

Figure 15.3: Fan flow and system resistance (adapted from Reference[4]).

Flow control is achieved by adjusting:
- Outlet dampers
- Inlet-box dampers
- Variable inlet vanes
- Variable pitch on propeller fans
- Variable speed
- The number of fans in parallel operation

Each of these controls affects specific output, stability, turndown ratio, startup and power savings.

Startup flow control for fans in parallel or even series operation is more complicated. It requires careful consideration because the differential pressure versus flow rate characteristics can have important variations from one type of fan to another, for example, forward leaning vanes, straight radial vanes and backward leaning vanes. In the past, insufficient consideration has been given to the different pressure versus flow rate characteristics. When only depending upon the system resistance curve to control the proper sharing of load between fans in parallel this method of flow control can result in the fans hunting and constantly varying the delivered flow rate. Reliable operation of this system – fans in parallel operation – requires careful understanding of fan characteristics and the associated load or system curve.

15.4.1 Monitoring After Initial Startup

Fans and blowers must be monitored after their initial startup. One problem that haunts this type of equipment is imbalance, a common phenomenon with all industrial fans including those in processing applications. Imbalance is a potentially dangerous condition and can result in breakdowns and costly startup delays. It is imperative that owner's engineers understand what it is, how to detect it, what causes it and how it could be addressed.

While some causes can be traced to manufacturing problems, the bulk of imbalance situations arises during startup and operation. Imbalance manifests itself by shaft vibration. However, bear in mind that vibration is a symptom only. It can have several causes such as mechanical looseness, coupling misalignment, defective bearings, insufficient flatness of bearing mounting surfaces, rotor cracks, driver vibration and belt slippage. Acceptable overall vibration levels are typically below 0.3 in./s peak or 5 mm/s rms.

A review of the causes of imbalance would most often reveal:
- Foreign matter built-up on blades
- Temperature differentials – if a fan rotor is left at rest, a differential temperature may develop between the top and the bottom of the housing leading to a rub at startup.
- Loose hub-to-shaft fit – during initial startup the fan hub may be securely held in place by setscrews after a period of time. However, the set screws may loosen due to fretting. The loosening of the set screws may allow the hub and the entire fan impeller to become displaced relative to the axis of rotation. The result would be extreme unbalance. For this reason, hub-to-shaft connections with an interference fit or with some type of tapered bushing are preferred.

In any event, the startup team should not "reinvent the wheel" by attempting to balance the fan as there are professionals available who balance fans and blowers routinely.

15.4.2 Progress Tracking

A large petrochemical project is liable to have a fair number of air-cooled heat exchangers – see Figure 15.2. Commissioning and startup progress must be tracked using a pre-startup and run-in log sheet as shown in Figure 15.4. A similar tracking sheet for FD fans is offered in Figure 15.5.

Air Fin Fan No. Designation	Pre-Startup Checks									Run-In				
	Moter Run-in	Gear Insoect.	Blade End Clear.		Blade Angles		Blade/Hub Tightness Checked	Motor/Gear Alignment Recorded		Run-In Start/Stop Time	Fan Rotation	Motor Amps	Vibration Recorded	Witness Signature/ Date
			Min	Max	Min	Max								

Figure 15.4: Typical air-fin fan pre-start and run-in log sheet.

Commissioning Status

Fan Number Designation	Driver Run-In Complete	Gear Run-In Complete	Align, Recorded	Control System Checked Out	Vibration Recorded

Performance Test*

Guide-Vane Setting	Discharge Pressure Inches. H$_2$O	Flow Meter Reading	Motor Ambs	Turbine Throttle Position, Inches	Maximum Bearing Temp. °F	Hot Alignment Complete	Fan Complete And Turned-Over To Process	Witness Signature/Date
10%								
10%								
10%								
25%								
25%								
25%								
50%								
50%								
50%								
75%								
75%								
75%								
100%								
100%								
100%								

*For each guide vane setting, agjust discharge dampers to provide:
First: Maximum Flow Rate
Second: Maximum Discharge Pressure
Third: Flow Rate Mean between "First" and "Second" Flow Rates

Figure 15.5: Typical forced draft fan run-in and performance log sheet.

References

1 FAN ENGINEERING, *An Engineer's Handbook on Fans and Their Applications*, edited by Robert Jorgensen, 1983, 8th Edition, published by BUFFALO FORGE COMPANY, Buffalo, N.Y.

2 API Standard 673, Centrifugal Fans for Petroleum, Chemical, and Gas Industry Services, THIRD EDITION | DECEMBER 2014.

3 The System Resistance Curve has several other equivalent terms such as "plant curve", "load curve" and "process head requirement".

4 Alberta Apprenticeship and Industry Training Material, http://tradesecrets.alberta.ca/ilm/

Bibliography

Bloch, H.P. & Geitner, F.K., Series *Practical Machinery Management for Process Plants*, Volume IV, *Major Process Equipment Maintenance and Repair*, Second Edition, an imprint of Gulf Professional Publishing., at www.elsevier.com 1997, and the preceding edition, Hardcover ISBN: 9780884156635, eBook ISBN: 9780080479002, 700 pages.

Bloch, H.P. & C. Soares, *"Process Plant Machinery"*, Butterworth-Heinemann, Woburn, Mass., USA, a member of the Elsevier Group, 2nd Edition, 1998.

Chapter 16
C&S of Process Pumps

16.1 Introduction

Chemical and most other processing cannot be imagined without pumping equipment. They are simple machines that lift, transfer or otherwise move liquids from one place to another. As we indicated in Figure 1.1, process pumps work according to two different principles. Displacement pumps are as old "as the hills" whereas centrifugal or kinetic pumps appeared on the scene as late as the middle of the nineteenth century.[1] Whereas niche development of positive displacement pumps continues, centrifugal pumps have plateaued in their development and standard versions[2] are today widely used in processing plants and facilities.

Many books and articles have been filled dealing with pump technology and their reliability enhancement.[3] We are, therefore, limiting our text to important issues connected to commissioning and startup of this type of equipment. Figure 16.1 depicts a typical heavy duty processing pump, the main subject of this chapter. Centrifugal pumps have one or multiple impellers. They are usually driven by electric motors, but also by general purpose steam turbines. A typical oil refinery, for example, has some 2,000 centrifugal pump sets within its confines ranging from fractional to 500 hp or <1 to 370 kW.

Figure 16.1: Typical heavy-duty horizontal end-suction process pump.

Horizontal pumps are usually more desirable than vertical pump configurations. But in some applications, special requirements such as low weight, compact configuration

https://doi.org/10.1515/9783110701074-016

and low net positive suction head (NPSH) might result in the selection of vertical pumps. Most should have a positive inlet pressure to prevent cavitation, which is caused by lack of enough inlet pressure to prevent liquid vaporization.

16.2 Prime Movers Used

Most small or medium centrifugal pumps are driven by constant-speed electric motors. Variable speed drives (VSDs) are become increasingly more common. They are used in medium and large pumps and are sometimes applied to small pumps to save energy and lower operating costs.

In the upstream business, in oil refineries and chemical intermediate product plants, steam turbines are also frequently used as pump drivers. They offer high speed, as well as direct-drive and flow control through their variable-speed capabilities, simple controls and non-sparking attributes for hazardous environments. Variable speed drivers can help the pump operate at or near its best efficiency point (BEP).

16.3 Commissioning

16.3.1 Risks Assessed

Figure 16.2 is to convey an understanding of centrifugal pump vulnerabilities. Figure 16.3 is showing an inherent-risk evaluation based on complexity considerations relative to other types of process machines – see Chapter 2.

Project related risks are presented in Table 16.1. As with other types of machinery dealt with in this text, we would like to discuss in the following text important issues relating to pump installation phases.[4]

Installation and pre-commissioning of pumps require more attention than is usually practiced. Key requirements for successful commissioning and operation include strong foundations, proper anchor bolts, high-quality grouting, minimum piping forces and good alignment. Pump installation concerns are:

– *Receiving and storage protection.* Because pumps usually represent the bulk of machinery on a new project, owners are well advised to plan for a suitable storage area with proper protection and management. The main feature of such a storage facility would be an oil mist lubrication system as described in many references[5,6]. Figure 16.4 illustrates the application of oil mist to a pump.
– *Foundations.* Foundation mass, for example, should be three to five times the mass of the pump package. Epoxy grout should always be used. Anchor bolts are best

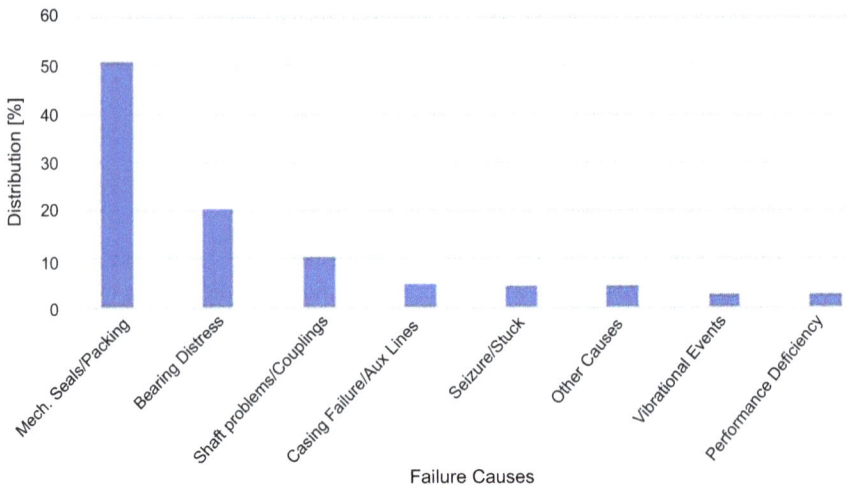

Figure 16.2: Risk profile – centrifugal process pump.

1	2
	150hp/112KW
	Motor driven
	CentPmp at 3600rpm
Rel. Risk #	**650**
Rel.Risk Level:	MEDIUM
Complexity	25
3.1	10
4.4	4
5.1	1
7.1	10
Severity	26
Table 2.4.1	8
Table 2.4.2	8
Table 2.4.3	8
Table 2.4.4	2

Figure 16.3: Relative inherent-risk assessment of a motor driven process pump set.

with a length to diameter ratio of 11 to 15. They should have sleeves to prevent entry of grout and accommodate relative thermal growth.

– *Baseplates.* Sufficient structural stiffness in the pump and baseplate minimizes pump – and driver shaft misalignment. Baseplates are available in a variety of configurations. Those fabricated from strong steel profiles or structures are preferred. Torsional stiffness, rigidity and flatness are important for a baseplate. Fabricated

Table 16.1: Project-related risk mitigation – motor driven centrifugal process pump set.

Phase	Risk	Most probable cause[1]	Suggested actions/mitigating measures
Rigging and lifting	**Low** – Careless treatment due to small size of pump	QP	– Plan to assign technician to supervise.
Handling, storage protection and staging	**Medium** – Careless handling – Inadequate protection	OR, HK	– The process starts with preparation for shipment by the OEM. Review plans for handling and storage protection – Ensure pump flanges are capped with plastic covers to protect flange serrations etc. – Use VPI or install desiccants inside the flanges to prevent moisture intrusion. – Use oil mist purge during storage whenever possible. Some pump vendors have oil mist connections already installed on the storage containers/boxes.
Foundation and grouting	**Low** – Improper procedures	QP	– Assure established procedures[2] are being followed.
Piping	**Medium** – Overlooked pipe elbows located too close to the pump causing vibration – Foreign objects lodged in pipe and not removed	OR, HK	– To be sure, final borescope inspection on large (process charge and boiler feedwater) pumps recommended. – Always have breakout spools installed at the discharge and suction flanges. Do not hydrotest the process piping along with the pump. – Piping alignment to the pump must follow guidelines given in Reference 2, below. – Use detailed checklist for pump piping.

Table 16.1 (continued)

Phase	Risk	Most probable cause[1]	Suggested actions/mitigating measures
Shaft alignment	**High** – Misalignment	QP	– Assure established procedures[2] are followed. – Establish general alignment tolerances for the project.
Lubrication system	**High** – Wrong oil/grease Specification – Mix-up – Contamination	QP, DO, CO	– Project management to assure checklists are being followed.
Startup and Initial Operation	**High** – Operating too far away from BEP – Incorrect procedure		– Train operators. – Consider by-pass loops if not already installed.

[1] Refer to Figure 2.12
[2] Procedures may be based on API RP 686, 2nd Edition, 2006, *Recommended Practice for Machinery Installation and Installation Design and / or* API RP 1FSC, 1st Edition, 2013, *Facilities Systems Completion Planning and Execution.*

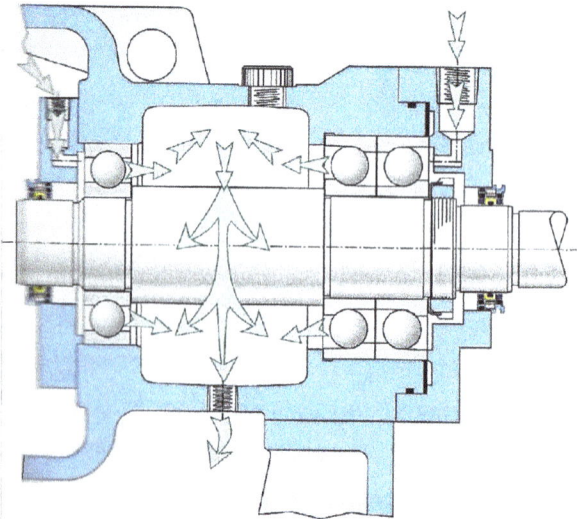

Figure 16.4: API 610 compliant new style of oil mist application method (source: AESSEAL Inc., Rotherham, UK and Rockford, Tennessee, USA).

steel baseplates should be sufficiently strong, heavy duty, properly welded, stress relieved and machined.

- *Piping.* All pumps require some auxiliary piping: for the lubrication system, seal systems, drains, vents and any cooling systems. Tubing and flexible hoses should be avoided as they have been responsible for many failure and reliability issues. The use of small-bore piping (SBCs) is a better option. Pipes should be properly supported, braced and gusseted to prevent vibration or stress as outlined before in Chapter 5. The configuration of piping should not cause excessive forces and loads on pump nozzles, and should limit misalignment between the pump and driver. Piping should only produce reaction loads within the allowable nozzle load at all operating conditions.
- *Piping field inspection.* Many flaws and oversights relating to pump piping can remain undetected and become the cause for repetitive failures and low pump reliability. We should have an eye for common mistakes such as:
  1. Geodetic oversights that will cause the pump to perform improperly.
  2. Lack of proper insulation for viscous hydrocarbon fluids.
  3. Are suction and discharge pipes properly dimensioned? For example, a rule of thumb would be 1.6–3.3 ft/s or 0.5–1 m/s entry velocity into displacement pumps, and maximum velocities of 8 ft/s or ~ 2.5 m/s into centrifugal pumps.
  4. Violation of basic rules such as:
     - Provide the suction side with a straight run of pipe in length equivalent to 5 to 10 times the diameter of that pipe between the suction nozzles in the first obstruction in the line.
     - The pipe diameter on both the inlet and the outlet sides of the pump should be at least one size larger than the nozzle itself. On the horizontal inlet side, an eccentric reducer is required to reduce the size of the pipe from the suction line to the inlet nozzle. By positioning the reducer with the flat side up, the potential problem of an air pocket in a high point in the suction line is eliminated. Concentric increaser should be used on the vertical discharge.
     - Avoid elbows close to the pump inlet. There is always uneven flow into a pump causing turbulence and vibration. Elbows tend to aggravate this situation.
     - Arrange piping in such a way that there is no strain on the pump casing. API 610 – reference[2] – identifies a maximum level of forces and moments on the pump flanges, but the real field operating forces might be much higher than expected. A visual examination during a walk-through is invaluable here because it might reveal thermal expansion constraints and other obvious piping support errors.
- *Shaft alignment.* Hot pumps require particular attention to a proper sequence of alignment because they encounter linear and volume growth during operation.

A final hot alignment check is essential. Please, again, refer to Chapter 4, Section 4.5.4.

There should be conscious agreement on how to bolt up pump suction and discharge piping. This is important as practices can vary – especially when different and multiple pipefitter teams are involved in the installation process. Table 16.2 represents a good procedure and detailed checklist for rotating equipment piping.

Table 16.2: Detailed checklist for rotating equipment piping.

Designation: _____

Location: _____

Service: _____

Note: Cross out items not applicable

1. Check the pump discharge flange for level in two directions 90° apart. Maximum out of level 0.002 in. per foot or 170 μm/m.
   _____
   (Yes or no)

2. Check parallelism of the suction and the discharge pipe flanges to pump flanges. Maximum out of parallelism 0.030 in. or 0.76 mm at the gasket surface.
   _____
   (Yes or no)

3. Check concentricity of the suction and the discharge piping to pump flanges. Flange bolts should slip in and out of the bolts-holes by hand.
   _____
   (Yes or no)

4. Check to see if proper piping supports have been installed and set to appropriate heights.
   _____
   (Yes or no)

5. Check to see that 3 to 5 pipe diameters of straight pipe exist before the suction and after the discharge. Straightening vanes can be used when conditions do not allow for straight runs of pipe.
   _____
   (Yes or no)

6. Tighten the piping bolts in a crisscross pattern to the appropriate torque value. Coat bolt threads with thread lubricant before tightening. Torque values are normally calculated for clean and lubricated bolts.
   _____
   (Yes or no)

7. Check pump shaft deflection using face and rim alignment set up, while tightening the flange bolts. Maximum allowable shaft deflection should not exceed 0.002 in. or 51 μm TIR.
   _____
   (Yes or no)

8. Provide a drop out spool piece at the suction flange for a conical strainer as shown in Figure 16.5.
   _____
   (Yes or no)

16.3.2 Problem Services

The C&S team must pay attention to problem services. They are:
- *Hot services* where the pumping temperature is above 350 °F or 175 °C. These are services requiring extra attention to pump mounting, pipe and pump thermal growth, high temperature seal arrangement and auxiliaries. Warm-up facilities will be needed to keep the spare pump near operating temperature and the piping hot during preparation for initial startup.
- *Low NPSH* services where suction is taken from a vessel containing fluid in liquid–vapor equilibrium. The concern here is:
  - Limited NPSH resulting in limited pump operating range,
  - Control of seal environment for reliable seal operation.
- *Cryogenic services* with the risk of vapor binding and lack of seal environment control.
- *Vacuum services*. Here the seal environment must be maintained above atmospheric pressure to keep the seal faces lubricated. Proper venting of the pump is an additional concern.
- *Low flash or toxic services*. The seal design and installation are critical to prevent the risk of a seal leak becoming the source of a fire, hygiene or environmental – SHE – hazard.
- *Auto-start services*. These pumps require attention to ensure reliable automatic starting capabilities.
- *Loading services*. The main concerns are water hammer from on – off valves located well downstream of the pump and reliable remote starting capabilities.
- *Emergency services*. This system design and installation must receive a final thorough review to verify the pumps will operate under the emergency conditions in which they will be needed.
- *Viscous services*. Warm-up facilities are important to make sure the liquid in the pump and the seal chamber are at a suitable viscosity before the pump is started.
- *Pump type*. It is important to single out services for which especially OEMs were selected, such as large pumps over 500 hp or 375 kW or high head or pressure, low flow services.

In the following pages, we are presenting a comprehensive checklist reflecting the tasks that have to be to be accomplished in the field during installation and initial startup – see Table 16.3.

Do not ever forget to install a suction strainer – Figure 16.4 – for obvious reasons.

16.4 Startup

Centrifugal pumps, belonging to the family of turbomachines, follow performance characteristics similar to turbocompressors. Centrifugal pump manufacturers offer

Table 16.3: Centrifugal pump installation and initial operation checklist.

How to use this checklist
- The underlying premise is that pump installation is according to recommended practices as outlined in, e.g., reference.[7]
- Place your initials behind the items you have personally witnessed to be correct.
- Attach an initialed copy of the machinery field installation checklist to cover items which project management personnel have reviewed for you.
- This checklist may also be used by maintenance.

1.0   Mechanical preparation
1.1   Safety
1.1.1 Complied with lock- and tag-out procedure
                                                                     _____
                                                                    (Yes or no)

1.1.2 Complied with rules for proper safety equipment and PPE
     _____
(Yes or no)

1.2   Preparing pump and baseplate for installation
1.2.1 Cleaned and removed all burrs from bottom of pump feet and baseplate pads _____
(Yes or no)

1.2.2 Cleaned pump and piping suction and discharge flanges _____
(Yes or no)

1.3   Pump installation on baseplate
1.3.1 Installed pump on base plate and checked the following:
    a)   Check nameplate against data sheet
    b)   Pump feet and/or supporting flanges are square with baseplate pads and seated solidly.
    c)   After tightening the base bolts, the bump was checked for level and leveled, if required.
    d)   The pump was checked to make sure that it was slightly higher than the driver so that shims could be placed under the driver's feet during final alignment.
    e)   Drain line provided _____
(Yes or no)

1.3.2 HiFi coupler pads or transducers have been bonded to equipment casing as listed below:
    ☐ driver inboard   ☐ driver outboard
    ☐ pump in board   ☐ pump outboard
    ☐ pump central   ☐ none _____
(Yes or no)

1.4   Pump and driver's coupling concentricity
1.4.1 Installed a coupling sweep and dial indicators and checked to make sure that the pump and driver's coupling halves outside diameters and faces were true with their bores (max. OD 0.002 in. or 50 μm; face 0.001 in. or 25 μm). Used pry to hold actual movement while checking the coupling-face run out on sleeve bearing machines. _____
(Yes or no)

Table 16.3 (continued)

1.4.2 Made first alignment check using laser alignment equipment.	_____ (Yes or no)
1.5 Installation of suction and discharge piping – see also Table 16.2	
1.5.1 a) Check-valves, drain-, vent, slopes and flare-lines installed as per OEM drawings, P&IDs and seal drawings.	
b) Multi-stage pump casing vent to flare/OWS provided as per drawing.	_____ (Yes or no)
1.5.1 Checked piping and pump flanges for parallelism. When the pipe was pushed in place without strain the bolts fit free in the flange holes	_____ (Yes or no)
1.5.2 Flange alignment is ☐ satisfactory ☐ unsatisfactory	_____ (Yes or no)
1.5.3 Placed and adjusted dial indicators on the top and side of the pump's coupling hub and/or flange observed. Observed the dial indicators while the pump's suction and discharge flange bolts were being tightened evenly to prevent pipe strain on the pump.	_____ (Yes or no)
1.5.4 Pump suction strainer – see Figure 16.5 – is ☐ installed ☐ not installed	_____ (Yes or no)
1.6 Auxiliary piping	
1.6.1 Cleaned and installed piping and plugs using thread compound. **Note:** All small-bore piping to be Schedule 160	_____ (Yes or no)
1.6.2 Checked small bore piping for adequate support – refer to project small-bore piping policy, if available.	_____ (Yes or no)
1.7 Lubrication – sleeve and ball bearings (AFBs)	
1.7.1 Cleaned and installed bottom side glass and snap drain on dry sump lubricated pumps adjusted lubricator assembly to oil level mark on bearing housing of wet-sump lubricated pumps. Used _____ on all pumps. _____ on steam turbines.	_____ (Yes or no)
1.7.2 Verified that vertical in-line pumps in hydrocarbon service have thrust bearing installed at motor outboard end.	_____ (Yes or no)
1.7.3 On dry-sump oil mist lubricated motors, verified that bearing identification code on nameplate shows thrust-loaded bearings <u>not</u> to be equipped with shields (no letter "P" on code sequence). Radial bearings are allowed one or no shield (one letter "P" or no letter "P" in code sequence).	_____ (Yes or no)

Table 16.3 (continued)

1.7.4 Cleaned oil-mist fittings at bearing housing, checked tubing for cracks and
      connected it to distribution block. Adjusted tubing so that is slopes from the
      manifold to the oil reservoir without sagging and preventing oil traps. Verified
      that drain tubing on horizontal motors is led to the drain rim of baseplates.   ————

                                                                       (Yes or no)

1.7.5 Observed oil-mist and escaping mist from bearings.
      Note: If mist was not being observed, check 1.7.6 below.   ————

                                                                       (Yes or no)

1.7.6 Checked oil-mist generator for proper. Check:
      ☐ Adequate oil in reservoir   ☐ Oil-mist pressure – 20 to 30" H_2O
      ☐ Oil temperature – 95 to 125 °F   ☐ Air pressure – 30 psig
      ☐ Air temperature – 130 to 155 °F   ————

                                                                       (Yes or no)

1.8   Hot-pump piping strain check
1.8.1   Placed and adjusted a dial indicator on the side of the pump's coupling hub
      and/or flange and observed the dial indicator while the pump was being
      heated as closely as possible to its normal operating temperature.
      Note: While making this check, do not tie down bearing bracket support on
      overhung-type pumps.
      ☐ Movement was more than 0.010 in. or 250 µm TIR. Made correction to
        piping support.
      ☐ Movement was less than 0.010 in. or 250 µm TIR.   ————

                                                                       (Yes or no)

1.9   Alignment of driver to pump
1.9.1   Reviewed thermal growth calculation and verified pump and driver were
      offset accordingly.   ————

                                                                       (Yes or no)

1.9.2   Connected coupling sweep (with less than 0.003 in. or 75 µm sag) to pump.
      Installed and adjusted dial indicators to the OD and face of the driver's
      coupling and recorded the coupling's misalignment. Calculated shim
      adjustments using traditional two-indicator alignment method.
      Note: Do not stack shims and use never more than three shims.   ————

                                                                       (Yes or no)

1.9.3   Alternative method of laser alignment was used. Record on file.   ————

                                                                       (Yes or no)

1.9.4   Final alignment after shim correction.
      a). Face: ———————— (max. 0.001 in. or 25 µm) b). OD ————————   ————

                                                                       (Yes or no)

1.10   Coupling and coupling guard
1.10.1 Checked driver sense of rotation.   ————

                                                                       (Yes or no)

1.10.2 Adjusted flexible disc coupling so that sleeve bearing motor driver will run its
      magnetic center and tighten set screws to lock it in please.   ————

                                                                       (Yes or no)

Table 16.3 (continued)

1.10.3 Greased gear-type coupling and limited its end float to prevent bearing and shaft shoulder contact on sleeve-bearing equipped motors. _____
(Yes or no)

1.10.4 Installed coupling guard. _____
(Yes or no)

STOP *Do not proceed unless all of the above items have been fully verified, properly initialed and any issues cleared.*

2.0 **Operating preparation**

2.1 Special procedures

2.1.1 Any additional process steps or checkout procedures pertaining to this pump or its driver have been reviewed and instructions issued to operating personnel.
□ supplements issued
□ no supplements required _____
(Yes or no)

2.1.2 Verified pump is suitable for run-in on water. _____
(Yes or no)

2.2 Machinery electrical and instrumentation specialist coverage

2.2.1 □ stand-by coverage needed and requested
□ stand-by coverage not requested. Operating personnel trained and equipped with instrumentation to take and record necessary surveillance data. _____
(Yes or no)

2.2.2 a) Vibration sensors connected and functional – if provided
b) Mechanical seal alarms checked and functional – if provided _____
(Yes or no)

2.3 Preparation for startup (process operators to verify asterisked items)

2.3.1*a) Checked seal flush system is according to specifications.
b) Ascertain that pumps with dual mechanical seals had their overhead reservoir (pots) and supply piping cleaned.
c) Verified seal pots were filled with specified fluid.
d) On steam quenched mechanical seals: verified orifices installed. _____
(Yes or no)

2.3.2*Turned on seal buffer/barrier fluid supply and seal flush. _____
(Yes or no)

2.3.3*Vented packing box of double seal to ensure that box is free of air or non-condensibles and full of fluid. _____
(Yes or no)

2.3.4 Adjusted packing-gland bolt nuts to finger-tight and observed slight leakage _____
(Yes or no)

2.3.5*Opened pump's suction valve and bled air/gas from pump. _____
(Yes or no)

2.3.6*Opened pump's discharge valve (when line is under pressure).
Cracked open discharge valve when line was not under pressure. _____
(Yes or no)

Table 16.3 (continued)

2.3.7 All vent and drain plugs installed with similar material of construction. _____
(Yes or no)

2.3.8 Double check location of high point vents, low point drains. _____
(Yes or no)

2.3.9 Identify any pump seal flange drain ports with quenches that should not be
plugged – where applicable _____
(Yes or no)

2.4   Initial startup data to be taken
2.4.1 □ data below refers to run-in on water
       □ data below refers to process fluid (other than water) _____
(Yes or no)

2.4.2 Suction pressure □ psi; □ bar _____ design; _____ actual _____
(Yes or no)

2.4.3 Suction temperature □ °F; □ °C _____ design; _____ actual _____
(Yes or no)

2.4.4 Discharge pressure □ psi; □ bar _____ design; _____ actual _____
(Yes or no)

2.4.5 Discharge temperature □ °F; □ °C _____ design; _____ actual _____
(Yes or no)

2.4.6 Flow rate □ gpm; □ m^3/h _____ design; _____ actual
2.4.6 Motor readings, amps _____ design; _____ actual _____
(Yes or no)

2.5   Test run – follow startup instructions, e.g., Table 16.4
2.5.1 Checked and adjusted packing for slight leak (packed pumps only). _____
(Yes or no)

2.5.2 Checked bearings for noise and acceptable temperature. _____
(Yes or no)

2.5.3 Checked for rubbing of seal throttle bushing and wear rings. _____
(Yes or no)

2.5.4 Verified that oil level in tandem seal pots remained unchanged. _____
(Yes or no)

2.5.5 Perform the following vibration check at conditions indicated.
       □ min. flow □ rated flow □ max. flow
       IB bearing – Horiz. disp. □mils □µm _____ | Vel. – □ in./s □ mm/s _____
       IB bearing – Vert. disp. □mils □ µm _____ |Vel. – □ in./s □ mm/s _____
       OB bearing – Horiz. disp. □mils □µm _____ | Vel. – □ in./s □ mm/s _____
       OB bearing – Vert. disp. □ mils □ µm _____|Vel. – □ in./s □ mm/s _____
       Axial. disp. □ mils □ µm _____ | Vel. ─ □ in./s □ mm/s _____
       (Max. velocity: 0.075 in./s or 1.91 mm/s peak unfiltered). _____
(Yes or no)

2.5.6 □ Initial vibration check showed excessive vibration indicating misalignment.
       □ A hot-check alignment adjustment was made.
       □ Vibration check was satisfactory.
       □ Doweled pump and driver _____
(Yes or no)

Strainer screen

Spool piece

Identification tab at
top

→ Preferred
Flow direction

← Acceptable,
2nd choice

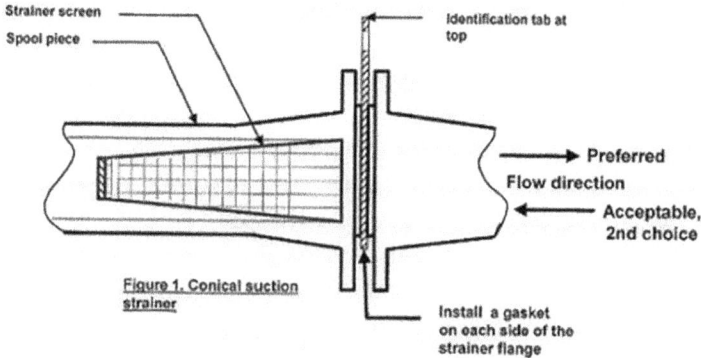

Figure 1. Conical suction
strainer

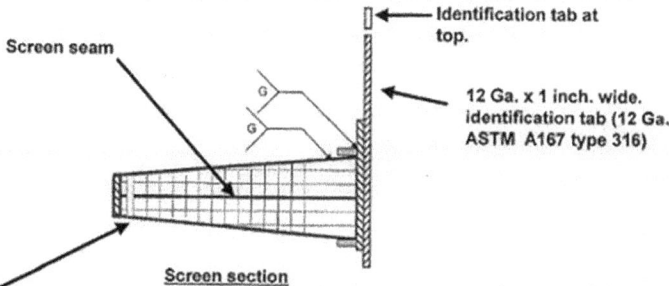

Install a gasket
on each side of the
strainer flange

Screen seam

← Identification tab at
top.

← 12 Ga. x 1 inch. wide.
identification tab (12 Ga.
ASTM A167 type 316)

Screen section

Wire mesh as specified.
install inside of a #3
mesh removable guard
screen.
Tacked in place as
indicated.

Figure 16.5: Suction strainer.

different performance curves for each standard pump. These curves show the head in terms of feet or meters versus flow rate or capacity and efficiency. There are usually several sets of curves for each pump model for various impellers and speeds.

Shutoff pressure or shutoff head is also important. It is the maximum pressure a pump can produce. It is the pressure at the left side of the performance curve equivalent to a near zero flow rate. This metric determines the pressure rating of downstream equipment and piping, particularly the rating of flanges and vessels, and the class of piping required. During startup, checking the shutoff pressure is an important activity which will verify the OEM's performance characteristic. This is accomplished by momentary closure of the discharge valve.

With the exception of speed variation, throttling is the only practical method of regulating the flow and head.

Many pumps use control valves at their discharge for control and operation. The difference between the head developed by the pump and the head required by the system head curve is usually lost energy. The back pressure produced by a control valve or a throttling valve imposes a variable amount of loss on the system. This can potentially shift the pump to operate away from its BEP.

There is a minimum continuous safe flow (MCSF) for each pump. Below that point, operation is unstable and therefore, not advisable. As the flow rate approaches minimum flow, pump vibration increases, and the temperature of the pumped liquid rises. A bypass line or similar facility may be used to recirculate some liquid from the discharge to suction to allow the pump to achieve its MCSF. It is important to know this value when starting up and operating a centrifugal pump. This will help to stay out of pump distress. Figure 16.5 presents an overview of centrifugal pump distress informing the reader that there is in fact only a relatively small flow range in which the pump will operate reliably.

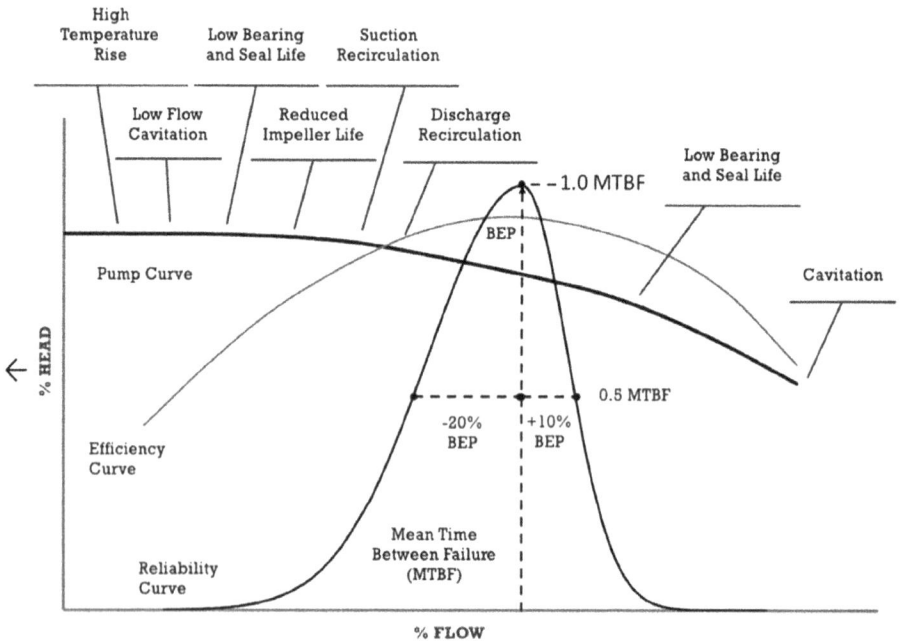

Figure 16.6: Staying near the center of the reliability curve is a wise course of action.[8,9,10]

Earlier, a comprehensive pump installation, commissioning and startup checklist was offered in Tables 16.2. and 16.3. Similar guidelines must be followed in order to achieve a successful startup of process pumps. This checklist as well as the procedure in Table 16.4 can only be of a general nature and may have to be modified for site- and machine specific services. Pertinent data has to be reviewed as applicable.

We must always remember to write in concise sentences if instructions are to become part of checklists which startup personnel or operators are asked to have on their person while on duty.

Table 16.4: Centrifugal pump startup and shutdown procedure.

PETRONTA *Chemicals* **OLEFINS LE-1 Unit**	*LEONBURGER*
Process:	Machine: (*Describe Function*)
Procedure no.:	*Example:* Startup Charge Pump P-502A&B
Purpose:	To safely start up, bring on line, operate and shutdown the pump set.
Safety, health environmental consideration	– Caution must be exercised when working near rotating equipment. – Ensure pump and piping has been purged with N_2 so that the system is air free. – Ensure RO is installed in MCSF by-pass valve line.
References:	– Posted (Plant Information System) or shirt pocket startup instructions – OEM O&M manual – Company sparing policy – Work permit manual – Task risk assessment document – P&D: LE-XXX
Responsibility and assistance	– Trained operator – Journeyman machinist if startup is first after commissioning or repair – Journeyman electrician if startup is first after commissioning or repair – Machinery FLS to coordinate work permit if required
Prerequisites:	– Work permit should be approved prior to startup – Standard PPE – if required
Materials/tools to be used	– Valve wrench – if required
Remarks A. **Starting up**	
1.	Close the discharge valve and open the suction valve. The closed discharge valve creates a minimum load on the driver when the pump is started. If experience has shown that the discharge valve is very difficult to open against pressure and that the motor will not kick off, the discharge valve may be just "cracked open" – about 1/8" or 3 mm open – before the pump is started.

Table 16.4 (continued)

	2. Be sure the pump is primed. Opening all valves between the product source and the pump suction should get product to the suction, but does not always ensure that the pump is primed. Open the bleeder valve from the pump casing – if available – until all vapor is exhausted and a steady stream of product flows from the bleeder. It may be necessary to open the bleeder again when the pump is started, or even to shutdown and again bleed off vapor if pump discharge pressure is erratic.
SHE risk!	**Note:** Priming of a cold service pump may have to be preceded by a "chill down." A cold service pump is one which handles a liquid that vaporizes at ambient temperatures when under operating pressures. Chilling down of a pump is similar to priming in that a casing bleeder or vent valve is opened with the suction line open. There are three additional factors to be considered for cold service pumps: – Chilling a pump requires time for the pump case to reach the temperature of the suction fluid. – Chill-down vents are always tied into a closed system. – On pumps with vents on the pump case and on the discharge line, open the vent on the discharge line first for chill down and then open the pump case vent to ensure that the pump is primed. Should it be necessary to have a cold service pump chilled down and ready for a quick start, e.g., refrigerant transfer pumps during unit startup, then the chill-down line can be left cracked open to get circulation of the suction fluid.
	3. If a minimum flow bypass line is provided, open the bypass. Be sure minimum flow bypass is also open on spare pump as it starts automatically.
SHE risk!	4. Never operate a centrifugal pump without liquid in it.
	5. Check lube oil and seal pot level.
	6. Start the pump. Confirm by observing the pressure gauge that the pump is operating. If the discharge pressure does not build up, stop the pump immediately and determine the cause. **Do not stay in this condition longer than 30 s!**
	7. Open the discharge valve slowly, watching the pressure gauge. The discharge pressure will probably drop somewhat, level off and remain steady. If it does not drop at all, there is probably a valve closed somewhere in the discharge line. In that case, close the discharge valve. **Do not continue operation for any length of time with discharge valve or line blocked.**

Table 16.4 (continued)

SHE risk!	8. If the discharge pressure drops to zero or fluctuates widely, the pump is not primed. Close the discharge valve and again open the bleeder from the casing to exhaust vapor. If the pump does not "pick up" at once, as shown by a steady stream of product from the bleeder and steady discharge pressure, shutdown the pump and driver and check for closed valves in the suction line. **A dry pump will rapidly destroy itself.** _____
	9. Carefully check the pump for abnormal noise, vibration (using hand-held vibration meter) or other unusual operating conditions. **An electrician, mechanical technician and machinery engineer should be present when pumps are initially commissioned.** _____
	10. Be careful not to allow the bearings to overheat. Recheck all lube oil levels. _____
	11. Observe whether or not the pump seal or stuffing box is leaking. _____
	12. Check the pump nozzle connections and piping for leaks. _____
	13. When steady pumping has been established, close the startup bypass and chill down line (if provided) and check that block valves in minimum flow bypass line (if provided) are open. _____
	B. Pump surveillance
	1. Especially during early stages of the pumping, but also on periodic checks, note any abnormal noises and vibration. If excessive, shutdown. _____
	2. Note any unusual drop or rise in discharge pressure. Some discharge pressure drops may be considered normal. When a line contains heavy, cold product, and the tank being pumped contains a lighter or warmer stock, the discharge pressure will drop when the line has been displaced. Also, discharge pressure will drop slowly and steadily to a certain point as the tank level is lowered. Any other changes in discharge pressure while pumping should be investigated. If not explainable under good operations conditions, shutdown and investigate thoroughly. Do not start up again until the trouble has been found and remedied. _____
	3. Periodically check oil mist bottom drain sight glass for water and drain if necessary. _____
	4. Regularly check both oil level and oil mist vapor flow from vents or labyrinths. _____
	5. Seal oil pots need to be checked regularly for correct level. Refill with fresh oil or methanol as required. _____

Table 16.4 (continued)

SHE risk!	6. Periodically check for excessive packing leaks, mechanical seal leaks, or other abnormal losses. Also, check for overheating of packing or bearings. Excessive heat will cause rapid failure of equipment and may result in fires. _____
	7. Operate all spare pumps once a week (plant wide policy for swinging pumps!) to prevent the bearings from seizing and to assure that the pump will be operable when needed. _____
	8. When a pump has been repaired, place it in service as soon as possible to check its operation. Arrange for a machinist to be present when the pump is started up. _____
	C. Shutting down centrifugal pumps
	1. Close the discharge valve. This takes the load off the motor and also may prevent reverse flow through the pump. _____
	2. Shutdown the driver. _____
	3. If pump is to be removed for mechanical work, close the suction valve and open vent lines to flare or drain as provided. Otherwise leave suction valve open to keep pump at correct operating temperature. _____
	4. Shut off steam tracing, if any. _____
	5. Shut off cooling water, sealing oil, etc. if pump is to be removed for mechanical work. _____
	6. At times, an emergency shutdown may be necessary. If you cannot reach the regular starter station – in case of fire, for example – stop the pump from the starter box, which is located some distance away, and is usually accessible. If neither the starter station nor the starter box can be reached, call the electricians. Do not, as a part of regular operations, stop pumps from the starter box. Use the regular starter station instead. _____

Finally, Table 16.5 allows us to track and manage the run-in of a large process pump population typically found on large process plant projects.

16.5 Positive Displacement Pumps

Figure 1.1 lists the most common type of displacement pumps. They exist most often in smaller, low energy level applications. Commissioning and startup would generally follow procedures similar to centrifugal pumps.

Table 16.5: Typical centrifugal pump run-in log sheet.

(1) Pump	(2) System	(3) Seals Flushed and Vented	(4) Suction Pressure psi/bar	(5) Shut-off Disch.Press. psi/bar	(6) Run-in Disch.Press. psi/bar	(7) Motor Amps - Start of Run	(8) Motor Amps - End of Run	(9) Turbine Speed - rpm	(10) Turbine Throttle Open - in./mm	(11) Seal Leak - Start of Run	(12) Seal Leak - End of Run	(13) Vibration - Start of Run	(14) Vibration - End of Run	(15) Max. Bearing Temperature - °F/°C	(16) Witness / Date / Duration of Run
P-501 A/B	GC														
P-502	GC														
P-503 A/B	GC														
P-504 A/B	GC														
P-505 A/B	GC														
P-510 A/B/C	GC														
P-511 A/B	GC														
P-512 A/B	GC														
P-513 A/B	GC														
P-514 A/B	GC														
P-515 A/B	GC														
P-516 A/B	GC														
P-517 A/B	GC														
P-518	GC														
P-520 A/B	GC														

References

1 A brief history of pumps, WORLD **PUMPS,** January 2009.
2 The term Standard Pump usually implies that the pump is designed and built according to one of several current standards. For example, API 610, ANSI, AVS, AFNOR, BSI and DIN.
3 Heinz Bloch, FLUID MACHINERY – LIFE EXTENSION OF PUMPS, GAS COMPRESSORS AND DRIVERS, HANDBOOK, 2020 Walter de Gruyter GmbH, Berlin/Boston, ISBN 978-3-11-067413-2.
4 API RP 686, 2nd Edition, 2006, *Recommended Practice for Machinery Installation and Installation Design and/or* API RP 1FSC, 1st Edition, 2013, *Facilities Systems Completion Planning and Execution.*
5 Heinz P. Bloch, Don Ehlert and Fred K. Geitner, *Optimized Equipment Lubrication, Oil Mist Technology and Storage Preservation,* ISBN HF012020, © 2020, Reliabilityweb, Inc., Reliabilityweb, Inc., www.reliabilityweb.com
6 Heinz. P. Bloch, *OIL-MIST LUBRICATION HANDBOOK – Systems and Application*, 1987, Gulf Publishing Co., Book Division, Houston, London, Tokyo. ISBN 0-87201-640-4.
7 API RP 686, 2nd Edition, 2009, *Recommended Practice for Machinery Installation and Installation Design.*
8 Paul Barringer, *API pump curve practices and effects on pump life from variability about BEP*, Weibull Analysis Course.
9 Heinz P. Bloch, PUMP WISDOM – PROBLEM SOLVING FOR OPERATORS AND SPECIALISTS, 2011, John Wiley and Sons, Inc., Hoboken, NJ, ISBN 978-1-118-04123-9.
10 Heinz P. Bloch, *Improving Machinery Reliability*, Volume I, Series *Practical Machinery Management for Process Plants*, 3rd Edition, 1998, Gulf Publishing Company, Houston TX, Pages 618 to 628.

Bibliography

H.P. Bloch and Allan R.B, pump user's HANDBOOK – LIFE EXTENSION, 2004, Fairmont Press, Inc., Lilburn GA, ISBN 0-88173-452-7.
H.P. Bloch & F.K. Geitner, Series *Practical Machinery Management for Process Plants*, Volume III, *Machinery Component Maintenance and Repair*, Fourth Edition, ELSEVIER, 2019, and preceding editions, 650 Pages, Chapter 3.

Chapter 17
Outlook and Conclusion

17.1 The Path Forward

The commissioning effort cannot end with the initial startup of machinery but it continues and assumes a different phase with a new cast. Process engineers will now work with the process designers. At this stage, problems with equipment and the process itself may become apparent. As part of the continuing commissioning effort, often an intense and hopefully short period of troubleshooting, problem solving, engineering correction and plant modification begins. Once the plant is fully operational, the final proving trial or performance run is undertaken to prove that the plant can do what it is supposed to do.

The values or range of values for each independent variable – flow, temperature, pressure, level, concentrations and so on to which the plant must be operated to are determined.

The plant is brought up to those conditions and the agreed trial period begins. The owner organization needs to ensure:
- Control of plant operating conditions has been achieved, that is, temperature, pressure levels, production rate and analyses are reasonably constant, or, in the case of a batch process, there is repeatability.
- Daily material and energy balances can be performed and confirmation obtained that they agree with "official" process design figures.
- Product specifications are being achieved consistently.

The verification process has to prove:
- Physical operation, capability and capacity of plant and equipment
- Energy and mass balance
- Process chemistry
- Efficiencies, yields and quality
- All to specification

17.2 Outlook

Hopefully, when the facility is working as intended, there will be time for reflection and constructive criticism while everything is still fresh in everybody's mind. A report that documents important insights gained during the project's C&S should be prepared. This report must capture practices or procedures in need of revision in terms of additions, deletion and modifications. The final report begins with the project review meeting. The goal of the project review meeting is to provide an open

https://doi.org/10.1515/9783110701074-017

forum where general aspects of the C&S effort and other associated issues can be discussed, as well as make specific assignments relative to written reports that will become the final report document and file. Specifically, the project review meeting should cover the following topics that are of general importance to the entire C&S effort:

- Safety, health and environmental performance and procedures
- Contractor performance
- OEM onsite technical support performance
- Schedule compliance
- Work assignments and execution
- Costs

When the individual reports are being prepared, a project file should be created. It should be included in the repository of all hardcopy generated in connection with this C&S. Further, it must include the following:

- All work orders initiated during the C&S phase of the project
- Contracts
- Insurance papers
- Equipment used in C&S progress reports
- Drawings and prints including as-built drawings and sketches.

The information gathered and assigned to the project review meeting should ultimately end up in a final report which includes the specific detailed recording at the meeting and any reports on equipment deficiencies and fixes within the company's management of change documentation system. The final report should even include the broad description of the C&S major focus. This report should be a comprehensive document and an accurate description of what was done, what needs to be done and what the cost impact was. It will become a basis for continuous improvement.

17.3 Conclusion

Commissioning is the final phase in project execution and the first step in setting up the operating team for success. Any cost savings accrued through the project life cycle can be completely lost if a facility is not commissioned efficiently, safely and on or ahead of schedule. Doing so effectively requires a comprehensive and uncompromising strategy. It must be centered on an optimized commissioning sequence that drives engineering, procurement and construction toward the quickest and safest start of production. Accomplishing this goal requires strategies that consider the transition from bulk progress to system completion in a process-based, logical order.

When properly executed by experienced partners, these commissioning strategies are the key to reducing startup costs, optimizing cash flow and getting a more rapid return on investment for stakeholders.

In conclusion, we hope to have helped to open our readers' minds to the extent where it is recognized that our topic covers an important asset management activity – even though one of many – but significant to the point where it requires attention to good organization, thorough planning and preparation before it can be transposed into reality. We demonstrated how commissioning and startup fit into the overall concept of a capital project. A seemingly unassuming project phase has been shown to be a key to successful new plant operation. We concluded that the thrust of a C&S effort should always be loss prevention by stressing safety, health and the protection of the environment.

Organizational goals and objectives of a capital project have to contain ways to promote and influence effective owner–contractor interfacing during the installation, commissioning and startup phases of new plant assets and particularly rotating equipment. Our intent was to demonstrate how an effort has to be made in the early phases of a project in terms of reviews and inspections as opportunities for reliability and maintainability input diminish with a project's progress in time. In numerous examples it was explained how mutually agreed procedures and checklists are part of a successful C&S effort ultimately maximizing new plant availability. Case studies were offered to illustrate practical issues in petrochemical plant commissioning.

We hope we presented a management-oriented general framework for plant commissioning that reflects the multidisciplinary and interdisciplinary nature of commissioning. The roles of mechanical, EIC and process specialists was explained as part of a concept that should be applied during the early phases of the project development to ensure adequate planning, staffing, budgeting and expectations for project success.

It is of utmost importance to have capable and experienced subject matter experts available and assist in the successful commissioning and startup of a facility. We do not see them being replaced any time soon. While industry has begun to talk about virtual commissioning of machinery,[1] we recognize that this endeavor is in its infancy to be suitable for application to the C&S effort of major process machinery. We think for a moment of the possibility of foreign matter ingress to the equipment and how to prevent is. However, one day in the not-too-distant future virtual commissioning will become another valuable tool to be used by subject matter experts to mitigate C&S risks and thus help in loss prevention.

References

1 BILL DAVIS, *How to validate machines with virtual commissioning*, Virtual commissioning begins with a vision of the desired machine behavior and sequence of operation, PLANT ENGINEERING, SEPTEMBER 3, 2020, plantengineering.com (21/03/04).

Bibliography

F.K. Geitner & H.P. Bloch, Series *Practical Machinery Management for Process Plants*, Volume II, *Machinery Failure Analysis and Troubleshooting*, Fourth Edition, Butterworth-Heinemann, an imprint of Elsevier, Oxford, UK, www.books.elsevier.com, 2012, and preceding editions, 743 Pages. ISBN 978-0-12-386045-3

Glossary

Actual capacity	This is the quantity of gas actually compressed and delivered to the discharge system by a compressor at rated speed and under rated suction (inlet) and discharge conditions. Actual capacity is expressed in cubic feet per minute or cubic meters per hour and is referred to the first-stage inlet flange temperature and pressure.
Ad hoc	The term "ad hoc" is a Latin phrase that literally means "to this" and is commonly understood as meaning "for this purpose." It can also be used to mean "as needed." It is commonly used in both business and government settings.
AFB	Anti-Friction Bearing
Agitator	This is a device typically consisting of a motor, gearbox, shaft and propeller used for mixing fluids and fluid suspensions in a tank or vessel.
Alignment	This is a condition where the centerline of two meeting machine shafts is brought into a colinear condition within standard accepted tolerances.
API	American Petroleum Institute (Standards).
Asset	This is an entity with monetary value. In the CAPEX project commissioning and startup context, an asset is considered to be any component of a plant or its equipment such as compressors, gearboxes and motors.
Availability general	The ratio of the time equipment is actually available to the process or to the total scheduled equipment time. The availability ratio is computed by adding the scheduled running time to the scheduled idle time and subtracting the unscheduled downtime, then dividing that value by the sum of the scheduled running time plus the scheduled downtime plus the scheduled idle time.
Availability machinery	"Time Unavailable" includes both the unscheduled or forced downtime and the scheduled downtime for maintenance, overhauls and uprates performed during process unit shutdowns. The machine is debited for the total time it is unavailable for operation regardless of whether it is needed by the process or not.
Availability factor machinery	This is also referred to as service factor (SF). It is defined as the percentage of time a machine or machinery train is available for operation whether or not it is needed by the process. In many plants, it is used as a KPI (key performance indicator).
Benchmarking	This is the process of measuring products, services and practices against the toughest competitors or those known as leaders in their field. The subjects that can be benchmarked include strategies, operations, processes and procedures. The objective of benchmarking is to identify and learn "best practices" and then to use those procedures to improve performance. This is a method to improve organizational process by the use of best practice in an industry and is also another term for a best practices study.
BEP	Best efficiency point.
BMS	Benchmarking study. See benchmarking.
Breakdown torque	This is a maximum torque an electric motor will produce. Frequently referred to as pull-out torque or maximum torque.
BiC	Best in class.
BOC	Best of class.
Capacity, actual	This is the quantity of gas actually compressed and delivered to the discharge system by a compressor at rated speed and under rated suction (inlet) and discharge

https://doi.org/10.1515/9783110701074-018

	conditions. Actual capacity is expressed in cubic feet per minute or cubic meters per hour and is referred to the first-stage inlet flange temperature and pressure.
CAPEX	Capital expenditures (CapEx) are funds used by a company to acquire, upgrade and maintain physical assets such as property, plants, buildings, technology or equipment. CapEx is often used to undertake new projects or investments by a company.
CBM	Condition-based maintenance. This is the maintenance initiated as a result of knowledge of the condition of an item. See also predictive maintenance.
CER	Cold eye review (e.g., critical machinery train data of a major petrochemical site).
CMMS	Computerized maintenance management system.
Compliance	Fulfilling local, state and federal government rules, for example, see SHE – safety, health and environmental rules.
Compression ratio	This is the ratio of absolute discharge to absolute suction pressure in a compression process, that is, P_2/P_1.
Continuous improvement	A systematic method of increasing equipment reliability. For example: "Each maintenance intervention should be considered an opportunity for improvement."
Condition-based maintenance (CBM)	This is the maintenance strategy initiated as a result of knowledge of the condition of an item. See also predictive maintenance.
Corrosive gas	This is a gas that attacks normal materials of construction. When water vapor gets mixed with most of the gases, it does not make them corrosive within the scope of this definition. In other cases, the presence of water initiates a corrosive action. Examples are carbon dioxide, hydrogen sulfide, chlorine and fluorine.
CSO	Car-sealed open.
CSS	Car-sealed shut.
CQI	Continuous quality improvement.
DBSE	Distance Between Shaft Ends.
DCS	Distributed control system.
DGS	Dry gas seal.
DEF	Diesel exhaust fluid. Also see SCR.
Design pressure	This is the pressure used to determine the stress levels in components that will either contain a fluid or gas under pressure at a corresponding temperature. The design pressure is always greater than the maximum allowable working pressure. See maximum allowable working pressure.
Design speed	Equal to maximum allowable speed.
DOD	Domestic object damage.
ECFT	Engineering contractor field team.
EE	Equipment Engineering (Department).
EFRC	European Forum for Reciprocating Compressors.
EHM	Equipment health monitoring.
EIC	Electrical Instrumentation and Control
EMS	Engine management system (IC engines).
EOTA	European Organization for Technical Approvals
EPC	Engineering, procurement and construction (contractor/company).
ERPM	Engineering resource planning.
ERV	Estimated replacement value. This is usually the total value of the plant as used for insurance purposes.
ESD	Emergency shutdown device (often used as collective, i.e., more than one device).

ESV	Emergency shutdown valve.
FAT	Factory acceptance test.
FEED	Front-end engineering and design (a stage in capital project development).
FLA	Full-load amperage (electrical device current).
FLS	First-line supervisor.
FOD	Foreign object damage.
Forced downtime	This is also known as "unscheduled outage," that is, it is not planned. The equipment is debited only for forced downtime.
FRACAS	Failure reporting analysis and corrective action system.
FSR	Field service representative (equipment vendor's/OEM's FSR)
FTP	Functional test procedure.
GMRC	Gas Machinery Research Conference
GPA	Gas Producers Association.
HAZOP	Hazard and operability analysis.
HAZID	Hazard identification.
HC	Hydrocarbon (processing industries).
HCPI	Hydrocarbon processing industries.
HDPE	High-density polyethylene.
HMI	Human–machine interface.
Hogging ejector	This is a steam turbine ejector that evacuates air and creates a vacuum on a condenser. It is a vacuum ejector that "pulls" vacuum and helps a thermal turbine unit to work. A hogging ejector is larger than normal ejector used when starting up and during heavy loads. Usually, they are a steam-operated venturi type. Some operators use vacuum pumps and would use several pumps during hogging operations and reduce to one pump when normal operation begins.
Horsepower (hp)	This is an engineering unit for measuring the power of prime movers. One hp equals 33,000 foot pounds of work per minute, which is used in the customary system of units. In the SI system, the standard is kW and MW.
HP	High pressure.
HRSG	Heat Recovery Steam Generator
I&E	Instrument and Electrical (Department).
IB	Inboard.
ICE	Internal combustion engine.
ICFM	Intake compressor flow measurement.
ID	Inside diameter.
IMO&R	Inspection, Maintenance, Overhaul and Repair.
Inert gas	This is a gas that does not combine chemically with itself or any other element. The four gases of this type are helium, argon, neon and krypton. In compressor terminology, it usually means a gas that does not supply or support any of the needs of combustion, such as nitrogen.
Inspection A.	This is a process of measuring, examining or testing a product or service against some requirement to identify nonconformance before it reaches a (internal/external) customer. **B.** This includes activities such as measuring, examining, testing, gauging one or more characteristics of a product or service and comparing these with special requirements to determine conformity.
IRN	Inspection Release Notice.
IoT	Internet of things.
ISO	International Standards Organization.
ISP	Inlet Steam Pressure

KO drum	Knock-out or suction drum
KPI	Key performance indicator.
LLDPE	Linear low-density polyethylene.
LDPE	Low-density polyethylene.
LHV	Lower heat value (fuel gas for gas turbines or internal combustion engines, in Btu or kJ.
LNG	Liquid (or liquified) natural gas.
LOTO	Lock out/tag out.
LP	Low pressure.
LPG	Liquid (or liquified) petroleum gas.
LST	Landing ship, tank.
Machinery string	A string of coupled machines. See machinery train.
Machinery train	MT is usually a unit or set consisting of driver, power transmission element(s), coupling(s)/gear box and driven machine(s). In a wider sense, definition of an MT could include appurtenances essential to the operation of such train, for example, trip and throttle valves, condensers, vacuum generation equipment, motor controls, transformers, surge controls, pulsation drums (reciprocating compressors) and check valves immediately up- or downstream of the set.
Machinery trip	A forced shutdown of a machinery train triggered by a malfunction resulting in unscheduled (process) outage results in statistics identifying causes, for example, process, maintenance mechanical, electrical, control and/or instrumentation and on another level, human error. A trip event typically becomes FRACAS history and data.
Maintainability	The ability of an item of plant equipment to be maintained under stated conditions of use.
Maximum allowable working pressure (MAWP)	This is the maximum continuous operating pressure for which a compressor has been designed when handling the specified gas at the specified temperature. It is not the design pressure of the compressor or compressor auxiliaries.
MCC	Motor control center.
MCSF	Minimum continuous stable flow (centrifugal process pumps).
MIG/MAG	Metal/Manual Inert Gas (welding process).
MIS	Management information system.
MOC	Management of change.
MOV	Motor Operated Valve.
MoC	Material of construction.
MR&O	Maintenance, repair and overhaul.
MQA	Machinery quality assessment. A process practiced by BiC companies to assure reliable equipment is acquisitioned.[1]
MTBF	Mean time between failure. A yardstick of equipment reliability.
MTBR	Mean time between repair. Also, a yardstick for maintainability.
MTS	Mechanical Technical Support (Department/Function).
MW	Molecular weight.
NDA	Non-disclosure agreement.
NDT	Non-destructive testing.
NEMA	National Electrical Manufacturers' Association.
NPSH$_a$	Net positive suction head available.
NPSH$_r$	Net positive suction head required.
OB	Outboard.

OD	Outside diameter. *Example*: Measuring the outside diameter (OD) of a pump shaft and the inside diameter (ID) of a hole.
ODE	Opposite drive end.
ODR	Operator-driven reliability.
ODP	Open drip proof (el. motor classification).
OEE	Overall equipment effectiveness – percent of actual operations against perfect production rates assuming no scheduled or unscheduled downtime, no defective product and no reduced production rates.
OEE	availability × process efficiency × rate of quality product.
OEM	Original equipment manufacturer.
OIMS	Operations integrity management system.
OMG	Oil mist generator.
OOD	Own object damage.
OPEX	Operating expense.
O/S	Overspeed.
OSHA	Occupational Safety and Health Act (USA). This deals with injuries/ 200,000 manhours. (Sometimes stated as OSHA injuries per 100 employees.) This is the standard indicator of safety program effectiveness.
Outage	(Process) outage. This results in statistics identifying causes, for example, process, maintenance or an overhaul scheduled in advance and performed while the process unit is shut down.
OWS	Oily water sewer (oil refinery and chemical plants).
PDT	Portable data terminal. This is used by process operators in "best-of-class" plants instead of log sheets. Collected field (process) data can be downloaded into computers for trending and analysis after completion of operator field rounds.
PdM	Predictive maintenance. This is an equipment maintenance strategy to discover incipient failures by monitoring the condition or health of the equipment. Using this data, predict the probable failure date, so as to enable corrective maintenance to be carried out in time.
PEEK	Polyetheretherketone.
PID	Piping and instrumentation diagram.
PIP standards	Process industry standards.
PIS	Plant information system.
PLC	Programmable logic controller.
PFD	Process flow diagram.
PHA	Process hazard analysis.
PHSA	Process hazard and safety analysis.
PLV	Pitch line velocity (gears).
PM	Preventive/periodic maintenance.
PMT	Process Mechanical Technical (Support Department – organizational concept).
PO	Purchase order.
Predictive maintenance (PdM)	an equipment maintenance strategy to discover incipient failures by monitoring the condition or health of the equipment. Using this data, predict the probable failure date, so as to enable corrective maintenance to be carried out in time.
PRV	Pressure relief valve. Usually, a spring-loaded valve is used to control or limit the pressure in a system by relieving fluid into a lower pressure stream or volume.
PSV	Pressure safety valve.
PSSR	Pre-startup safety review.
PTO	Power take-off.

QA	See quality assurance.
QC	See quality control.
Quality assurance	**A.** A program that is intended, by its actions, to guarantee a standard level of quality. **B.** QA is the planned and systematic activities implemented within the quality system and demonstrated as needed to provide adequate confidence that an entity will fulfill requirements for quality (source: ANSI/ISO/ASQ A8402-1994. Quality Management and Quality Assurance – Vocabulary). *Example*: Designing and providing a (quality control) checklist to maintenance millwrights, a procedure, a "best practice."
Quality control	**A.** The process of measuring quality performance, comparing it with the standard and acting on the difference. The operational techniques and the activities used to fulfill requirements of quality. **B.** A system of control meant to guarantee, by periodic inspections (hands-on measurements, comparison to standards, using gauges or tools) that a certain amount of quality is being maintained during the production or repair of the product in question. Materials, procedures, tools, and other things, as well as the product itself are inspected. **C.** Quality control is operational techniques and activities used to fulfill requirements for quality. It involves techniques that monitor a process and eliminate causes of unsatisfactory performance at all stages of the quality loop (source: ANSI/ISO/ASQ A8402-1994. Quality Management and Quality Assurance – Vocabulary).
R&M	Run and maintain.
Ranking	This is used in benchmarking processes: A benchmark is established by stating indices or parameters observed with best of class performers and then ranking the competitors' parameters against this best of class benchmark. Comparison (ranking) elements impact (machinery) reliability against those found in the "best of class."
RFQ	Request for quotation.
RCFA	Root cause failure analysis.
RCM	Reliability-centered maintenance.
Reliability	Rotating equipment: Time when a machine is not operating due to a problem with any component or subsystem is defined as "forced" or "unscheduled" downtime. Forced downtime is an "unscheduled outage" in that it is not planned or scheduled in advance and performed while the process unit is shut down.
Reliability factor	Rotating equipment: RF is defined as the percentage of time the machine is operating during process run-time. The machine is debited only for forced downtime or unscheduled work required during process run-time and not for work performed during a scheduled unit shutdown.
ROI	Return on investment.
RSC	Reliability stewardship committee.
SBC	Small bore piping connection.
SC	System completion.
SCADA	Supervisory control and data acquisition.
SBC	Small Piping Connections.
SCADA	Supervisory control and data acquisition.
SCP	System commissioning procedure. See also SOP.
SCR	Selective catalytic reduction. An SCR system is a system installed on diesel vehicles to reduce harmful nitrous oxide (NO_x) emissions. It works by injecting an automotive-grade urea, or diesel exhaust fluid (DEF) through a specially designed catalyst into the exhaust stream of a diesel engine. The DEF sets off a chemical

	reaction which converts NO_x into nitrogen, water and very small amounts of CO_2. SCR technology can achieve up to 90% NO_x reduction.
Service factor (SF)	See availability.
Settling-out pressure	This is the pressure within the compressor system when a compressor is shut down without de-pressuring the system, or the maximum pressure the system can reach under static conditions.
SHE	Safety, health, environment (issues, problems, risk).
SIT	Site integration test.
SIS	Safety instrumented system.
SME	Subject matter expert.
SOP	Standard operating Procedure.
SPC	Spare parts coordinator.
SRV	Safety relief valve.
SPIR	Spare parts interchangeability record.
SPO	Spare parts order.
SPOC	Single point of contact.
Stonewall	Stonewall is the point on a turbocompressor performance curve (characteristic) in which the maximum flow (capacity) and minimum pressure rise is encountered, beyond which further decreases in system resistance will not increase the flow rate. This is not particularly damaging to a single-stage compressor but can cause serious damage to the rotors and blades of all multistage centrifugal and axial compressors. *Stonewall* is also known as *Choke point*.
Surge	A damaging flow condition that occurs whenever a turbocompressor's outlet pressure is too high in relation to the flow through it. Surge manifests itself by repeated cycles of abrupt flow reversals which are often audible. If the surge is not controlled, the compressor can be severely damaged.
Surge control	A surge control and prevention system is designed to protect a turbocompressor from surge. Typical methods of surge control are blow-off to atmosphere or recirculation from the outlet to the inlet of the machine. An anti-surge control strategy must be closely integrated with the compressor load control strategy.
Surge point	An indication by a point, triangle, line or end of a turbocompressor performance curve showing where surge is expected to occur under stated conditions or a flow value provided by the machine's OEM where surge is expected to occur under stated operating conditions.
Train	See machinery train.
TI	Temperature Indicator.
Technician	Any maintenance worker performing direct maintenance labor is a technician. This includes mechanics, electricians, millwrights, artisans and other tradespeople, and is often called "direct maintenance personnel", "hourly personnel" or simply "mechanics."
TEFC	Totally enclosed fan cooled (el. motor classification).
TEWAC	Totally enclosed water-air cooled (el. motor classification).
TIR	Total indicator run-out (using a dial indicator or "clock").
TMR	Triple modular redundant (system as applied to machinery controllers).
Trip equipment	This is usually a forced shutdown of a machinery train, furnace and so on, resulting in unscheduled (process) outage, which is used in statistics identifying

	causes, for example, design, maintenance, process (human error), mechanical, electrical, control and/or instrumentation.
TTV	Trip and throttle valve of the steam turbine system.
TIG	Tungsten Inert Gas (welding process).
TYOSP	Two-year operational spares.
Uptime	See availability. This is the ratio of actual production hours to scheduled production hours, which is a measure of unscheduled interruptions to production (machinery trips). Indicator of effectiveness of preventive and predictive, or condition-based maintenance (PM and PdM/CBM) in combination with other factors.
VDDR	Vendor data and drawing requirements.
Viscosity	This is the measure of a fluid's resistance to flow, which is typically measured as the time required for a standard quantity of fluid at a certain temperature to flow through a standard orifice; the higher the value the more viscous the fluid.
Vision	**A.** An over-arching statement or decision on the way an organization wants to be. An ideal state of being at a future point. **B.** A statement of the desired end state of the organization articulated and deployed by the executive leadership. Organizational visions are inspiring; clear, challenging, reasonable and empowering. Effective visions honor the past while they prepare for the future. A broad statement founded on current trends and issues of where the organization will be in the next decade but not how it will get there.
VOC	Volatile organic compounds
VPI	Vacuum pressure impregnation (El motor windings)
VVP	Variable volume pockets (in reciprocating compressor cylinders)
Winding RTD	This is a device used to measure temperature changes in the motor windings to detect a possible overheating condition. These probes are imbedded into the winding slot, and their resistance varies with temperature.
WO	Work order

References

1 MQA is explained in detail: Bloch, H.P. & Geitner, F.K., *COMPRESSORS – How to Achieve High Reliability & Availability*, The McGraw-Hill Companies, email: bulksales@mcgraw-hill.com, New York, NY and other global cities, 2012, 268 pages. Bibliography.

Bibliography

[1] Robert X. Perez, Kane's ROTATING MACHINERY Dictionary, 2019, Third Coast Publishing Group L.L.C., ISBN: 978-1-7330413-0-0

Index

https://doi.org/10.1515/9783110701074-019

www.ingramcontent.com/pod-product-compliance
Lightning Source LLC
Chambersburg PA
CBHW060957210326
41598CB00031B/4853